Stored Product Protection and Postharvest Technology

Edited by **Fernando Plath**

SYRAWOOD
PUBLISHING HOUSE

New York

Published by Syrawood Publishing House,
750 Third Avenue, 9th Floor,
New York, NY 10017, USA
www.syrawoodpublishinghouse.com

Stored Product Protection and Postharvest Technology
Edited by Fernando Plath

International Standard Book Number: 978-1-68286-128-8 (Hardback)

Contents

Preface

Over the recent decade, advancements and applications have progressed exponentially. This has led to the increased interest in this field and projects are being conducted to enhance knowledge. The main objective of this book is to present some of the critical challenges and provide insights into possible solutions. This book will answer the varied questions that arise in the field and also provide an increased scope for furthering studies.

This book discusses the technologies related to stored product protection and postharvest crop handling. This field studies the preservation and safety of crops and food stocks including processing, storage related problems, etc. The chapters included herein discuss some of the important topics such as postharvest shelf life, techniques to avoid spoilage, postharvest physiology, etc. This book, with its detailed analyses and data, will prove immensely beneficial to professionals and students involved in this area at various levels.

I hope that this book, with its visionary approach, will be a valuable addition and will promote interest among readers. Each of the authors has provided their extraordinary competence in their specific fields by providing different perspectives as they come from diverse nations and regions. I thank them for their contributions.

Editor

Shelf life of Karonda jams (*Carissa carandas* L.) under ambient temperature

R. A. Wani, V. M. Prasad, S. A. Hakeem, S. Sheema, S. Angchuk and A. Dixit

Department of Horticulture, SHIATS Allahabad U.P-211007, India.

The studies were based on variations of sugar and to find out the best treatment for maximum storage period. The experiment comprised of 5 levels of addition of sugar and data obtained was analyzed by completely randomized design. Results obtained from study showed that Treatment 4 (1000 g pulp + 1150 g sugar) possessed an ideal value of total soluble solids (TSS), pH, acidity, moisture, ascorbic acid, iron, and overall acceptability at 0, 20, 40 and 80 days of storage. These seven parameters show that the quality of Karonda jam obtained by incorporating 1150 g of sugar was of good texture and quality. Based on the experimental study it was concluded that among all the treatments, treatment 4 was the best with regard to physical, chemical and sensory parameters of jam.

Key words: Karonda jams, sugar concentrations, preservation, shelf life.

INTRODUCTION

Fruits are amongst the first food items known consumed prehistorically by human beings. Fruits whether fresh, dried or processed have always formed a part of the staple diet of human beings because they are rich in nutrients and provide some of the essential minerals, vitamins, and the like, apart from that, they also help in curing a number of diseases. Preservation is a way to keep fruits for longer duration as it prevents the food from decay and spoilage. Karonda is a fruit of dry areas containing fair amount of vitamin C and minerals. The Karonda fruit is an astringent, antiscorbutic and as a remedy for biliousness and useful for cure of anemia. In traditional medicine the fruit is used to improve female libido and to remove worms from the intestinal tract. The fruits have anti-microbial and antifungal properties and its juice used to clean old wounds which have become infected. The fruit have an analgesic action as well as an anti- inflammatory one. The juice can be applied to the skin to relieve any skin problems. Fruits are generally harvested at immature stage for vegetable purpose, fully ripen fruits are consumed fresh or processed. Traditional healers of Chhattisgarh, India have expertise in treatment of different types of cancer from Karonda.

Carissa Carandas Linn. (Karonda) is a widely used medicinal plant by tribals throughout India and popular in various indigenous system of medicine like ayurveda, unani and homoeopathy. Karonda is good appetizer. Usually the fruit is pickled before it gets ripened. Ripe Karonda fruit contains high amount of pectin therefore it is also used in making jelly, jam, squash, syrup, tarts and chutney which are of great demand in international market. According to FPO specifications, a jam should contain a minimum of 68% TSS in the final product and the fruit content in the final product should not be more than 45% (w/w).

Preservation of Karonda jam in glass jars, which cannot be hermetically sealed is rather difficult, as the surface of the jam in the jar is susceptible to mould growth and after moisture evaporation from the jam resulting in surface graining and also shrinkage of jam. Developing countries are being encouraged to diversify their food exports by developing new products and adding more value to

Table 1. Chemical parameters of Karonda jam during storage at room temperature.

Treatment	pH Days after storage			Total soluble solids (%) days after storage			Acidity (%) days after storage			Moisture (%) days after storage			Ascorbic acid (mg/100 g) days after storage			Iron content (mg/100 g) days after storage		
	0	40	80	0	40	80	0	40	80	0	40	80	0	40	80	0	40	80
T1	3.6	3.5	3.4	64	67	68	0.76	0.79	0.80	20	17	14.5	3.6	3.4	3.1	36.7	35.2	32.6
T2	3.6	3.5	3.6	66	68	70	0.75	0.77	0.79	21	19	18	3.5	3.2	2.9	36.0	35.3	32.7
T3	3.7	3.6	3.6	66	68	70	0.75	0.77	0.79	21	19	18	3.5	3.2	2.9	36.0	35.3	32.7
T4	3.7	3.6	3.5	67	68	70	0.75	0.77	0.79	22	21	19	4.0	3.7	3.5	37.0	35.4	30.8
T5	3.8	3.6	3.5	69	71	73	0.72	0.75	0.77	23	22	20	3.8	3.5	3.4	37.0	35.4	32.7
SEM ±	0.08	0.06	0.12	1.31	1.23	0.65	0.01	0.01	0.01	0.63	0.60	0.75	0.10	0.06	0.09	0.07	0.20	0.11
CD at 5%	0.20	0.15	0.28	3.03	2.85	1.51	0.02	0.02	0.27	1.45	1.39	1.73	0.24	0.14	0.21	0.17	0.20	0.11

existing products. High concentration of sugar facilitates storage (Tarr and Baker, 1985; Bhandari, 2004) as such it was suggested to determine the best treatment of sugar variation in Karonda jam for maximum storage.

MATERIALS AND METHODS

The experimental work was conducted in the laboratory of Horticulture Department, SHIATS Allahabad during 2008 to 2009. There were five treatment combinations and experiment was laid out in completely randomized block design with three replications. The fruits selected for processing purpose were crushed in Karonda Grater and the pulp collected was subjected to boiling and concentrated by adding sugar till the end point was judged through drop test, TSS (68 to 70%) and by sheet test.

Different concentrations of sugar like 850, 950, 1050, 1150 and 1250 g were added to 1.0 Kg of fruit pulp as 1, 2, 3, 4 and Treatment 5, respectively. The observations were recorded on physical characteristics like pH, TSS, acidity, moisture, ascorbic acid, iron content and physical parameters like texture, flavor, color, appearance, taste, after taste and over all acceptability.

RESULTS AND DISCUSSION

All the treatments in the present investigation had significant impact for all observed traits. However, treatments differed significantly from one another at various time intervals (Tables 1 and 2). Among the chemical parameters observed pH, moisture content, ascorbic acid content and iron showed a gradual decline where as total soluble solids, total acidity registered a subsequent increase. Similar results were also obtained by Karhasushenko (1998) and Pino et al. (2004). pH of all the treatments underwent a decrease during preservation because of an increase in overall acidity of jam during preservation. All the treatments showed better values of pH during storage but the Treatment 4 showed an ideal value of pH during storage thus indicating 1150 g of sugar may be recommended for 1 kg of Karonda pulp as jam at this pH possesses a good setting property. The hydrogen ion concentration indicates strength of jam. There was a regular decrease of pH value of all the treatments during storage. Similar findings were also obtained by some other scientists Baker (1989); Joseph (1994); Lal et al. (1998) and Dheeraj et al. (2008). TSS of all treatments underwent an increase because of breakdown of complex sugar in to simple sugar during the period of preservation. TSS of Treatment 4 was found to be ideal during the period of storage (Table 1). Jam at this TSS possesses a firm texture, excellent body and sweet taste. Observations showed a

subsequent increase in TSS values for all the treatments during storage.

Ashraf (1987) and Manivasagan et al. (2006) also reported similar results in different studies. Total titrable acidity determines the strength of jam and there was a subsequent increase in total acidity of jam during preservation and acidity of Treatment 4 was found in accurate range (see Table 1). Jam at this titrable acidity possesses a firm texture and good setting property. Similar results were reported by Wang (1999) and Singh and Kumar (2000). Moisture content of all the treatments decreased during preservation and increased with enhancement of sugar concentration in Karonda pulp. All the treatments registered better water activity values and the best water activity values were recorded for 4 and 5 Treatment. The water content of jam directly controls chemical reaction rates and microbial activity. These findings are supported by different research works reported by Gordon et al. (2000) and Nayak et al. (2011).

Ascorbic acid content showed a gradual decrease during preservation because of breakdown of ascorbic acid by anti ascorbic acid compounds. The maximum ascorbic acid content was found in Treatment 4 as shown in Table 1 indicating ideal sugar concentration. Similar results were obtained

Table 2. Sensory parameters (storage period marks according to 9 point Hedonic scale) of Karonda jam during storage at room temperature.

Treatment	Texture says after storage			Flavor says after storage			Color and appearance says after storage			Taste days after storage			After taste days after storage			Overall acceptability days after storage		
	0	40	80	0	40	80	0	40	80	0	40	80	0	40	80	0	40	80
T1	6.0	7.33	7.33	6.33	6.66	7.33	7.00	7.33	7.00	8.33	6.00	6.66	6.66	6.66	7.00	5	6	7
T2	6.83	7.00	7.00	6.33	7.00	7.33	7.66	6.66	7.33	6.33	7.00	7.66	5.66	6.66	7.33	5	5	7
T3	6.33	6.66	6.66	7.00	7.66	7.33	6.00	7.33	7.66	7.33	7.66	7.33	6.33	6.66	7.66	6	7	8
T4	6.66	7.66	8.66	7.66	8.33	9.00	8.33	9.00	9.00	7.33	8.33	9.00	7.33	8.00	9.00	7	8	9
T5	6.00	5.33	6.00	5.00	6.00	5.66	7.33	6.66	6.33	5.00	5.66	6.33	5.66	5.33	5.66	4	5	5
SEM±	0.53	0.34	0.59	0.50	0.39	0.59	0.59	0.93	0.34	0.55	0.39	0.42	0.36	0.42	0.39	0.51	0.65	0.60
CD at 5%	0.23	0.80	1.37	1.16	0.90	1.05	1.37	1.02	0.80	1.28	0.90	0.02	0.84	0.97	0.90	1.19	1.51	1.39

by Upasana and Bhatia, (1985) and Joshi et al., (1986). There was regular decrease in the iron content (mg/100g) of all the treatments and an ideal value of iron content (mg/100g) was shown by Treatment 4. Similar results were reported by Nayak et al (2011) and Gordon et al (2000).

The sensory parameters like texture, color, flavor, appearance, taste, after taste and overall acceptability showed significant increase during preservation. All the treatments were awarded better score by a panel of 7 judges but the Treatment 4 got the best score (see Table 2). The score card is based on the 9 point Hedonic Scale. Thus the sensory quality of the jam increased during the period of storage. Overall acceptability of jam incorporated with 1150 g of sugar was found to be liked extremely on Hedonic scale in comparison to other treatments during preservation. Similar results were reported by Pino et al (2004); Shashi and Badiyala, (2002); Singh et al (2005) and Tandon et al (2003).

Conclusion

From the present findings it is concluded that Karonda, which is the minor fruit of India can be utilized for making jam and this jam can be stored for at least three months without undergoing any deterioration and evidently Treatment 4 (1.0 kg Karonda pulp + 1150 g sugar) showed the best results with regard to physical, chemical and sensory parameters of jam.

REFERENCES

Ashraf SN (1987). Studies on post harvest technology of Jamun Fruit. PhD Thesis N.D. University of technology Faizabad (U.P.)

Baker GL (1989). High polymer pectin's and their desertification. Adv. Food Res. 1:395.

Bhandari SP (2004). Studies on physico-chemical composition, grading, storage and processing of jamun (Syzygium cuminii (Linn.) Skeels). M.Sc. (Agri.) Thesis, Dr. Balasaheb Sawant Konkan Krishi Vidyapeeth, Dapoli, Dist. Ratnagiri, Maharashtra.

Dheeraj S, Lobsang W, Prahalad VC (2008). Processing and marketing feasibility of underutilized fruit species of Rajasthan. Contributed paper presented at IAMO fórum 2008.

Gordon DL, Schwenn KS, Ryan AL, Roy S (2000). Gel products fortified with calcium and method of preparation. United States Patent. 6077557.

Joseph GL (1994). Better pectin. Food Eng. 23:71.

Joshi GD, Prabhudesai VG, Salvi MJ (1986). Physico-chemical characteristics of Karonda (Carissa carandas L.) fruits. Maharashtra J. Hort. 3(1):39-44.

Karhasushenko (1998). Changes in the chemical properties of

jam during storage. KhelbopekarnayaiKanditerskaya, Promyshlennost 7:26-27.

Lal G, Siddappa GS, Tandon GL (1998). Preservation of fruits and vegetables. ICAR Publication, New Delhi, India.

Manivasagan S, Rana GS, Kumar S, Joon MS (2006). Qualitative changes in Karonda (Carissa carandas Linn.) candy during storage at room temperature. Haryana J. Hortic. Sci. 35(1):19-21.

Nayak P, Bhatt DK, Shukla DK, Tandon DK (2011). Evaluation of aonla (Emblica Officinalis G.) segments-in-syrup prepared from stored fruits. Res. J. Agric. Sci. 43(2).

Pino J, Marbot R, Vazques C (2004). Volatile flavour constituents of Karnda (Carissa carandas L.) Fruit. J. Ess. Oil Res. 16(5):432-4.

Shashi KS, Badiyala SD (2002). "Karonda, fruit plant for marginal land.

Singh BP, Pandey G, Pandey MK, Pathak RK (2005). Shelf life evaluation of aonla cultivars. Indian J. Hort. 62(2):137-140.

Singh R, Kumar S (2000). Studies on the effect of post-harvest treatments on decay loss and biochemical changes during storage of aonla (Emblica officinalis G.) fruit cv. Chakaiya. Haryana J. Hortic. Sci. 29(3):178-179.

Tandon DK, Yadav RC, Sood S, Kumar S, Dixit A (2003). Effect of blanching and lye peeling on the quality of aonla candy. Indian Food Packer. 57(6):147-152.

Tarr LW, Baker GL (1985). Fruit jellies, the role of sugar. University Delaware Agriculture Experimental Station, Tech. Bull. 11:3.

Upasana R, Bhatia BS (1985). Studies on pear Candy processing. India Food Pack. 39:40-46.

Wang CY (1999). Post harvest quality decline, quality maintenance and quality evaluation. Acta. Hort. ISHS, 485:389-392.

A review on the role of packaging in securing food system: Adding value to food products and reducing losses and waste

Umezuruike Linus Opara and Asanda Mditshwa

South African Research Chair in Postharvest Technology, Faculty of AgriSciences, Stellenbosch University, Stellenbosch 7600, South Africa.

Packaging is an essential component of the food system, assuring the safe handling and delivery of fresh and processed food products from the point of production to the end user. Technological developments in packaging offer new prospects to reduce losses, maintain quality, add value and extend shelf-life of agricultural produce and consequently secure the food system. The objective of this review is to highlight the contributions of packaging in securing the food system by maintain quality and reducing food losses and waste. The review also discusses some of the novel and emerging packaging technologies that have revolutionized the way we handle and package food to meet the increasing consumer demand for consistent supply of high quality, safe and nutritious products.

Key words: Food system, food products, packaging, waste, plastic.

INTRODUCTION

Maintaining food quality and improving safety, and reducing postharvest losses waste are key objectives of a sustainable food system. High incidence of postharvest losses and waste pose a major problem in the food industry and world at large. An estimated 1.3 billion tonnes of food is wasted annually in production, distribution, and homes (Quested et al., 2011). Reports from developed countries such as Britain, Sweden and USA have indicated that almost one third of purchased food is wasted at food service institutions and households (Wikström and Williams, 2010). In addition to the effects of a wide range of socio-economic, climatic and environmental factors, the loss and wastage of already harvested food is a major contributor to food and nutritional insecurity. Moreover, reducing food loss and preventing waste also has environmental benefits given that each tonne of prevented food waste contributes to avoiding 4.2 tonnes of carbon dioxide emissions that would have been associated with the waste (Quested et al., 2011).

In the early days of agriculture, leaves and animal skin were used as packaging materials to carry food over short distances and to secure them for later use. In modern food systems, the principal functions of packaging have widened to include containment, protection, communication and convenience. Paine and Paine (1992) noted that "to ensure delivery, the package must at least provide information as to the address of recipient, describe the product and perhaps describe how to handle the package and use the product." Despite the overriding importance of packaging in maintaining quality and wholesomeness and facilitating the movement of food along the value chain, there is continuing debate on the amount of packaging used in the food industry in relation to packaging waste the environmental impacts, as well as the role of packaging in reducing food losses and waste (Opara, 2011).

Inappropriate processing and packaging (or lack of these) can contribute to 25 to 50% food loss, especially in developing countries. About 10% of fruit and vegetables shipped to European Union are discarded due to unacceptable quality and spoilage (World Packaging Organization, 2008). These high levels of postharvest loss and waste suggest that food production is only half the battle to feed the world (Opara, 2011). Examining the role of packaging in reducing postharvest food losses and waste is particularly important given that packaging also contributes to municipal waste after completing its function of protecting the contents. The need to handling and dispose large quantities of packaging after utilising the food contents, therefore, constantly puts packaging waste in bad light in public discussion about waste, often ignoring the critical role that packaging plays in securing the food system.

The objective of this paper, therefore, is to highlight the role of packaging in the food industry with particular attention to the impacts of packaging in maintaining product quality and safety and reducing the incidence of postharvest food losses and waste. Recent advances in smart and intelligent packaging designed to minimise some of the negative impacts of packaging on the environment and food waste are highlighted. The environmental impacts of food packaging are examined and measures to reduce packaging waste are discussed. In the next section, we highlight the different types of materials and formats used in food packaging.

Types of packaging materials and formats used in the food industry

A wide range of packaging materials and packaging formats are used in the fresh and processed food industry to handle, store, and distribute fresh and processed food products, from farm to the consumer. Different types of materials such as glass, plastic, metal, cardboard are used for making packaging containers and the material used depends on the nature of the food product because different packaging materials possess a range of performance characteristics that exert significant impacts on shelf-life (Robertson, 2011). Bottles and glass jars are often used for packaging liquid food stuff while solid food products are mostly packed on plastics and cardboards. Processed fruit and vegetables are usually packed in airtight metal containers to prevent oxygen transmission that might lead to spoilage of the product through microbial growth and oxidation of lipids (Robertson, 2010). According to the World Packaging Organization (2008), the most important consumer packaging are made of paper and board (38%), followed by plastic (30%) with rigid plastics alone taking an 18% share, metal (19%), glass (8%), and others (5%). Moreover, approximately 70% of overall consumer packaging are used in food industry where 48% of all the

packaging are made from paperboard.

Plastic

Historically, packaging was used primarily to prevent food contamination with unwanted objects. However, consumer demand for desirable food quality has led to a surge in packaging innovation. For instance, Cha and Chinnan (2004) noted the increasing use of plastic films in food packaging, which combines the biophysical properties of plastic films with biopolymer coatings to maintain the nutritional and sensory quality of the product. Using plastic as packaging material also offers marketing advantage. Unlike metal and aluminium packaging materials, harnessing the transparency of film packaging for product visibility is now widely practised, enabling consumers to assess the visual quality of the product prior to purchase. However, the variable permeability to light, gases and vapours of plastics is a major drawback. The various kinds of plastic films include low density polyethylene (LDPE), laminated aluminium foil (LAF), high density polyethylene (HDPE), polypropylene (PP), polyethylene (PE).

Paper and cardboard

Paper and cardboard are made from cellulose fibres derived from wood and plant fibres using sulphate and sulphite (Robertson, 2011). The poor barrier properties of plain paper makes it unsuitable for long time storage. Protective properties of paper are usually improved by coating, laminating or filled with waxes and resins. Paper and cardboard are widely used in corrugated boxes, milk cartons, sacks, and paper plates. Packaging material based on paper has an advantage due to its high recyclability at relatively low cost. Paperboard packaging such as carton are the most widely used packaging in the horticultural industry. For horticultural food products such as fruit and vegetables which remain alive after harvest (Figure 1), the use of ventilated packaging is essential to facilitate the delivery of cold air to produce inside the packaging during precooling and refrigerated storage. The design challenge is to balance the cold chain requirements for optimum airflow while maintaining the mechanical integrity of the package and produce. Given that the marketability of fresh produce is reduced when precooling is delayed (Figure 2), resource-efficient package design for optimum cooling without adverse effects on produce quality are essential for cost-effective postharvest handling and marketing of fresh horticultural foods.

Metal

The good physical protection and recyclability of metal is

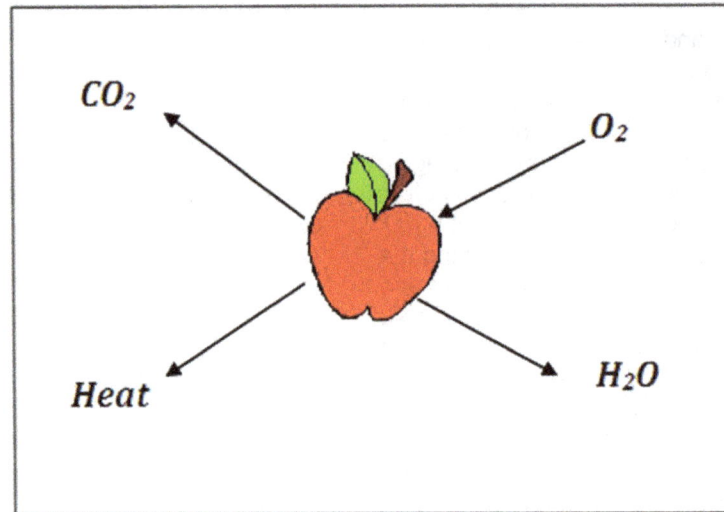

Figure 1. Fresh foods such as fruit and vegetables are alive and continue to respire after harvest. Reducing the respiration rate and reducing the heat produced through efficient airflow inside ventilated packaging is important in maintaining product quality.

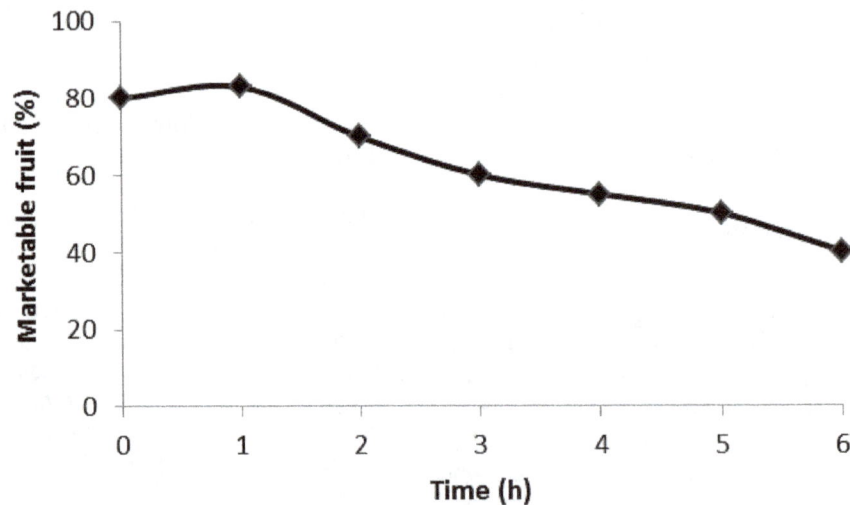

Figure 2. The effect of delay on precooling of horticultural produce (Adapted from Brosnan and Sun, 2001).

widely preferred in many food applications. Aluminium and steel are 2 metals predominately used in packaging (Marsh and Bugusu, 2007). Aluminium is commonly used in making cans, foil, and laminated paper. Carbonated beverages and seafood are often packed on aluminium packaging material. The high cost of aluminium compared to other metals is the main disadvantage of using it in food packaging systems. Steel packaging material is often used to make cans for drinks and processed foods such as beans and peas. The high mechanical strength and low weight of steel makes it

relatively easy to store and ship food (Marsh and Bugusu, 2007). Steel can be recycled many times without quality loss and its cost is significantly lower than aluminium hence it's highly used in packaging systems.

Glass

Glass is another common packaging material which dates back to 3000 BC (Marsh and Bugusu, 2007) and is used mostly for packaging processed foods especially

Table 1. Common packaging formats used for different products.

Packaging formats	Example of produce
Paperboard cartons	Fresh produce (apple, strawberry)
Polyethylene-laminated cartons	Processed produce (orange juice)
Wooden box	Fresh produce (strawberry)
Tetra recart carton	Processed produce (meat)
Tetra wedge package	Processed produce (meat)
Can	Processed food (minimally processed tomato pulp)
Glass bottle	Minimally processed food (tomato sauce, orange juice)
Plastic bottle	Processed food (citrus juice)

where moisture and oxygen barrier is of great importance. Carbonated beverage drinks contain dissolved carbon dioxide creating pressure within the package, and glass is often the suitable packaging capable of withstanding carbon dioxide pressure. Moreover, the odourless and static chemical property of glass that ensures unimpaired taste and flavour of the contents makes it advantageous for food packaging (Marsh and Bugusu, 2007). The reusability and recyclability of glass-based packaging material contribute to less negative impacts on the environment; the heavy weight of glass adds to the transportation costs of food products.

OTHER TYPES OF PACKAGING MATERIALS

Packaging can also be of mixed material. This kind of packaging may be resource and energy efficient than using a single material. However, the drawback of such packaging is the difficulty to recycle, which is attributed to the lack of infrastructure to separate the materials. New biodegradable, plant-based packaging materials are needed to combat environmental problems associated with such mixed packaging. Identification of biodegradable packaging materials and development of innovative methods to degrade plastic are thus needed.

Packaging formats used for different food products

Choosing the right format of packaging is important to meet the functions of packaging. There are some considerations for selecting appropriate packaging, including suitable structure and form, efficiency and disposal after use. While engineering and economic aspects of packaging performance are important, the environmental issues associated with packaging also need to be addressed when choosing packaging. Common packaging formats used in the food industry include paperboard cartons, wooden boxes, metal cans, glass, and plastic bottles (Table 1). For horticultural produce such as fruit, the packaging format may be

paperboard produced as single layer or multi-layer cartons and stacked into pallets or bulk bins made of wood or plastic (Figure 3).

Developments in package-food-environment interaction

Developments in sensors and information and communication technologies have enabled designers to impart desirable packaging attributes which promote greater interactions between the product and package as well as enable the consumer to make decisions about the quality and safety of the product contained in the package. The use of these highly instrumental packaging systems have various functions that are important in maintaining produce quality and safety as well providing other value-added services to the consumer.

Smart packaging

Smart packaging refers to an improved packaging system with functional attributes that add benefits to the food product and subsequently the consumers. Smart packaging uses an integrated approach with mechanical, chemical, and electrical driven-functions to ensure an improved usability of food products. Some of the prominent facets of smart packaging include use-by dates, usage of self-heating or self-cooling containers with electronic displays storage temperature, and nutritional information of the product (Mahalik and Nambiar, 2010).

Active packaging

Active packaging is categorized into active scavenging systems (absorbers) and active releasing systems (emitters) (de Kruift et al., 2002). Under scavenging packaging system, unwanted compounds such as oxygen, excessive moisture and ethylene which accelerate the spoilage process in foods are removed from the product. For instance, oxygen may cause off-

Layered package   Bulk package

A pallet of packages

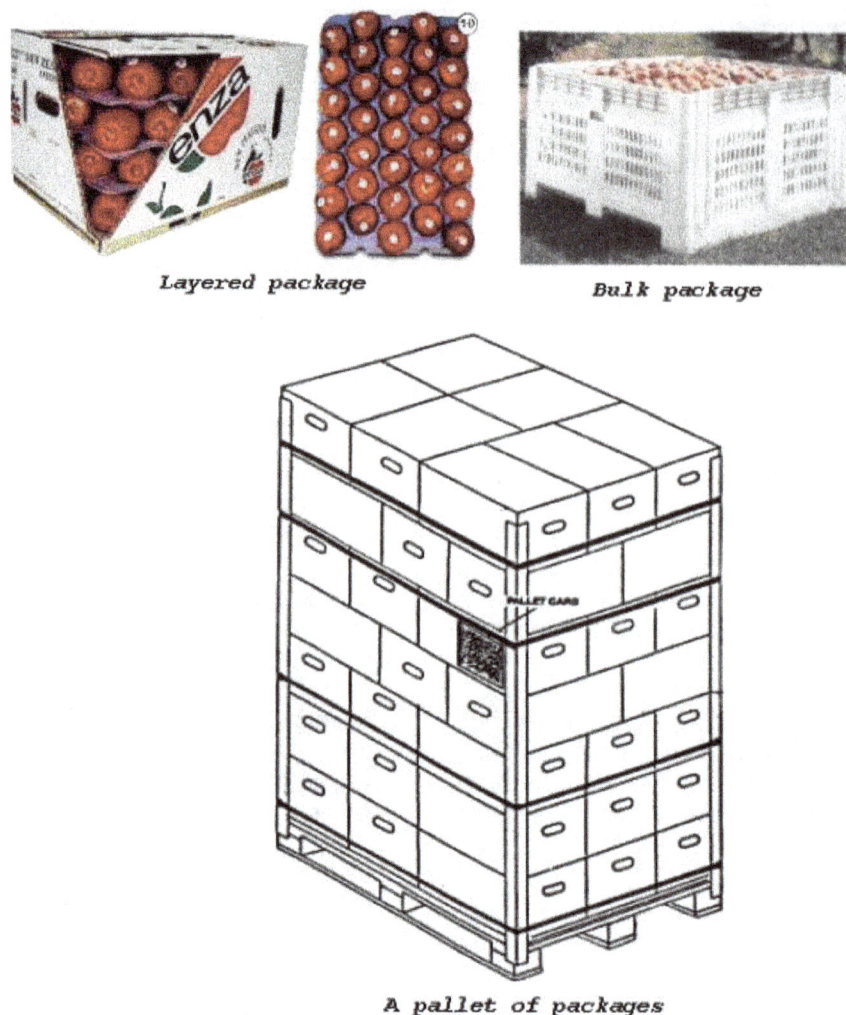

Figure 3. Different packaging formats used for handling fresh fruit such as apple (Opara, 2011).

flavours, nutrient loss (through oxidation) and colour changes; hence the usage of oxygen scavengers to maintain quality and extend shelf life of some food products (Berenzon and Saguy, 1998) (Figure 4). The moisture content of packed horticultural products should be controlled because high moisture content favours microbial growth. The softening of dry crispy food products such as biscuits and caking of coffee result from unregulated moisture content. Moisture controlling systems are often used to scavenge excess moisture that contributes to product quality loss. However, it is worth noting that excess moisture loss might impose lipid peroxidation and desiccation of packed products. It is therefore imperative to have a good understanding of product physiology, structure and composition when designing the packaging as food stability is closely linked to water activity. Active releasing packaging system is another aspect of active packaging and this involves the addition of beneficial agents to the package to preserve the quality of the content. Releasing packaging system

favours the addition of compounds such as carbon dioxide, moisture, preservatives and antioxidants into the package. Carbon dioxide releasing systems are also used to retard respiration of horticultural crops and subsequently prolong shelf-life. The main objective of active packaging, with both scavenging and releasing systems, is ensuring exceptional food quality and extended shelf-life.

Intelligent packaging

Intelligent packaging refers to the use of packaging as an intelligent messenger to monitor the condition and provide quality information of packed foods to the consumers (de Kruift et al., 2002). Indicators such as temperature, microbial growth, product authenticity, and pack integrity are used in intelligent packaging. At the moment, freshness (Figures 5 and 6) and leakage indicators are commercially available for monitoring food

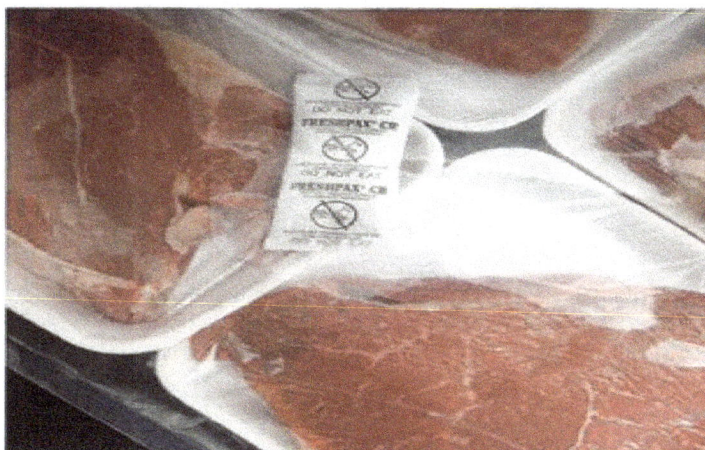

Figure 4. Oxygen absorber in polyethylene tray packed meat (Packaging Europe, 2013).

Figure 5. Meat packed in polyethylene trays with fresh label monitors detecting expiration date through colour-changing as a response to the ammonia level emitted by aging food (Marlin, 2012).

Figure 6. Packaged golden drop fruit with food spoilage indicator label (Green = fresh; orange = warning) (Nopwinyuwong et al., 2010).

Table 2. Examples of the effects of packaging on quality and shelf-life of horticultural products.

Product	Packaging material	Effects on quality attributes	Reference
Plum	Cardboard box (compared to unpackaged).	Fruit firmness was retained 55-60% compared to 36-47% in unpackaged produce. High Chroma value (good colour) in packaged fruit compared to unpackaged fruit was recorded	Valero et al. (2004)
Blueberries	Polylactide containers (compared to clamshell containers)	Polylactide containers had 4% weight loss compared to 48% for clamshell containers after 9 days of storage at 10°C.	Almenar et al. (2008)
Pomegranate juice	Glass bottle (Compared to paperboard cartons with polyethylene layers)	High juice quality was obtained in glass bottle package. Anthocyanin degradation was 78% in juice packed inside glass bottle compared to 95% for paperboard cartons.	Pérez-Vicente et al. (2004)

quality. High temperatures are often correlated with food deterioration as result of irreversible biochemical reactions combined with microbial growth (de Kruift et al., 2002). The time-temperature indicator therefore measures the change that mimics the targeted quality attribute with the same behaviour under the same time-temperature exposure. The pH and enzymatic changes of the product might also give information about the quality of food. The Vitsab TTI indicator (Vitsab Sweden AB, Sweden) measures the enzymatic reactions that subsequently cause pH change of the product. The package contains the Vitsab TTI indicator window indicating the difference between acceptable and distasteful.

IMPACTS OF PACKAGING ON FOOD QUALITY, SHELF-LIFE AND SAFETY

Sensory and nutritional quality

The type of packaging exerts considerable effect on the sensory quality of produce. For instance, litchi (cv. 'Mauritius') packed in biorientated polypropylene (BOPP-3) were found to be of exceptional nutritional and sensory quality compared to fruit packed in BOPP-1 and BOPP-2 with less polypropylene layer (Sivakumar and Korsten, 2006). In addition, non-perforated polypropylene plastic bags were found to be more suitable for table grapes than perforated plastic bags based on higher sensory scores for crunchiness, juiciness and overall fruit quality. Previous studies have demonstrated the potential of packaging to either negatively or positively influence the nutrient composition of food. Some packaging materials and forms promote nutrient loss during storage whilst some can preserve nutrients (Table 2). For instance, high

losses of aroma compounds have been reported in citrus juices packed in low density polyethylene paperboard than other packaging (Ebbesen et al., 1998). Mexis et al. (2009) studied the effects of different packaging materials and found considerable variability in product shelf life, with maintenance of nutrient content ranging from 2 to 12 months depending on type of package.

Shelf life

Packaging is often used as a tool to extend shelf life by preventing or reducing water loss, especially in fresh produce. Studies by Miller and Krochta (1997) showed that polyethylene bags reduced water loss and extended storability of various fruit and vegetables. Unpacked foods are often exposed to a range of microorganisms which have the potential to reduce shelf-life (Paine and Paine, 1992). The choice of packaging type and material has also effects shelf-life. For instance, Lee et al. (2002) reported that red pepper paste packed on polyethylene plastic had prolonged shelf-life compared to other forms of plastics, while Mexis et al. (2009) reported prolonged shelf-life and reduced microbial growth of shelled-walnuts packed on polyethylene terephthalate//polyethylene compared to polyethylene pouches.

Food safety

Harmful microorganisms feeding on unpacked food which are later consumed by humans can result in food poisoning, sickness or even death (Paine and Paine, 1992). Maintaining hygiene during food handling is important to assure the safety of consumers as well as promote longer shelf-life of food products. While effective

Table 3. Examples of the effects of packaging on horticultural food product losses and waste.

Product	Packaging	Effect on quality attributes	Reference
Sweet corn	Polystyrene trays wrapped (Polyolefin film) versus polystyrene trays (PVC film)	Decay of polyolefin film wrapped cobs was 1.5% compared to 45.2% decay for PCV film	Aharoni et al. (1996)
Sweet corn	Cardboard trays (Polyolefin film) versus Cardboard trays (PVC film)	Decay of polyolefin film wrapped cobs was 1.5% compared to 51.4% decay for PCV film	Aharoni et al. (1996)
Blueberries	Polylactide containers versus clamshell containers	Berries packed on clamshell containers were unmarketable after 3 days of storage at 10°C unlike Polylactide packed fruit that was still marketable after 18 days of storage	Almenar et al. (2008)
Cabbage	Monooriented polypropylene (OPP) trays versus PVC-PE trays	OPP trays prolonged shelf-life to 10 days unlike PVC-PE packed vegetable with 7 days shelf-life	Pirovani et al. (1997)
Celery	Perforated polypropylene (PP) film versus unpackaged	PP film allowed a shelf-life of 31 days while unpacked was unacceptable after 20 days	Rizzo and Muratore (2009)
Tomato	Plastic container versus cartons	Plastic container had 39.8% fruit loss while 80.6% was lost in carton stored fruit after 21 days storage at 10°C	Linke and Geyer (2002)
Red pepper	Polyethylene bags versus unpacked	Fruit packed in polyethylene bags had no decay, whereas fruit inside polyethylene bags had 11.7% decay after 14 days at 3°C	Meir et al. (1995)

packaging contributes to reducing spoilage and maintaining food quality, studies have also shown that packaging (and its related components) is a potential source of food contamination (Muncke, 2009). Some substances used in food packaging such as bisphenol have been found to contain endocrine disrupting compounds that are highly detrimental in biological systems (Vom Saal et al., 2007). Muncke (2009) described the contamination of food by packaging as being regulated by diffusion-controlled processes which depend on temperature and storage time of the product. This process leads to the leaching of food contaminants compounds from packaging to foodstuff.

ROLE OF PACKAGING IN REDUCING FOOD LOSSES AND WASTE

Roughly 30 to 40% of food produced in both developed and developing countries are lost or wasted, with more losses occurring in developing countries (Godfray et al., 2010). The lack of proper postharvest technologies and cold-chain infrastructure are often cited as the principal factors aggravating food losses and waste in developing countries. The use of cost-effective and resource-efficient packaging technologies can contribute to reducing food losses and waste during postharvest handling (Opara, 2011). Almost one-third of rice grain produce in Asia may be lost due to pests and spoilage related to poor packaging equipment, and 10 to 15% postharvest losses of cereals and grain legumes are commonly recorded in developing countries (FAO, 1997). Some regions in Africa and Latin America experience postharvest food losses as high as 50%.

Several researchers (Table 3) have reported the potential of applying appropriate packaging to reduce postharvest losses and waste in a wide range of products

(Marsh and Bugusu 2007; Quested et al., 2011). For example, García et al. (1998) reported decay incidence of 86.5% in unpackaged strawberries compared to only 33.8% in fruit packaged with polypropylene film. Rizzo and Muratore (2009) prolonged the shelf-life of celery by 31 days using perforated polyethylene film while celery that was not packaged decayed before 20 days of storage. In another study demonstrating the importance of packaging in relation to food waste, Meir et al. (1995) reported 11% decay incidence of unpackaged red pepper stored at 3°C for 14 days compared with no decay in packaged produce. As shown in Table 3, selecting appropriate packaging material is a critical factor in realising the potential of packaging to reduce postharvest food losses and waste.

THE IMPACT OF PACKAGING ON FOOD PRICE

While packaging contributes to reducing postharvest food losses and waste and maintaining product quality and safety, it also affects the cost of the product to the consumer. On the other hand, like other energy intensive industries, the price of packaging material is also affected by energy costs which are often linked to crude oil price. Generally, the cost of packaging material represents 17% of the total cost of product (Lange and Wyser, 2003). The type of packaging used also influences the price of food products. Products packed in glass bottles generally cost higher than those on plastic bottles (Lange and Wyser, 2003) and this is commonly attributed to higher transportation costs associated with higher weight of glass packages. Packaging exerts influence on food prices in developed countries than in developing countries, with packaging and marketing accounting far greater proportion of food prices in developed than developing countries (Elobeid and Hart, 2007). Consumer willingness to pay high prices for products is closely related to packaging style and material used.

FUTURE PROSPECTS AND CONCLUSIONS

Packaging is an essential component of the food system and plays a critical role in containing, protecting and preservation food and other agro-industrial raw materials from field to the end user. Researchers have shown that the use of appropriate packaging can contribute to reducing food losses and waste, and maintenance of product quality and safety. However, packaging is a major contributor to the cost of food, and packaging waste has been implicated as a major cause of municipal waste stream. The issue of food contaminants associated with packaging particularly due to use of recyclable paper needs to be addressed. To address these safety and sustainability challenges, the role of cost-effective and resource-efficient packaging design is crucial. The

application of emerging technologies in packaging design offers new prospects for advanced quality monitoring using electronic devices that monitor and report real time information on nutritional quality and safety of food.

The synergy of recent advances in biotechnology, nanotechnology and material science offer new opportunity to develop new packaging materials and design to address some of the changes facing the industry, including product safety, environmental impacts and sustainability of packaging. With increasing power and lower cost of information and communication technologies, the development of highly advanced packages incorporating nano-sensors to capture and analyse environmental signals and adjust stress response treatments on fresh foods through series of controllers to maintain storage quality and subsequently prevent food spoilage have become more of a reality than science fiction. Recent developments and applications of nanotechnology have produced antimicrobial packaging in response to the problem of food spoilage and losses. For fresh horticultural produce which continue to be alive after harvest, balancing the cold chain requirements through optimal ventilation design without compromising the mechanical integrity of the package will remain a packaging design challenge for engineers and food scientists.

ACKNOWLEDGEMENT

This work is based upon research supported by the South African Research Chairs Initiative of the Department of Science and Technology and National Research Foundation. The authors are grateful to the South African Postharvest Innovation Programme and Fruitgro[Science] for the award of the project on "Packaging of the Future".

REFERENCES

Aharoni Y, Copel A, Gil M, Fallik E (1996). Polyolefin stretch films maintain the quality of sweet corn during storage and shelf-life. Postharvest Biol. Tech. 7(2):171-176.

Almenar E, Samsudin H, Auras R, Harte B, Rubino M (2008). Postharvest shelf life extension of blueberries using a biodegradable package. Food Chem. 110(1):120-127.

Berenzon S, Saguy IS (1998). Oxygen absorbers for extension of crackers shelf-life. Food Sci. Tech. 31:1-5.

Brosnan T, Sun DW (2001). Precooling techniques and applications for horticultural products: a review. Int. J. Refrig. 24(2):154-170.

Cha DS, Chinnan MS (2004). Biopolymer-based antimicrobial packaging: A review. Crit. Rev. Food Sci. Nutr. 44(2):223-237.

de Kruift N, Van Beest M, Rijk R, Sipilainen-Malm T, Paseiro Losada P, De Meulenaer B (2002). Active and intelligent packaging: Application and regulatory aspects. Food Addit. Contam. 19(1):144-162.

Ebbesen A, Rysstad G, Baxter A (1998). Effect of temperature, oxygen and packaging material on orange juice quality during storage. Fruit Proc. 8:446-455.

Elobeid A, Hart C (2007). Ethanol expansion in the food versus fuel debate: How will developing countries fare? J. Agric. Food Ind. Org. 5(2). DOI: 10.2202/1542-0485.1201.

A review on the role of packaging in securing food system: Adding value to food products and reducing losses...

13

FAO (1997). www.fao.org/english/newsroom/factfile/IMG/FF9712-e.pdf [Accessed on 25.10.2012].

García JM, Medina RJ, Olías JM (1998). Quality of strawberries automatically packed in different plastic films. J. Food Sci. 63(6):1037-1041.

Godfray HCJ, Beddington JR, Crute IR, Haddad L, Lawrence D, Muir, JF, Pretty J, Robinson S, Thomas SM, Toulmin C (2010). Food Security: the challenge of feeding 9 billion people. Science 327:812-818 [DOI: 10.1126/science.1185383].

Lange J, Wyser Y (2003). Recent innovations in barrier technologies for plastic packaging? A review. Pack. Tech. Sci. 16(4):149-158.

Lee DS, Jang JD, Hwang YI (2002). The effects of using packaging films with different permeabilities on the quality of korean fermented red pepper paste. Int. J. Food Sci. Technol. 37(3):255-261.

Linke M, Geyer M (2002). Postharvest behavior of tomatoes in different transport packaging units. Acta Hortic. 599:115-122.

Mahalik NP, Nambiar AN (2010). Trends in food packaging and manufacturing systems and technology. Trends Food Sci. Tech. 21(3):117-128.

Marlin B (2012). Packaging innovation: Exploring active and intelligent packaging. http://www.onemarlin.com/2012/08/20/packaging-innovation-exploring-active-and-intelligent-packaging [Accessed on 11/01/13].

Marsh K, Bugusu B (2007). Food packaging? Roles, materials, and environmental issues. J. Food Sci. 72(3):R39-R55.

Meir S, Rosenberger I, Aharon Z, Grinberg S, Fallik E (1995). Improvement of the postharvest keeping quality and colour development of bell pepper (cv. 'Maor') by packaging with polyethylene bags at a reduced temperature. Postharvest Biol. Tech. 5(4):303-309.

Mexis SF, Badeka AV, Riganakos KA, Karakostas KX, Kontominas MG (2009). Effect of packaging and storage conditions on quality of shelled walnuts. Food Control 20(8):743-751.

Miller KS, Krochta JM (1997). Oxygen and aroma barrier properties of edible films: A review. Trends Food Sci. Technol. 8(7):228-237.

Muncke J (2009). Exposure to endocrine disrupting compounds via the food chain: Is packaging a relevant source? Sci. Total Environ. 407(16):4549-4559.

Nopwinyuwong A, Trevanich S, Suppakul P (2010). Development of a novel colorimetric indicator label for monitoring freshness of intermediate-moisture dessert spoilage. Talanta 81(3):1126-1132.

Opara UL (2011). From hand holes to vent holes: What's next in innovative horticultural packaging? Inaugural Lecture, Stellenbosch University, South Africa.

Packaging Europe (2013). Multisorb technologies introduces maplox program for map low-oxygen case-ready meat packaging. http://www.packagingeurope.com/News/21938 [Accessed on 11/01/13].

Paine FA, Paine HY (1992). A handbook of food packaging. 2nd Edition. Blackie Academic and Professional.

Pérez-Vicente A, Serrano P, Abellán P, García-Viguera C (2004). Influence of packaging material on pomegranate juice colour and bioactive compounds, during storage. J. Sci. Food Agric. 84(7):639-644.

Pirovani ME, Güemes DR, Piagentini A, Di Pentima JH (1997). Storage quality of minimally processed cabbage packaged in plastic films. J. Food Qual. 20(5):381-389.

Quested T, Parry A, Easteal S, Swannell R (2011). Food and drink waste from households in the UK. Nutr. Bull. 36(4):460-467.

Rizzo V, Muratore G (2009). Effects of packaging on shelf life of fresh celery. J. Food Eng. 90(1):124-128.

Robertson G (2010). Food packaging and shelf life: A Practical Guide. Taylor and Francis Group, Boca Raton, USA.

Robertson GL (2011). Paper-based packaging of frozen foods. In Handbook of Frozen Food Processing and Packaging Edited by D. (Sun Second), CRC Press.

Sivakumar D, Korsten L (2006). Influence of modified atmosphere packaging and postharvest treatments on quality retention of litchi cv. Mauritius. Postharvest. Biol. Techn. 41(2):135-142.

Valero D, Martínez-Romero D, Valverde JM, Guillén F, Castillo S, Serrano M (2004). Could the 1-MCP treatment effectiveness in plum be affected by packaging? Postharvest Biol. Tech. 34(3):295-303.

Vom Saal FS, Akingbemi BT, Belcher SM, Birnbaum LS, Crain DA, Eriksen M, Farabollini F, Guillette LJ Jr, Hauser R, Heindel JJ, Ho SM, Hunt PA, Iguchi T, Jobling S, Kanno J, Keri RA, Knudsen KE, Laufer H, LeBlanc GA, Marcus M, McLachlan JA, Myers JP, Nadal A, Newbold RR, Olea N, Prins GS, Richter CA, Rubin BS, Sonnenschein C, Soto AM, Talsness CE, Vandenbergh JG, Vandenberg LN, Walser-Kuntz DR, Watson CS, Welshons WV, Wetherill Y, Zoeller RT (2007). Chapel hill bisphenol A expert panel consensus statement: Integration of mechanisms, effects in animals and potential to impact human health at current levels of exposure. Reprod. Toxicol. 24(2):131-138.

Wikström F, Williams H (2010). Potential environmental gains from reducing food losses through development of new packaging – a life-cycle model. Packaging Tech. Sci. 23:403-411.

World Packaging Organization (2008). Market statistics and future trends in global packaging. World Packaging Organisation/PIRA International Ltda. P. http://www.worldpackaging.org/publications/documents/market-statistics.pdf (Accessed 25.10.2012).

3

Post-harvest losses in mandarin orange: A case study of Dhankuta District, Nepal

Rewati Raman Bhattarai[1], Raj Kumar Rijal[2] and Pashupati Mishra[1]

[1]Tribhuvan University, Central Campus of Technology, Hattisar, Dharan, Nepal.
[2]Food Research Officer, Regional Food Technology and Quality Control, Hetauda, Nepal.

Worldwide postharvest fruit and vegetables losses are as high as 30 to 40% and even much higher in developing countries like Nepal. A systematic survey was conducted to assess the extent of loss due to post harvest conditions in oranges at field, transport, storage and market levels during October to January, 2011. The survey data were collected using oral questionnaires, personal interviews, group discussions and informal observation in the field and *Krishi Bazar,* Dharan. The production of oranges in Dhankuta this year was found to be reduced by 40 to 50% than previous year which was observed to be followed by alternate pattern. Consequently, the price was doubled this year. The post harvest loss was found to be 46% from harvesting to distribution. The losses during harvesting, transportation, grading, packaging and marketing were found to be 7, 25, 3, 1 and 5% maximum, respectively. The storage losses were found to be 5% during 2 to 4 days in *Krish Bazar* while 40.1% during 21 days experimental condition in room. The losses in experimental condition comprised 15.02% evaporation loss, 14.34% pathological loss and 10.74% other losses. The most observed disease was fungal attack in oranges. Reducing postharvest losses is very important; ensuring that sufficient food, both in quantity and in quality is available to every inhabitant in our planet. Postharvest horticulturists need to coordinate their efforts with those of production horticulturists, agricultural marketing economists, engineers, food technologists, and others who may be involved in various aspects of the production and marketing system.

Key words: Orange, survey, post harvest loss, storage, loss reduction, Nepal.

INTRODUCTION

Fruits constitute an important item of our food and they play a significant role in the human diet through the supply of vitamins and minerals (Prabhakar et al., 2004). There is a growing need of fruit and vegetables on world market. Therefore, fruit and vegetables have a fast growing market share (Kabas, 2010). Post-harvest losses during handling, transport, storage and distribution are the major problems in agrarian economy, especially in perishable fruits and vegetables. Besides resulting in low

per capita availability and huge monetary losses, these increase transport and marketing costs also (Subrahmanyam, 1986). In Nepal, post harvest loss for fruits is 20 to 35% (Kaini, 2000). The mid-hill region (1000 to 1500 m altitude) has a comparative advantage in the cultivation of citrus fruits, especially mandarin and sweet orange. These are grown in almost all mid-hill areas (900 to 1400 m) of the country between 26° 45' and 29° 40' latitude and 80° 15' and 88° 12' longitude. The mid-hill region of Nepal, which accounts about 1.5 million ha is quite suitable for citrus cultivation (Shrestha and Verma, 1998). Mandarin orange contributes to augmenting food availability, improvements in nutrition, generation of employment and income and also helps in maintaining the environment (Anonymous[1], 2009).

Citrus, particularly the mandarin orange is the most important and highly commercial fruit crop in the hills of Nepal Anon[1], 2009. Mandarin is a group name for a class of oranges with thin, loose peel. These are treated as members of a distinct species, *Citrus reticulata Blanco*. Mandarins include a diverse group of citrus fruits that are characterized by bright coloured peel and pulp, excellent flavor, easy-to-peel rind and segments that separate easily (Parashar, 2010). In Dhankuta district, the area under oranges production is 652 ha and the major VDCs producing oranges are *Khoku, Chhintang*, Dhankuta, *Khuaphok, Belhara, Maunabudhuk* (*Statistical information on Nepalese Agriculture, MOAC, 2006-2007*). Significant damages are being occurred in the process of growing, harvesting, and post-harvesting on oranges. According to Kabas (2010) this causes losses on farmers and also on country's economy. This also causes decrease in food availability. Dhankuta is a major citrus producing area with superior quality mandarin oranges. To the best of our knowledge, the recent data on post harvest loss of mandarin orange of Dhankuta district is not available. Similarly, the survey data on production and quality of mandarins produced now is not available. No information regarding the post harvest losses during harvesting, transportation and storage is available for mandarin grown in Dhankuta. Similarly, losses due to disease and pests are not known till date. The present study thus was to estimate the post harvest loss of oranges particularly in trade route of Dhankuta-Dharan.

MATERIALS AND METHODS

A systematic survey was conducted to assess the extent of loss due to post harvest conditions in oranges at field, transport, storage and market levels during October to January, 2011. The survey data were collected using oral questionnaires, personal interviews, group discussions and informal observation in the field and *Krishi Bazar,* Dharan. The questions and discussions dealt about losses during harvesting, transportation, storage and distribution of oranges. It also included information on past and present orange production and price fluctuation. The loss due to evaporation and disease was assessed at 3 days intervals for a month (December to January 2012) at room. The fruits were bought and kept in storage conditions for a month in room and accessed for storage losses. As regards phyto-pathological disorders, the Laboratory examination was not done. Only visual examination of infected surface was carried out.

The nature and extent of postharvest losses due to insect, mechanical damage and spoilage were quantified by obtaining on the samples from harvesting, collection centers and retail market. Lots of 30 to 40 kg was randomly sampled in three replications at collection center. Similarly samples were collected in triplicate from retail market.

RESULTS AND DISCUSSION

Survey findings

The data on losses during harvesting, transportation, storage and distribution of oranges were collected. Most orange producers in Dhankuta district reported that the production of oranges this year was about 40 to 50% less than previous year. Consequently, the price of orange was increased by two times. The reason for less orange production this year was reported by most farmers to be due to climatic changes like environment, soil and rainfall. They also reported that oranges are generally produced in alternate years. They expected that more oranges will be produced next year compared to this year. The most preferable orange from survey was found to be from *Khoku* and it was highly priced. These oranges in *Khoku* are pesticide free. Generally orange tree gives fruit in five years and in *Khoku*, orange tree of 200 years old was also found. About 50% of surveyed producers reported that the qualities of oranges are declining. According to survey, this was due to over farming, use of grafting to produce new variety and diseases like scaling, *kharane* (Power Mildew), etc. The survey findings on losses during harvesting, transportation, storage and distribution are given subsequently.

Losses during harvesting

The nature, causes and percentage of losses in field visit during harvesting are given in Table 1.

Losses during transportation

The oranges were found to be transported from field to *Krishi Bazar* in mini trucks or tractors or *doko* (a basket made of bamboo) covering with plant leaves for cushioning. The dealers in *Krishi Bazar* reported that each day they receive 4 to 5 trucks of oranges having 15 to 25 quintals from different areas. From the survey, the losses in transportation of Mandarins from field to *Krishi Bazar* was observed to be around 25% from all areas and from *Krishi Bazar* to consuming markets/ terminal markets was negligible (about 1%). The main nature of transportation losses were reported to be damage, bursting, peeling, shrinkage, etc.

Losses during packaging

Improper handling, overloading and dropping of the fruits during weighment are the main factors of losses during packaging. The dealers generally sell the produce after packing in polythene bags in loose condition and overload the bags. So losses of upto 0.2 to 1% occurred.

Table 1. Nature, causes and percentage of losses.

S/N	Area	Losses during harvesting		Percentage
		Nature/type	Cause	
1	Khoku	Bruises, injuries, rottening	Improper handling, falling of fruits, fungal attack	2.5
2	Chhintang	Scratches, puncture of fruits, insect infestation	Improper harvesting, infestation	2-3
3	Dhankuta	Physical damage, bruises, cuts, over ripen, immature	Improper harvesting, falling of fruits, improper handling, monkey attack	4-5
4	Budhimorang and Khuaphok	Bruises, scratches	Improper handling	1-2
5	Bhirgaon	Damaged, insect infestation, bruises	Due to fall of fruits, cracked and spoiled fruits	4-7
6	Maunabudhuk	Bruises, softness, puncturing of fruits	Falling from height, open stacking, Improper plucking	5.5

Losses in the markets

The sellers generally do not undertake any special preparation for markets. They simply keep the fruits for sale in the markets. Most of the losses in the markets are at wholesaler, commission agents and retailers level. This may be due to the fact of over ripening of fruits, improper handling during packaging, falling of the fruits; shrinkage and rottening of fruits etc. according to market survey, losses upto 5% were reported.

Losses during grading in Krishi Bazar, Dharan

Scientific grading of Mandarins is not done by any of the dealers. At the most, the fruits are sorted out according to size, shape and color. There are hardly any losses in the process of grading, but due to improper handling of the fruits or due to over ripened fruits, there are likely losses of the fruits at the time of grading. The percentage of losses may be 0.5 to 3.

Evaluation of losses in oranges during storage

Storage losses were evaluated in two ways. In one way, survey was performed in *Krishi Bazar* and in other; simulation of storage condition was done in room. During survey, no cold storage facility was reported. The dealers in *Krishi Bazar* store oranges in cemented floor on plastics, straw bed, *doko*, crates, and sacks and on plant leaves bed. The oranges do not remain there for long time. Only storage for 2 to 4 days was observed. During this time, losses upto 5% was reported by dealers. This loss was due to evaporation loss, rottening, breakage, etc.For the assessment of post harvest storage losses in room condition, oranges were evaluated according to the following categories:

a. Oranges without shrinkage or diseases and disorders.
b. Oranges without pathological diseases.
c. Oranges without disorders or evaporative loss.

In every case, losses were evaluated after having divided oranges visually. The results obtained are shown in the Figure 1. They represent the average of some measurements carried out in December to January 2012 in storage. The total loss during storage for 21 days was found to be 40.1%. Paudel et al. (2004) verified that the maximum loss in mandarin oranges stored in improved cellar stores was 23% on a weight basis and 15% in number for a 120-day storage period. The pattern of evaporative losses during storage is given in Figure 2.

Losses due to diseases in oranges during storage

No significant disease attack was found during storage in *Krishi Bazar*. This may be due to short storage period in that area. But the dealers reported that sometimes if storage is for long period due to strike and low demand, blackening and rottening occurred. In our experimental study, a pathological loss of 14.34% was observed which was due to surface infection by fungus. Prabhakar et al. (2004) reported that the extent of loss in local mandarin varied from 15.1 to 22.1% in which spoilage due to green mold ranged from 8.5 to12.0% followed by blue mold of 4.3 to 7.0%. Other diseases of lesser importance were sour rot, anthracnose and stem end rot. The lower value in present study may be due to storage performed in small batch.

During storage, *Penicillium digitatum* and

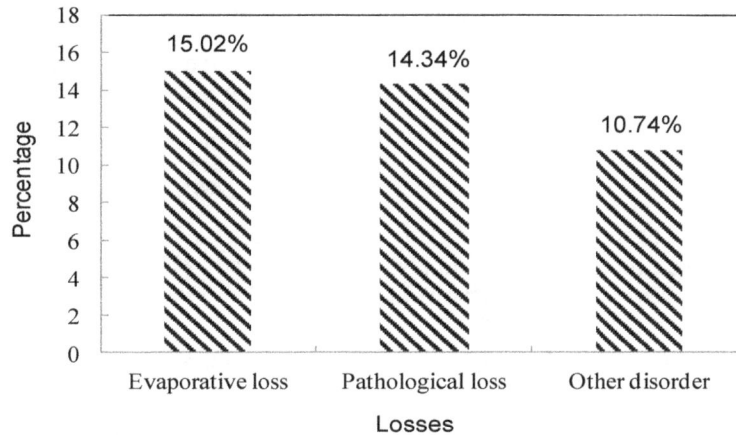

Figure 1. Losses in oranges during storage. The values are means of three batches.

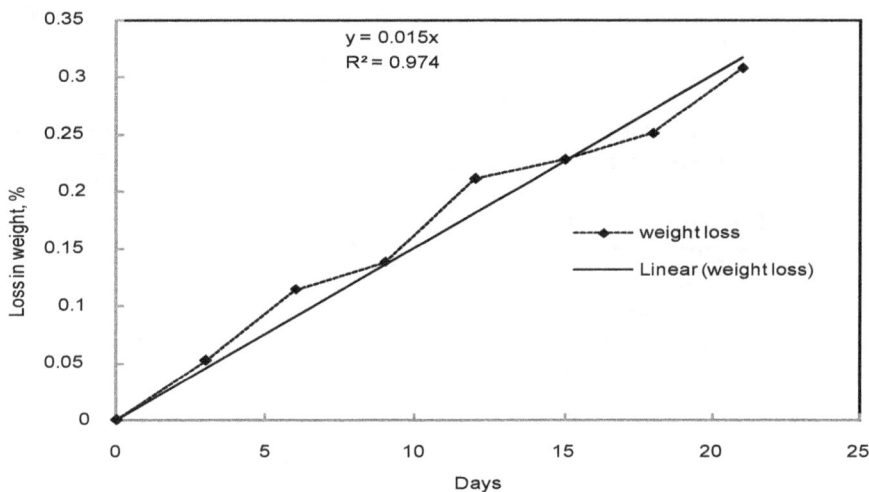

Figure 2. Trend in evaporative loss during storage. Values are means of three batches.

Penicillium italicum account for severe losses in mandarin worldwide (Prabhakar et al., 2004). Mandal (1981) has reported from West Bengal as high as 35% loss due to *Penicillium* spp. alone in mandarin orange. Losses from post-harvest diseases in oranges can be both quantitative and qualitative. These diseases are mainly caused by fungi and bacteria. Initially, only a few pathogens may invade and break down the tissue systems, followed by subsequent attack of weak pathogens. High temperature and humidity accelerate the process of post harvest decay by microorganisms.

Technology for reducing post-harvest losses

A systematic analysis of each commodity production and handling system is the logical first step in identifying an appropriate strategy for reducing postharvest losses. Also, a cost-benefit analysis to determine the return on investment in the recommended postharvest technologies is essential (Kader, 2005).

Post-harvest losses can be minimized by adopting certain pre-harvest strategy and post-harvest management/technology. The principal pre-harvest strategy and postharvest technology for reducing the post-harvest losses are as follows (Parashar, 2010):

(i) Pre-harvest treatment
(ii) Correct stage of harvesting
(iii) Proper harvesting method
(iv) Proper curing
(iii) Washing, cleaning and grading;
(iv) Scientific packing
(v) Pre-cooling

(vi) Cold storage
(vii) Suitable means of transport and
(viii) Efficient marketing.

Conclusions

Minimizing postharvest losses of oranges is a very effective way of reducing the area needed for production and/or increasing food availability. The production of oranges in this year was reduced by 40 to 50% than previous year. Consequently, the price was doubled. Production follows alternate cycle each year.

The losses during harvesting, transportation, grading, packaging and marketing were found to be 7, 25, 3, 1 and 5% maximum respectively. The storage losses were found to be 5% during 2 to 4 days in *Krish Bazar* while 40.1% during 21 days experimental condition in room. The losses in experimental condition comprised 15.02% evaporation loss, 14.34% pathological loss and 10.74% other losses. The most observed disease was fungal attack in oranges.

During the process of distribution and marketing, substantial losses are incurred which range from a slight loss of quality to total spoilage. Post-harvest losses may occur at any point in the marketing process, from the initial harvest through assembly and distribution to the final consumer. The causes of losses are many: Physical damage during handling and transport, physiological decay, water loss, pathogens, etc. Reduction of post-harvest losses reduces cost of production, trade and distribution, lowers the price for the consumer and increases the farmer's income.

Reducing postharvest losses in oranges is very important; ensuring that sufficient food, both in quantity and in quality is available to every inhabitant in our planet. Postharvest horticulturists need to coordinate their efforts with those of production horticulturists, agricultural marketing economists, engineers, food technologists, and others who may be involved in various aspects of the production and marketing system.

REFERENCES

Anonymous[1](2009).   www.agribiz.gov.np/publications/feasibilty-final-report.pdf. Accessed on 9th September, 2011.

Kabas O (2010). Post-Harvest Handling of Orange. Available in www.batem.gov.tr/yayinlar/bilimsel_makaleler/mekanizasyon/8.pdf. Accessed on 13th September, 2011.

Kader AA (2005). Increasing Food Availability by Reducing Postharvest Losses of Fresh Produce. Available in http://ucce.ucdavis.edu/files/datastore/234-528.pdf. Accessed on 11th September, 2011.

Kaini BR (2000). Country Paper on Post-harvest Techniques for Horticultural Crops in Nepal, In Report of the APO Seminar on appropriate Technologies for Horticultural Crops, held in Bangkok from 5–9 July 1999. Asian Productivity Organization, 2000, pp. 11–218.

Mandal NC (1981). Post harvest diseases of fruits and vegetables in West Bengal, Ph. D. Thesis, *Visva Bharati Bhidan Chandra Krishi Vishwavidyalaya, Kalyani*, West Bengal, pp. 210.

MOAC (2006-2007). Agri-Business Promotion and Statistics Division, MOAC, Statistical Information on Nepalese Agriculture.

Parashar MP (2010). Post'-Harvest Profile of Mandarin. Available in http://agmarknet.nic.in/preface-mandarin.pdf. Accessed on 1st October, 2011.

Prabhakar K, Raguchander T, Parthiban VK, Muthulakshmi P, Prakasam V (2004). Post harvest fungal spoilage in mandarin at different levels of Market. Madras Agric. J. 91(7-12):470-474.

Shrestha PP, Verma SK (1999). Development and Outlook of Citrus industry in Nepal. In: Proceedings of National Horticulture Workshop. Nepal Horticulture Society, Kathmandu, Nepal, pp. 48-57.

Subrahmanyam KV (1986) Post-harvest losses in horticultural crops: An appprisal. Agricultural Situation India 41:339-343.

4

Pollen spectra of honeys produced in Algeria

Samira Nair[1], Boumedienne Meddah[1] and Abdelkader Aoues[2]

[1]Laboratory Research on Biological Systems and Geomatics, Faculty of Nature and Life, University of Mascara, Algeria.
[2]Laboratory of Experimental Biotoxicology, Biodepollution and Phytoremediation, University of Es-Senia, Oran, Algeria.

The objective of this study was to evaluate the quality of 10 honey samples produced in Algeria. The samples were prepared using the methodology described by Louveaux and co-workers. Honeys were considered to be monofloral whenever the dominant pollen was found to be over 45% of total pollen. The results obtained in the present study show the variability of the honey samples. The botanical families Myrtaceae, Rutaceae and Lamiaceae are most frequently found. The identified pollen spectrum of honey confirmed their botanical origin.

Key words: Honey, Algeria, pollen spectrum, botanical origin.

INTRODUCTION

Codex Alimentarius Commission (2001) defines honey as "the natural sweet substance produced by honey bees from nectar of blossoms or from secretions of living parts of plants or excretions of plant sucking insects on the living part of plants, which honey bees collect, transform and combine with specific substances of their own, store and leave in the honey comb to ripen and mature".

Diversity of vegetation in Algeria makes possible diversification of apicultural production. Beekeeping is practiced mainly in the north of the country, where the floral diversity is ensured almost all the year (Hussein, 2001).

The botanical origin of honey is one of the most important parameters of honey quality (Tucak et al. 1998, 2000, 2004). The quality of honey depends on the multifarious plants that bees use in their nourishment. The honey obtained from different multifarious plants has different characteristics and applications, both in medicine and in food industry. Melissopalynology is an important tool in determining the floral sources upon which the bees foraged to produce honey (Ohe et al.,2004; Lieux 1975, 1977; Louveaux et al. 1970; Sawyer 1988). Each flower species has a unique pollen grain which, using proper techniques, may be studied to determine the geographical origin and major floral sources of the honey (Lieux, 1972; Jones et al., 2001).

Pollen analysis provides some important information about honey extraction and filtration, fermentation (Russmann, 1998), some kinds of adulteration (Kerkvliet et al., 1995) and hygienic aspects such as contamination with mineral dust, soot, or starch grains (Louveaux et al., 1978).

The present study was carried out to determine the critical analysis of different honey samples to identify the pollen types in honey samples of Algeria.

MATERIALS AND METHODS

Sample collection

Ten samples of honeys produced in various regions of Algerian North were collected from beekeepers in 2006. The samples were stored in a refrigerator in airtight plastic containers until analysis.

Pollen analysis

A microscopic pollen analysis of the honey samples was performed according to the method described previously (Louveaux et al., 1978) and (Ohe et al., 2004). Ten grams of honey were dissolved in

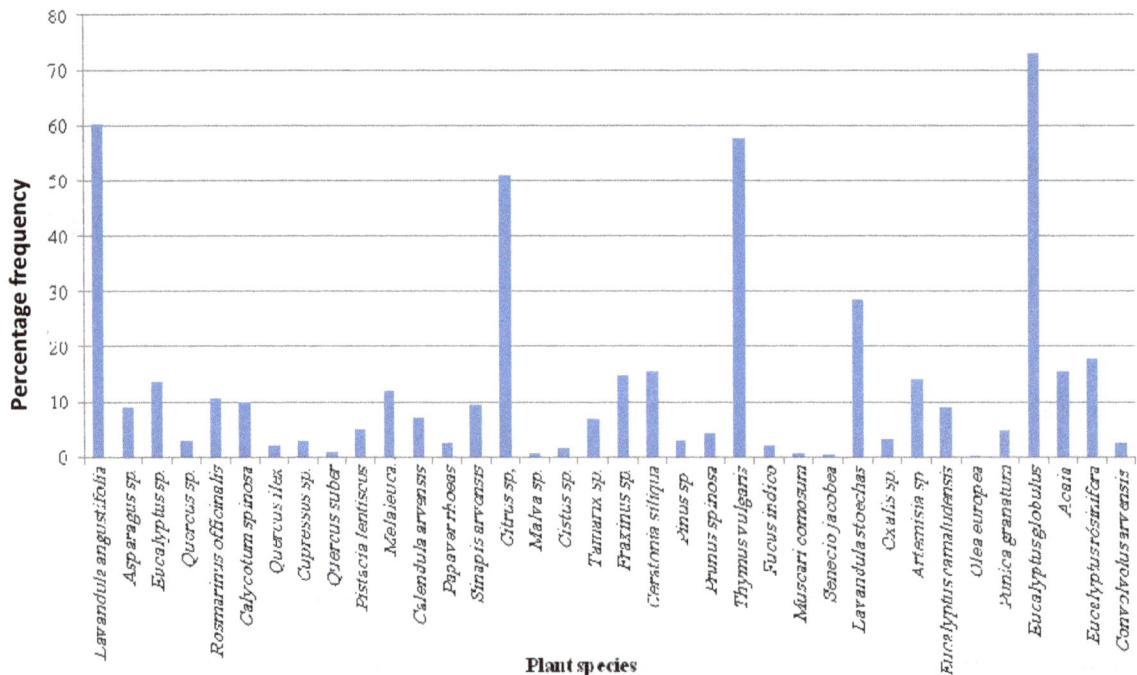

Figure 1. Pollen spectrum of Algerian honey samples honeys.

20 ml of distilled water. This mixture was divided into two centrifuge tubes of 15 ml, and centrifuged for about 5 min, at low speed. Distilled water was again added to the sediment, repeating the previous operation. Approximately 5 ml of glycerine–water 1:1 were added to the sediment, and it was left to rest for 30 min. After this time, the sample was centrifuged. The sediment was removed with aid of a stylet, embedded in glycerine jelley and deposited on a microscopic slide, sealing with paraffin wax.

For the identification of pollen types and the interpretation of pollen spectra, specific training and extensive experience are required. A collection of reference pollen slides and photographic atlas are very helpful (Sawyer, 1988; Maurizio and Louveaux, 1965; Ricciardelli, 1998).

Statistical analysis

To classify honeys, statistical method such as cluster analysis was applied. This method is frequently used to screen data for clustering of samples. The main goal of the hierarchical agglomerative cluster analysis was to spontaneously classify the data into groups of similarity (clusters), searching objects in the n-dimensional space located in the closest neighborhood and to separate a stable cluster from other clusters (Simeonov et al., 2007).

Cluster analysis was displayed in order to find similarities between the honey samples. In order to do that, Ward's hierarchical cluster method for pattern recognition was used.

RESULTS AND DISCUSSION

For the identification of pollen types and the interpretation of pollen spectra, specific training and extensive experience are required. A collection of reference pollen slides and photographic atlas are very helpful (Maurizio and Louveaux, 1965; Sawyer, 1988; Ricciardelli d'Albore,

1997, 1998).

The results of microscopical analysis of the sediment from the honeys used in this work are briefly summarized. Percentages are always referred to pollen from nectar plants.

A total of 36 pollen taxa were discovered and identified in the analyzed honey samples, from the 10 studied samples, 70% were unifloral honeys from *Eucalyptus globulus*, *Thymus vulgaris*, *Citrus* sp. and *Lavandula angustifolia*, 30% multifloral honeys with a high percentage of *Lavandula stoechas* (28,49%) (Figure 1). Predominant pollen is found in 7 samples and honey samples of Makda (S3), Hacine (S4) and Ain faress (S6) are polyfloral honeys. More than 9 pollen types are found in honey samples of Ménaouer (S1), 8 in samples of Makda (S3) and Freguig (S5) and 5 to 7 in the others (Table 1).

Quantitative analysis has shown low pollen concentrations in the studied honey samples, 5 samples belonged to the class I of representatives (under-represented honeys, with less than 20,000 pollen grains in 10 g honey, 5 to the II class (normal honeys, with 20,000 to 100,000 PG/10 g). Our results are quite in agreement with Ouchemoukh et al. (2007), these authors found in their study of 11 Algerian samples lower PG/10 g values, ranging from 20×10^3 till 40×10^3. Their samples were collected in various regions of the province Bejaia. The results of a study of Makhloufi et al. (2010) on the pollen richness of 66 Algerian honeys in which the values for the PG/10 g for the classes I, II, III and V were 33, 40.9, 22.7 and 3%, respectively.

The results of qualitative pollen analysis indicate the

Table 1. Pollen content in the honey samples (%).

Pollen taxa	Samples									
	S1	S2	S3	S4	S5	S6	S7	S8	S9	S10
Lavandula angustifolia					60.33					
Asparagus sp.			9.37			8.99				
Eucalyptus sp.	15			20		22.96	9.88	0.57		
Quercus sp.						3.02				
Rosmarinus officinalis						18.95			2.51	
Calycotum spinosa						10.1				
Quercus ilex					2.05					
Cupressus sp.					3.08					
Quercus suber					1.02					
Pistacia lentiscus	5.98				9.26			0.14		
Melaleuca Leucadendron	11.98			15	15			5.97		7.78
Calendula arvensis				7.02	7.5					
Papaver rhoeas				2.26	1.76					3.5
Sinapis arvensis		9		9.52			10.44			
Citrus sp,		46		45			62.5			
Malva sp.	0.43			1.19						
Cistus sp.			1.5							
Tamarix sp.			7.03							
Fraxinus sp.			14.84							
Ceratonia siliqua			15.62							
Pinus sp.		6.5	1.95					0.11		
Prunus spinosa	4.44									
Thymus vulgaris	57.91									
Fucus indico	1.98									
Muscari comosum	0.78									
Senecio jacobea	0.59									
Lavandula stoechas		21				35.98				
Oxalis sp.		3.5								
Artemisia sp.		14								
Eucalyptus camaludensis							12	6.98	12	5
Olea europea							0.18			
Punica granatum							5			
Eucalyptus globulus								86.2	52.39	80.22
Acaia									15.61	
Eucalyptus résinifera									18	
Convolvolus arvensis										2.5

diversity of resources utilized by honeybees in the region of investigation. The botanical families Myrtaceae, Rutaceae and Lamiaceae were most frequently found in the samples. Out of 66 Algerian honey analyzed by Makhloufi et al. (2007) showed that the main botanical species for honey production in Algeria are found to be *Eucalyptus* spp., Umbelliferae (above all *Pimpinella*), *Hedysarum*, Cruciferae, Compositae (mainly *Carduus*), *Trifolium* spp. and, to a lesser extent, *Echium, Rubus* and *Citrus*.

Cluster analysis is comprised of a series of multivariate methods that are used to find true groups of data orstations. In clustering, the objects are grouped such that similar objects fall into the same class.

Figure 2 shows the dendrogram that corresponds to clusters of the observations corresponding to each geographical origin of the honey samples. It was possible to distinguish two different groups. The first group is composed of honey produced in Ain fares; the second cluster clearly creates two separate subgroups; The first subgroup includes the stations: Guetna, Hacine and Bouguirat; however the most common plant species pollen in the samples of these stations was *Citrus* sp. The second includes the stations Sidi Ali, Hadjadj and Sirat, a representative of the *Eucalyptus golobulus* occurred in these samples.

Trees 10 variables
Minimum jump
Euclidean distance

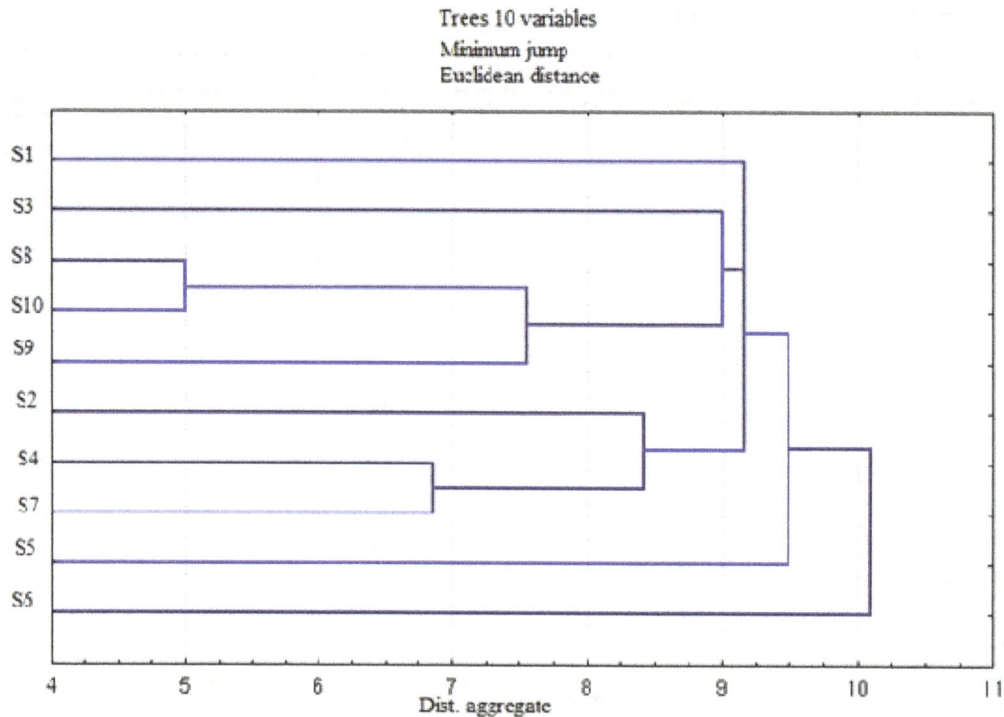

Figure 2. Dendrogram of cluster analysis.

The botanical composition of regional honey depends on the climatic conditions during the apicultural period. Based on the vegetation of each of the areas that the ten samples were obtained, it could be said that these samples were actually produced from beehives around that region because they showed pollen indicative of species indigenous and characteristic of those areas.

Conclusion

Algerian North is characterized by diversified flora which constitutes a valuable source of nectar flow for honeybees. In total, 36 pollen species were identified and the main pollen forms were Eucalyptus globulus, Thymus vulgaris, Citrus sp. and Lavandula angustifolia. Multifloral honeys comprised 30% of the honey samples.

Based on cluster analysis, two different groups of honey were observed according to different pollen types found in the samples. The identified pollen spectrum of honey confirmed their botanical origin.

REFERENCES

Codex Alimentarius Commission (2001). Codex standard 12, Revised Codex Standard

Hussein MH (2001). Beekeeping in Africa. 1- North, East, North-East and West African countries. Apiacta. 36(1-2):32-48, and 81-92.

Jones GD, Bryant B, Goodman DK, Clarke RT (2001). Alcohol dilution of honey. 9th, Inter. Palynol. Cong. Houston. 1996. pp. 453-458.

Kerkvliet JD, Shrestha M, Tuladhar K, Manandhar H (1995). Micro-

scopic detection of adulteration of honey with cane sugar and cane sugar products, Apidologie 26:131-139.

Lieux MH (1972). A melissopalynological study of 54 Louisiana (U.S.A.) honeys. Rev. Palaebot. Palynol. 13:95-124.

Lieux MH (1975). Dominant pollen types recovered from commercial Louisiana honeys. Econ. Botany 29:78-96.

Lieux MH (1977). Secondary pollen types characteristic of Louisiana honeys. Econ. Botany 31:111-119.

Louveaux J, Maurizio A, Vorwohl G (1970). Methods of melissopalynology. Bee World. 51:125-131.

Louveaux J, Maurizio A, Vorwohl G (1978). Methods of Melissopalynology. Bee World. 59:139-153.

Makhloufi C, Kerkvliet D, Ricciardelli D'albore G, Choukri A, Samar R (2010). Characterization of Algerian honeys by palynological and physico-chemical methods. Apidologie 41:509-521.

Makhloufi C, Schweizer P, Azouzi C, Persano OL, Choukri A, Hocine L, Ricciardelli D'Albore G (2007). Some properties of Algerian honey, Apiacta 42:73-80.

Maurizio A, Louveaux J (1965). Pollens de plantes mellifères d'Europe, Union des groupements apicoles français, Paris.

Ohe von der W, Persano oddo L, Piana M, Morlot M, Martin P (2004). Harmonized methods of melissopalynology. Apidologie 35:18-25.

Ouchemoukh S, Louaileche H, Schweizer P (2007). Physicochemical characteristics and pollen spectrum of some Algerian honeys, Food Control. 18:52-58.

Ricciardelli d'Albore G (1997). Textbook of Melissopalynology, Apimondia, Bucharest. Apimondia Publishing House, 1997. 308 pp.

Ricciardelli D'Albore G (1998). Mediterranean Melissopalynology. Perugia: Institute of Agricultural Entomology, University of Perugia.

Russmann H (1998). Hefen und Glycerin in Blütenhonigen – Nachweis einer Gärung oder einer abgestoppten Gärung. Lebensmittelchemie. 52:116-117.

Sawyer R (1988). Honey identification. Cardiff Academic. Press, Cardiff. p. 350.

Simeonov V, Wolska L, Kuczynska A, Gurwin J, Tsakovski S, Protasowicki M, Namiesnik J (2007). Sediment-quality assessment by intelligent data analysis Trends in Analytical Chemistry. 26(4):323-

331.

Stawiarz E, Wroblewska A (2010). Melissopalynological analysis of multifloral honeys from the sandomierska upland aera of Poland. J. Agric. Sci. 54(1).

Tucak Z, Periškić M, Bešlo D, Tucak I (2004). Influence of the Beehive Type on the Quality of Honey. Coll Antropol. 28(1):463-467.

Tucak Z, Puškadija Z, Bešlo D, Bukvić Ž, Milanković Z (1998). Chemical organicleptic honey determination in honey-herbs in The Region Slavonia and Baranja. Sup. 30, Biotehniške fak., Univ. u Ljubljani. pp. 299-302.

Tucak Z, Tucak A, Puškadija Z, Tucak M (2000). Nutritious healing composition of some kinds of honey in Eastern Croatia. Agriculture 6(1):129-132.

5

Changing demographics, expanding urban areas and modified agricultural extents and their impacts on water availability and water quality in Jordan

Khaled A. Alqadi[1], Lalit Kumar[2] and Al-Zu'bi Jarrah[3]

[1]Ecosystem Management, School of Environmental and Rural Science, University of New England, Armidale, NSW, 2351, Australia.
[2]Water Resources and Environment Management, Al Balqa Applied University, Amman, Jordan.

Current water use in Jordan is unsustainable in terms of both supply and quality. The growth in population, primarily as a consequence of pulse immigration stemming from regional conflicts, has led to serious water shortages in urban centers, which is expected to worsen in the future. The agricultural sector is moving towards intensification and a high reliance on irrigation, which is unsustainable in the face of dwindling supplies and rising contamination, principally due to salinity. The decline in field crops is a consequence of climatic fluctuations such as rainfall, while the nature of the plots, often small and isolated, make economies of scale problematic. Unless there is a significant shift in the trend of population growth, and controls on the use of irrigation, Jordan faces inevitable social conflict and irrevocable loss of agricultural land.

Key words: Water contamination, population growth, agriculture, climate fluctuation, salinity.

INTRODUCTION

There has been a significant rise in the population and standard of living in Jordan and this has placed pressure on the domestic water management. The understanding for the need to conserve water resources and manage urban waste has become an issue for large cities and the agricultural lands that surround them (Daigger, 2009). Jordan is suffering significant water pressure on demand and supply with the total available water per capita at less than 142 m^3 per annum (Ministry of Water, 2011). The pressure on the water supply has risen in the urban environment driven by population and industrial growth, while poor agricultural practices and increasing salinity has affected water availability to the agricultural sector. The agricultural sector accounts for the consumption of 60% of available water in Jordan (Ministry of Water, 2011). Increased efficiencies in irrigation as part of reform to on-farm water management has been identified as key to resolving the water issues in Jordan (Table 1). Irrigation is critical to agriculture in Jordan, with less than 4% of the total area under production able to be sustained by average rainfalls, and water is the primary regulator of the size of the agricultural sector.

There are three distinct agricultural regions in Jordan: the rangelands where gazing forms the basis for primary agricultural practice; the highlands with rain-fed dependant field crops and forestry projects; and the Jordan Valley with a high dependence on irrigation (Al-Rahahleh et al., 2007). The use of inappropriate

Table 1. Current water consumption in Jordan by source (million m^3 per year) (Hussein et al., 2010:185).

Area	Surface water	Groundwater	Waste water	Total
Municipal	46	170	-	216
Irrigation	350	313	52	715
Industrial	2.5	22	-	24.5
Others	51.5	3	-	54.5
Total	450	508	52	1010

agricultural practices for the topography of Jordan, and the increase in population, has led to continued long term degradation of land, with overgrazing demonstrated as a significant negative agricultural practice that affects agricultural water budgets primarily through the loss of rainfall to runoff (Raddad, 2005). The use of waste-water for agricultural production is problematic, primarily due to the treatment processes and potential for contamination (InWEnt, 2005).

Jordan is a net importer of food and the Jordanian government has moved to address this issue through intensification of the agricultural sector, particularly in the Jordan Valley. Poor resource management and the lack of consistency in policy have impacted on agricultural production nationally. The Jordanian agricultural sector has been in relative decline with GDP falling from 14.1 percent in 1971 to 3% of GDP in 2006. This decline is mirrored in absolute value declining from a peak of JD 223 million on 1991 to 115 million in 2006 (Royal Commission for Water, 2009).

There have been significant historical events in Jordan that have caused the population to swell, leading to increased pressure on state resources. The population of Jordan has risen from just over 500,000 at independence in 1946 to over 6 million by 2010, causing significant shifts in water use. This increase in population has strained water resources and land use in Jordan. Farm management practices have been modified with changes in land availability towards a more intensive high input set of farm practices. This shift in intensification has led to increased production, but at the cost of increased water demand and significant declines in water quality (Menzel et al., 2009). The quality of water has also been affected by change in climate with droughts leading to irreversible hyper-salinity. This paper will examine the impact of the increase in population, changes in aspects of agricultural practice and the decline in water availability and quality to all sectors through time. This paper will demonstrate that the increased population and change in agricultural practices have led to a significant degradation of Jordanian water assets.

WATER RESOURCES

Jordan is highly dependent on groundwater resources as it represents a viable supply for 80% of the country (Royal Commission for Water, 2009). The physical boundaries of groundwater differ from the administrative boundaries that have been designated by the various water-resources institutions in the region and this discrepancy makes protection of water resources problematic as regional bodies have differing policies on the draw-down of this groundwater (Royal Commission for Water, 2009). The groundwater in Jordan can be divided into two forms, renewable groundwater which accounts for 4% of total available groundwater and is fed primarily by rain infiltration, and fossil water which constitutes 96% of all groundwater (Hussein et al., 2010). Whereas groundwater is one of the primary water resources available in Jordan, current usage rates are unsustainable or unusable as the underlying bedrock contributes to the high rate of salinity of the water (Dottridge and Jaber, 1999).

The assessment of groundwater needs to be considered on an aquifer by aquifer basis to achieve effective management and reduce the impact of salinity due to depletion (Dottridge and Jaber, 1999; Salameh and Bannayan, 2004). The national approach however, unites all aquifers under one set of policy guidelines for their use and fails to address the particular issues for each aquifer which has led to significant environmental problems such as hyper saline water being drawn to the surface, leading to salinity problems faced in many regions of Jordan (Royal Commission for Water, 2009).

Surface flows from springs occur where the water table intersects the surface topography with concealed discharge, including seepages and it offers localised sources for fresh water. Under natural conditions, the aquifers discharge water proportionally to total annual infiltration that they receive. Recharge into aquifers is derived naturally from infiltration of precipitation, streams, wadis (valleys), lakes, ponds, or other impoundments that seep through the soil profile. Recharge is induced by anthropogenic activities such as injection wells, wastewater pond seepage, or irrigation seepage, or distribution pipe leakage (Al Khandak, 2002).

Springs are valuable sources of fresh water in Jordan and vary greatly in the volume of water they discharge and their salinity (Raddad, 2005). Springs flowing from water-table aquifers are extremely variable and are influenced significantly by climatic conditions with some

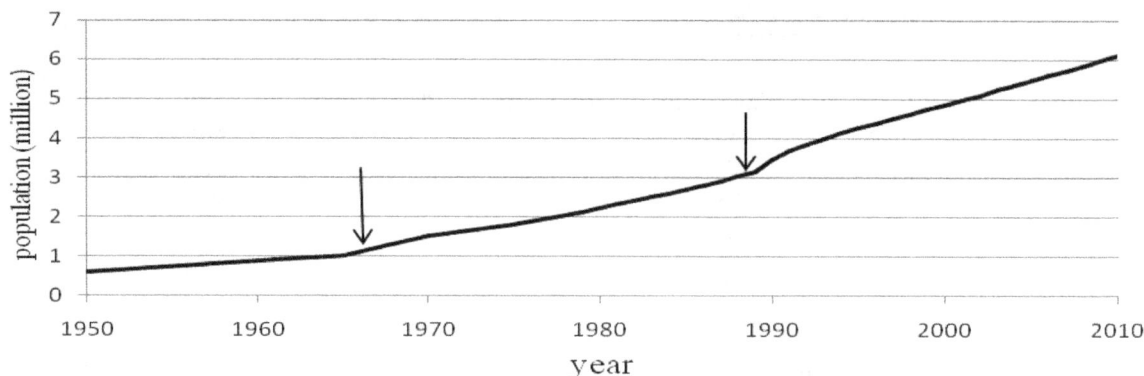

Figure 1. Population trends in Jordan since independence (Source: Department of Statistics, 2011).

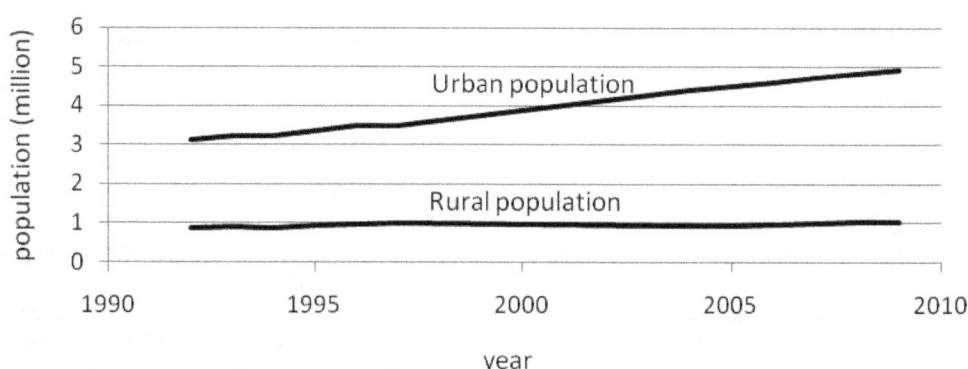

Figure 2. Growth of urban and rural population in Jordan from 1994 to 2008.

springs ceasing to flow during periods of low precipitation. Springs issuing from confined aquifers have larger and more consistent flow rates, and these springs tend to show less influence from climate than do the springs which are derived from the water table.

Surface water in most regions of Jordan drains to the Mediterranean, Red, or Dead Seas and is highly utilised (Carr et al., 2010). There has been an increasing problem of contamination of water by agriculture while greater draw-down of this water resource has led to increasing salinity as a consequence of over exploitation (Royal Commission into Water, 2009).

POPULATION TRENDS

Jordan has the highest population growth rate in the Middle East, averaging near 5% annual growth driven by increasing immigration (Martin, 1999); however this has declined in recent years to 2.8% with localised political stability (Jordan Department of Statistics, 2011) (Figure 1). Figure 1 illustrates two significant periods of rapid population growth: The first commenced in 1966 when the Middle East was in conflict involving by the 1967 Israeli wars, and the second was from 1990 during the

Palestinian uprising and increasing tensions between Israel and Jordan. The Iraq war saw another wave of war refugees entering Jordan with between 300000 and 500000 Iraqis seeking shelter between 2003-2005 (Ababsa, 2010). Therefore, the population of Jordan's urban centres has been artificially swollen by people displaced by regional conflicts. These refugees have tended to aggregate in larger urban areas (Figure 2) placing pressure on existing water infrastructure (Ababsa, 2010). This is particularly relevant given the current political instability in the region, particularly in Syria, associated with the Arab spring uprisings. The population of Jordan has reached critical levels at which existing water infrastructure is currently having difficulties in meeting the needs of the urban centres with demand for sweet water rising by 6.5% annually (Magiera et al., 2006; Martin, 1999).

There has also been a rapid growth in the urban population, primarily a reflection of immigration rather than mass movement of the rural population within Jordan (Figure 2). While the rural population has remained fundamentally constant through time at approximately one million, the urban population has increased from about 3.1 million in 1994 to around 5 million in 2008. This urban growth has not been constant

Table 2. Changes in the population of Jordan through time represented by the 12 governorates) (in '000).

Governorates	Capital	1994	2008	% change
Ajlun	Ajlun	95	134	41
Al-Aqabah	Al-Aqabah	80	127	59
As-Balqa	As-Salt	276	392	42
Al-Karak	Al-Karak	170	228	34
Al-Mafraq	Al-Mafraq	179	275	54
Amman	Amman	1576	2265	44
At Tafilah	At Tafilah	63	82	30
Az-Zarqa	Az-Zarqa	639	871	36
Irbis	Irbis	752	1041	38
Jarash	Jarash	123	176	43
Maan	Maan	77	111	44
Madaba	Madaba	107	146	36

across all urban areas, with some areas having much higher growth rates than others (Table 2). For example, Al-Karak has grown from 170 thousand in 1994 to 228 thousand in 2008 (34% growth over 14 years), while the population of Al-Aqabah has grown from 80 thousand in 1994 to 127 thousand in 2008 (59% growth over 14 years) (Table 2). Amman, the capital city and the largest urban area in Jordan, has grown from 1.58 million in 1994 to 2.3 million in 2008 (a growth rate of 44% over 14 years). This growth has occurred without significant investment in the level of water infrastructure. Maladministration has been identified as one of the main causes for water shortages in urban areas and, although this problem is not recent, it illustrates systemic regulatory failure (Cooley, 1984). The supply of water in urban areas is often of such unreliable nature that many households only have mains-water for as little as one day a week, having to buy water from the market to meet domestic needs (Potter et al., 2010). The spatial extent of Amman has also grown significantly from 1918 to 2010; with most of the change post-1956. Figure 3 represents the growth of Amman from 0.32 km^2 in 1918 to over 162.94 km2 in 2010, and this rise in urban areas is mirrored in declines in fertile land from 383.86 km^2 in 1918 to over 297.41 km^2 in 2010, and vacant land falling from 232.42 km^2 in 1918 to over 158.62 km^2 in 2010.

Amman is at the centre of unregulated urban growth that has been precipitated by mass immigration and movement of refugees. This unplanned development has led to increased pressure on green areas and water resources (Ta'any et al., 2009). Prior to the arrival of immigrants from neighboring countries, the land surrounding Amman was considered fertile and suitable for cereal production (Al-Rawashdeh and Saleh, 2006). Prior to 1950, the majority of urbanisation occurred over vacant lands in close proximity to existing dwellings; however recent migration has encroached onto agricultural lands and it is estimated that 86 km^2 of fertile

land has been lost to urban development around Amman from 1918 to 2010, accounting for 23% of the total arable land in the Amman region. Figure 3 clearly illustrates the extent of urban growth around Amman, as compared to 1918.

Table 3 illustrates the change in land use in the greater Amman region from 1918 to 2002. There have been two significant waves of Palestinian migration: The first in 1948 which saw the capital expand 1,284% in size (from 1918 to 1953); the second major population growth spur came in 1967 when the city limits expanded to 101 km^2, representing a rise of 2,280% (from 1953 to 1983). The close of the twentieth century saw the Gulf wars in the Middle East and this lead to the displacement of a large number of Iraqis which swelled the urban area of Amman by 45 km^2.

Amman, at the turn of the last century, housed 5,000 residents, but Amman city and surrounds today covers an area of 618 km^2 with 2.3 million residents with a long historical tradition dating to the Ammonites. This rapid increase in population came with little or no regulation and has placed significant pressures on the water resources in the areas that have been urbanised, particularly as domestic wells become contaminated from infiltration of waste products (Salameh et al., 2002). Coupled with the increase in water demand associated with population growth, Jordan, and Amman in particular, has seen declines in quality as a consequence of inadequate industrial and urban treatment of waste water (Hadadin and Tarawneh, 2007). The water table has risen in areas where water treatment occurs and stabilisation ponds provide seepage which raised the water table while reducing net quality of available water. In areas with rising water tables, there have not been problems with increasing salinity, but the nitrate levels and acidity have risen (Ta'any et al., 2009). This increase in pollutants exacerbates the water quality issues created by the removal of green areas, and a higher rate of runoff

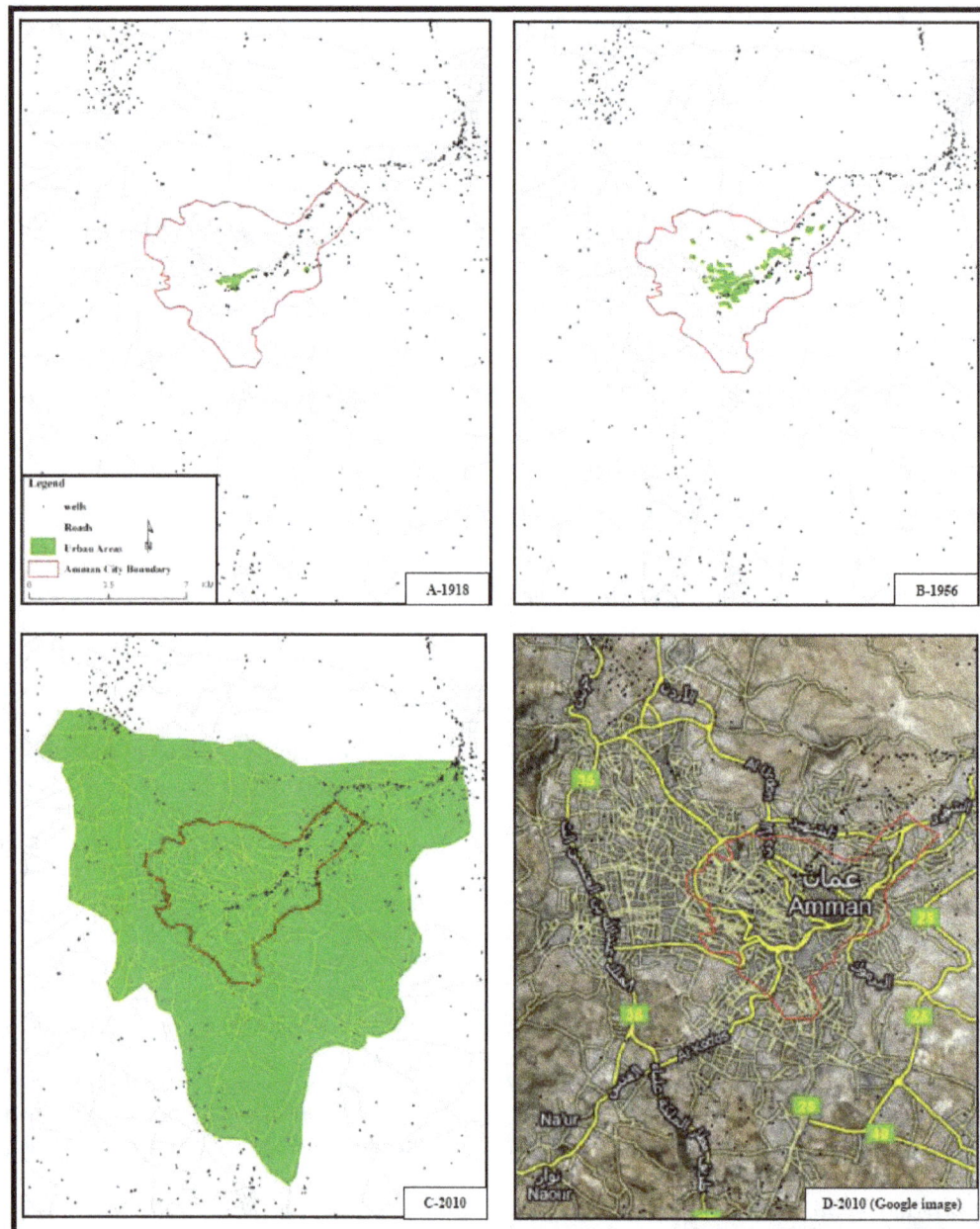

Figure 3. Changes in the spatial extent of Amman between 1918 to 2010 with the consequence of loss of agricultural land as the urban footprint rose from 0.3 km^2 in 1918, to 4.5 km^2 in 1953, and over 170 km^2 in 2010. All maps are to the same scale. Black dots show locations of bore holes and wells.

being captured. Also, one of the most significant threats to urban water supplies is the loss to the system through leakage, theft, and overflows which accounts for 50% of the total domestic supply (Mohsen, 2007; Potter et al., 2007; Rosenberg and Lund, 2009).

AGRICULTURAL LAND USE

Historically, agriculture in Jordan had a seasonal focus and was highly dependent on the annual rain. Land holdings for crops were small and self sufficient. Nomadic grazing accounted for one third of all farming (Baer, 1957; Issawi and Dadezies, 1951). The period immediately after independence in 1946 saw the development of large scale agricultural production through the provision of long term water security for those farmers on small holdings of 20 to 30 dunums (Hazleton, 1979). These reforms often did not take into consideration the significant implications for long term water security but rather focused on long term production growth. However, increasing salinity coupled with

Table 3. Land use in the greater Amman Region 1918-2002 (area in km^2).

Class	1918	1953	983	1996	2002
Urban area	0.321	4.444	105.764	150.764	162.924
Fertile lands	383.856	383.593	331.693	301.346	297.413
Unoccupied area	234.425	230.576	181.237	166.492	158.267

Source: Al Rawashdeh and Saleh, 2006:214).

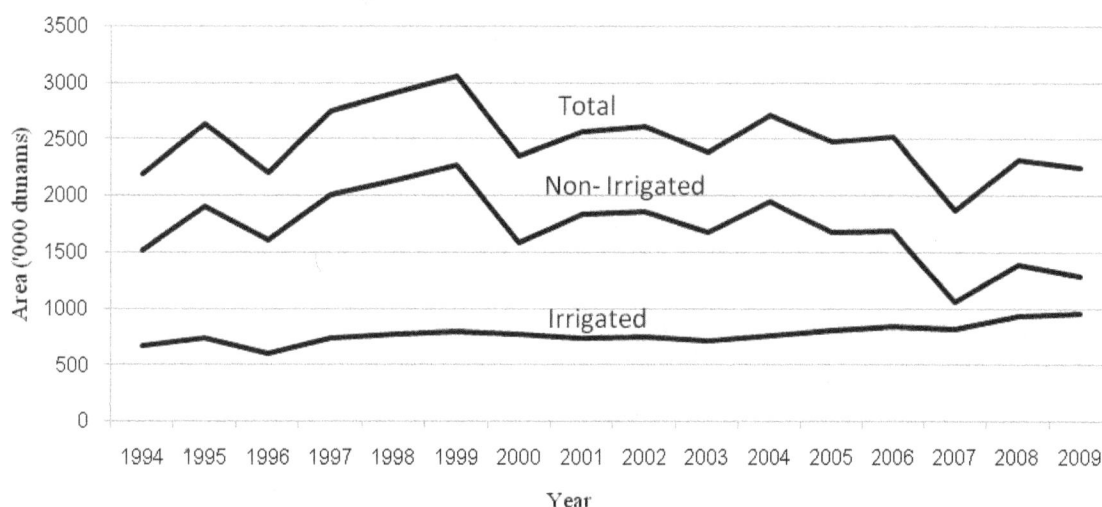

Figure 4. Agricultural land use in Jordan showing trends in irrigated and non irrigated sectors from 1994-2009 (Source Department of Statistics, 2011).

changes in the climate in Jordan, particularly frequent droughts towards the end of the last century, has seen a move away from small rural holdings to larger more intensive farming operations. This has impacted on the demographics within Jordan, with a trend for the rural youth population to migrate to urban centres as the agriculture sector is unable to sustain employment. Access to better education and a perceived higher quality of life are also causes of youth movement to urban areas.

The agricultural sector has also shifted away from non-irrigated, rain-fed agricultural practice, leading to a minor decline in the total area under production. While this decline was occurring, the period 1994 to 2004 (Figure 4) showed considerable fluctuations in the total agricultural production area, and this can be directly attributed to shifts in farm practices such as increasing salinity, irrigation and the use of fertilisers, and the onset of agricultural population pressures.

There has been a long-term trend in the decline in semi-arid and arid crop production, falling to 26.55% in 2008 of the area under field crops in 1961 (WTO). This decline represents a shift from field crops to irrigated tree crops with an accompanying movement from non-irrigated production, which reflects the government agricultural policy direction towards irrigation based food security given the level of climate variability (Al-Adamat et

al., 2007; Davis, 1958). While the production of vegetables has not changed significantly, this is not mirrored by the increasing use of irrigated water for tree crops, particularly after the end of the drought in 2007. Tree crops have become more favored in Jordanian agriculture as a consequence of soil problems and the need to reduce water in irrigation systems for vegetables and other water dependant crops. There is a clear indication that agricultural production is moving away from low water rain-fed cropping systems, such as wheat and barely, to irrigated systems, such as chickpeas and bananas (Al-Adamat et al., 2007).

The total area that is actually cultivated has declined as a consequence of two independent pressures. The first is the increase in population which has usurped available land for urbanization, including dwellings, roads and other infrastructure. The second is the withdrawal from agricultural production marginal lands due to reduced viability as water allocations is diminished and economies of scale leads to reduced profitability driven by fragmented cropping systems and increasing open markets to cheaper products (Jabarin and Epplin, 1994). Intensification of irrigated agricultural land-use near urbanised areas experiencing population growth is being fueled by treated waste water. Intensification of Jordanian agriculture will continue to place pressures on water

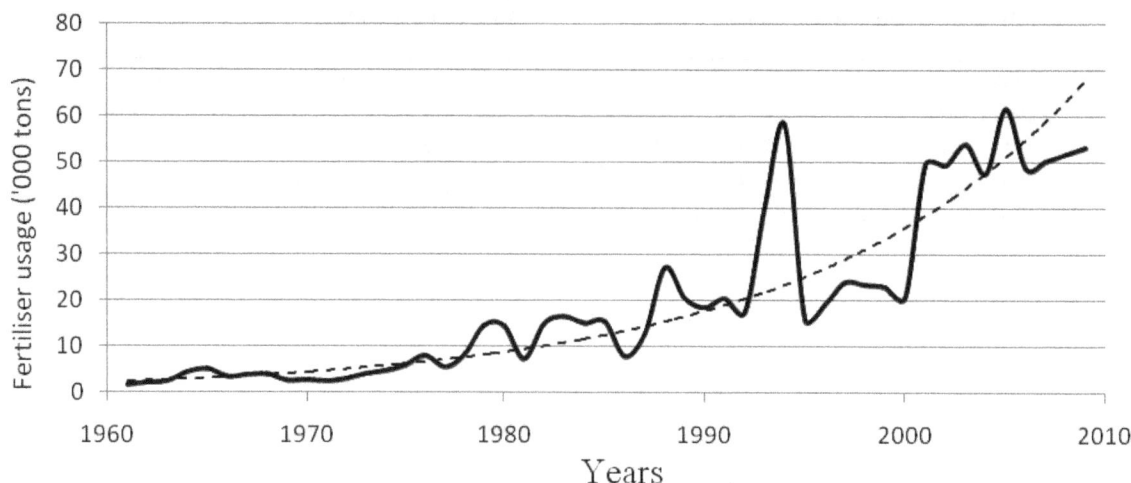

Figure 5. Agricultural fertiliser use in Jordan 1961-2009 (Adapted from Assi and Ajjour, 2009).

allocation, availability and quality, particularly as suitability, climate variability and increased domestic demand all affect water insecurity and the uptake of farming on marginal lands (Millington et al., 1999). Improvements in varietal use has also allowed for the continuation of agricultural production under rain fed conditions in limited parts of the country such as Mashagar (Badarneh and Ghawi, 1994).

One of the most significant changes in farm practices has been in the use of fertilisers. The use of fertiliser is dependent on water availability, with a peak in fertiliser imports prior to the drought of 2003 to 2004, and the long term growth of fertiliser use with intensive and irrigated farming (Figure 5). Fertiliser use has risen steadily from an annual average of less than 5 thousand tons per annum in the 1960s to an average of over 50 thousand tons by 2000. Jordanian fertiliser use is highly variable over time, and the periodic pulse rise in application of fertiliser is indicative of the uptake of improved agricultural farming practices and a rise in intensification. There are four distinct periods in the development of fertiliser use in Jordan during the last half century (Assi and Ajjour, 2009). The pre Green Revolution (per 1970s), saw limited fertiliser application and highlights the simple, low input farming practices of that period. The period of the green revolution during the seventies led to the doubling of fertiliser inputs during that period, much of which was in association with the introduction of drip irrigation systems. The third modernisation period, 1988 to 2003, saw a rise in the total inputs through time almost doubling for the previous period and representing an increasing internal production of fertiliser in Jordan. This new production enabled the overcoming of the supply problems associated with importation. The fourth period, from 2004 to 2010, represents a saturation and plateauing of fertiliser use in agriculture as farmers faced increasing production issues such as water supply and land availability pressures, and increased competition as

a consequence of increasing international trade. There is significant deviation within each period and these can be attributed to rainfall and modernisation, such as the 1992 use of fertilisers through the suction pipes attached to the irrigation system. More recent applications are informed and targeted on achieving effective soil nutrition and avoiding significant pollution from excessive application of fertilisers.

It is estimated that the restrictions on agricultural expansion will occur based on declines in land availability and the ongoing decline in available sweet water and this will continue without significant mitigation works to adjust for increasing salinity. Notwithstanding peace and regional stability, cooperation is the key to effective management of the agricultural sector in Jordan (Menzel et al., 2009).

CLIMATE

Intensive and irrigation agriculture, along with the sweet water resources in Jordan, are highly concentrated in the 4% of the country that is considered semi-arid to semi-humid (Table 4). There are three major regions associated with a semi-arid to semi-humid climate zones in Jordan, and these are located in the Jordan Valley, the Dead Sea, and in the south part of the country near the Red Sea. In contrast, the major urban centres are located in the climate zone predominantly responsible for the desert and arid regions which dominates over 96% of the country, primarily allowing grazing and sporadic cropping (Table 4). The immediate agricultural surrounds of urban centres have benefited in recent years through improved water recycling and treatment centres.

Rainfall in Jordan is seasonal and variable, with the desert regions receiving less than 200 mm of rain annually. In contrast, the semi-humid regions receive more than 500 mm annually. The fundamental climatic

Table 4. Climate and regional classification in Jordan.

Region	Rainfall (mm) (annual average)	Temperature (°C) (annual average range)	Area		Primary agricultural use
			'000 ha	% of Total	
Desert	<200	1-44	8080	90.5	Sporadic grazing
Arid	200-350	3-40	490	5.5	Grazing and limited grains
Semi-arid	350-500	3-33	170	1.9	Grazing, barley and wheat
Semi-humid	>500	3-33	2.1	2.1	Tree crops, wheat, tobacco, sorghum, and summer crops

problem faced in the water crises is that the semi- humid regions account for less than 2.1% of the total country (Table 4). Drought is a significant agricultural and environmental problem, and is an inevitable result of the irregular climate. During a drought, there is reduced infiltration into ground water and a lack of flow in natural water ecosystems; both reductions exacerbate the issue of salinity due to the national soil mineralogy.

WATER CONTAMINATION

There are four primary vectors for water contamination from agricultural production: First, fertilisers, pesticides and herbicides that are leached from farm lands; second, the use of saline and untreated water which accumulates in the soil or is washed into waterways; third, return flow from irrigation which is one of the leading causes of hyper-salinity with the salinity of irrigation water ranging from 250 to 8000 mg/L depending of the aquifer where the water is sourced with the salinity of ground water highly variable between regions; and fourth, the transportation of contaminants with run-off water both into the river systems and through infiltration (Royal Commission for Water, 2009).

The river systems in Jordan have a natural saline character resulting from the underlying soil chemistry. The Jordan River is saline and is not suitable for irrigation or domestic use without treatment. The Zarqa River is only usable for irrigation and drinking outside of dry periods as the contamination from agriculture and waste water renders it unusable during the dry periods. The Yarmouk River is still a sound source for water but is increasingly becoming more stressed as municipal waste water is added. The diversion of inflows by Syria and Israel has exacerbated the problem of reduced flows in Jordanian water courses (Royal Commission for Water, 2009).

In Jordan, landfill leachate is a significant threat to groundwater and aquifers. Such leachate has led to a reduction in irrigation in areas surrounding landfill sites, primarily because of high concentrations of chloride, bicarbonate and nitrate, with nitrate levels rising from 4 mgL^{-1} in 1978 to 423 mgL^{-1} in 1996 (Abu-Rukah and Al-Kofahi, 2001).

CONCLUSION

Jordan is facing significant continued water shortages in the near future without significant reform. The fundamental problems are the rising levels of contamination; draw down of non-renewable aquifers, the intensification of agriculture, as well as a high population growth rate. The decline in the agricultural sector is primarily restricted to the arid and semi arid regions and in field crops. This trend is reversible if climatic conditions improve, such as good rainfall. The long term trend, however, is a move towards irrigated trees and vegetables crops. The use of irrigation water, with high salinity as a consequence of the underlying soil chemistry, will continue to exacerbate the salinity problem which significantly impacts on the agricultural sector (Möller et al., 2007). The increase in urbanisation, driven by pulse immigration and the rural urban shift, has led to water shortages in Jordan's major cities, particularly the capital Amman. Amman has undergone signification growth in both population and spatial extent, and this has led to increased pressure on fresh water supplies and created declines in water quality. As the population increases, the urban footprint increases, and the land available for agriculture declines. The level of water pollution from urban areas is also increasing as treatment infrastructure is unable to process wastewater. The water quantity is also declining as more intensive farming systems and inputs such as fertilisers exacerbate problems with water quality through contaminated runoff and increasing salinity. Without a significant shift in policy, the current population growth and agricultural policies are unsustainable in the context of the currentwater situation and land use.

REFERENCES

Ababsa M (2010). The evolution of upgrading policies in Amman. *Second* International Conference on Sustainable Architecture and Urban Development; Amman.

Abu-Rukah, Y, Al-Kofahi O (2001). The assessment of the effect of landfill leachate on ground-water quality- a case study El-Akader landfill site-north Jordan. J. Arid Environ. 49:615-630.

Al Khandak H (2002). Water Demand Management, Conservation and Pollution Control in Jordan. *Regional Conference* on Water Demand Management, Conservation and Control; Amman.

Al-Adamat, R, Rawajfih, Z, Easter M, Paustian K, Coleman K, Milne E, Falloon P, Powlson D S, Batjes N H (2007). Predicted soil organic carbon stocks and changes in Jordan between 2000 and 2030 made using GEFSOC Modelling System. Agric. Ecosyst. Environ, 122:35-45.

Al- Rahahleh M, Rakabat S, Saraheen W, Awaedeh M, Khalifeh M, Sharma RK (2007). Jordan Accession to WTO: Impact on Agricultural Sector; Afro-Asian Rural Development Organization: Chanakyapuri.

Al-Rawashdeh S, Saleh B (2006). Satellite monitoring of Urban spacial growth in the Amman area. Jordan. J. Urban Plan. Develop, 132(4):211-216.

Assi R, Ajjour R, (2009). *Jordan's second* National Communication to the United Nations Framework Convention on Climte Change (UNFCCC): The Hashemite Kingdom of Jordan, Amman.

Badarneh DMD, Ghawi IO (1994). Effectiveness of inoculation on biological nitrogen fixation and water consumption by lentil under rainfed conditions. Soil Biol. Biochem. 26(1):1-5.

Baer G (1957). Land tenure in the Hashemite Kingdom of Jordan. *Land Econ.* 33(3):187-197.

Carr G, Potter RB, Nortcliff S (2010). Water reuse for irrigation in Jordan:perceptions of water quality among farmers. *Agricultural Water Management,* DOI; 10.1016/j.agwat.2010.12.011, in press.

Cooley JK (1984). The War over Water. Foreign Policy 54:3-26.

Daigger G (2009). Evolving urban water and residuals management paradigms: water reclamation and reuse, decentralisation and resource recovery. Water Environ. Res. 81(9):809-823.

Davis HRJ (1958). Irrigation in Jordan. Economic Geography 34(3):264-271.

Department of Statistics (2011). Population of the Kingdom by Sex: The Hashemite Kingdom of Jordan, Amman.

Dottridge J, Jaber NA (1999). Groundwater resources and quality in northeastern Jordan: safe yield and sustainability. Appl. Geography 19:313-323.

Hadadin NA, Tarawneh ZS (2007). Environmental Issues in Jordan, Solutions and Recommendations. Am. J. Environ. Sci. 3(1):30-36.

Hazleton JE (1979). Land reform in Jordan: The East Ghor Canal Project. Middle Eastern Stud. 15(2):258-269.

Hussein K, Nawaz K, Majeed A, Khan F, Lin F, Ghani A, Raza G, Afghan S, Zia-ul-Hussnain S, Ali K, Shahazad A (2010). Alleviation of salinity effects by exogenous applications of salicylic acid in pearl millet (*Pennistum glaucum* (L.) R.Br.) seedlings. Afr. J. Biotechnol. 9(50):8602-8607.

InWEnt (2005). *Prospects of Efficient Wastewater Management and Water Reuse* in Jordan. InWEnt, Amman Office, Jordan.

Issawi C, Dabezies C (1951). Population movement and population pressure in Jordan, Lebanon, and Syria. *Milbank* Memorial Fund Q. 29(4):385-403.

Jabarin AS, Epplin FM (1994). Impacts if land fragmentation on the cost of producing wheat in the rain-fed region of northern Jordan. Agric. Econ. 11:191-196.

Magiera P, Taha S, Nolte L (2006). Water demand management in the Middle East and North Africa. Management of Environmental: An Int. J. 17(3):289-298.

Martin N (1999). *Population, Households and Domestic Water use in Countries of the Mediterranean* Middle East (Jordan, Lebanon, Syria, the West Bank, Gaza and Israel). International Studies for Applied Systems Analysis, Interim Report, IR-99-032.

Menzel L, Koch J, Onigkeit J, Schaldach R (2009). Modelling the effects of land-use and land-cover change on water availability in the Jordan River region. Adv. Geosci. 21:73-80.

Millington A, Al-Hussein S, Dutton R (1999). Population dynamics, socioeconomic change and land colonisation in northern Jordan. Appl. Geography 19:363-384.

Ministry of Water and Irrigation (2011). Annual Report: The Hashemite Kingdom of Jordan, Amman.

Mohsen MS (2007). Water strategies and potential of desalination in Jordan. *Desalination* 203:27-46.

Möller P, Rosenthal E, Geyer S, Flexer A (2007). Chemical evolution of saline waters in the Jordan-Dead Sea transform and in adjoining area. Int. J. Earth Sci. 96(3):541-566.

Potter RB, Darmame K, Barham K, Nortcliff S (2007). An Introduction to Urban Geography of Amman, Jordan: University of Reading, Geographical. P. 182.

Potter RB, Darmame K, Nortcliff S (2010). Issues of water supply and contemporary urban society: the case of Greater Amman, Jordan. Philosophical Trans. Royal Society, A, 368:5299-5313.

Raddad K (2005). Water Supply and Water Use Statistics in Jordan; IWG-Env International Work Session on Water Statistics: Vienna, June 20-22.

Rosenberg DE, Lund JR (2009). Modelling integrated decisions for a municipal water system with recourse and uncertainties: Amman, Jordan. Water Resour Manage. 23:85-115.

Royal Commission for Water (2009). *Water for life; Jordan's Water strategy, 2008-2022:* Royal Commission for Water the Hashemite Kingdom of Jordan, Amman.

Salameh E, Alawi, M, Batarseh M, Jiries A (2002). Determination of trihalomethanes and the ionic composition of groundwater at Amman City, Jordan. Hydrogeol. J. 10:332-339.

Ta'any A, Tahboub AB, Saffarini GA (2009). Geostatistical analysis of spactiotemporal variability of groundwater level fluctuations in Amman-Zarqa basin, basin Jordan: A case study. Environ. Geol. 57:525-535.

6

Efficacy of plant leaf extracts on the mycelial growth of kolanuts storage pathogens, *Lasiodioplodia theobromae* and *Fusarium pallidoroseum*

S. O. Agbeniyi[1] and M. S. Ayodele [2]

[1]Cocoa Research Institute of Nigeria, P. M. B. 5244, Ibadan, Oyo State, Nigeria.
[2]Department of Biological Sciences, University of Agriculture, Abeokuta, Ogun State, Nigeria.

The efficacy of leaf extracts of five plant species namely: *Glyricidia sepium* (Jacq.) Linn, *Tectona grandis* Linn. *Ocimum gratissimum* Linn. *Anacardium occidentales* Linn. and *Carica papaya* Linn. against storage fungi *Lasiodioplodia theobromae* and *Fusarium pallidoroseum* was evaluated. The potency of these leaf extracts after storage at ambient temperature for 15 and 30 days, respectively was also tested on the radial growth of *L. theobromae* and *F. pallidoroseum*. The results indicate that leaf extracts from *O. gratissimum* and *A. occidentales* are effective in inhibiting the radial growth of *L. theobromae* and *F. pallidoroseum,* respectively. *O. gratissimum* even at 2.5% concentration gave 35.89% mycelial growth inhibition of *L. theobromae* and 10% concentration gave 50.3% mycelial growth inhibition after five days. The extract of *C. papaya* exhibited less antifungal activity than either *O. gratissimum* or *A. occidentales*. Generally, with the exception of *C. papaya* leaf extract, there was no significant difference (P = 0.05) between the fresh leaf extract and the stored extracts in the inhibition of the mycelial growth of either *L. theobromae* or *F. pallidoroseum and the* potency of the leaf extracts was retained even after 30 days of storage at ambient temperature.

Key words: Leaf extracts, kolanuts, storage pathogens, mycelial growth.

INTRODUCTION

Kolanuts are widely cultivated in West Africa, where they are used as stimulants to counteract fatigue, suppress thirst and hunger, and are believed to enhance intellectual activity (Nickalls, 1986). In addition the nuts are exported to Europe and North America, where they are used chiefly as flavouring agents (Oludemokun, 1982). Disease incidence during storage is a major post harvest problem that farmers and kolanut traders seek to solve. The major post harvest pathogens in west Africa for the nut are *Lasiodiplodia theobromae* and *Fusarium* pallidoroseum (Agbeniyi, 2004). *L. theobromae* is a ubiquitous pathogen of tropical woody trees reported to cause shoot blight and dieback of many plant species including black branch and dieback disease of cashew. It has also been reported to cause gummosis of *Jatropha podagrica* (Fu et al., 2007).

F. pallidoroseum has also been implicated as causative pathogen of brown rot disease of kolanuts. Presently, the only control strategy practice by farmers is to remove diseased nut at intervals during the storage period. The use of chemical fungicides is not desirable due to health hazard on the consumers. Plants extracts have previously been used successfully to control other plants diseases in plants (Alkhail, 2005) and they could as well be employed in the control of kolanuts storage rot. This strategy has however not been explored. The present study was initiated to elucidate the efficacy of *Glyricidia sepium* (Jacq.) Linn, *Tectona grandis* Linn. *Ocimum gratissimum* Linn. *Anacardium occidentales* Linn. and *Carica papaya* Linn. leaf extracts on suppression of the the mycelial growth of kolanut storage pathogens, *L. theobromae* and *F. pallidoroseum in vitro.*

MATERIALS AND METHODS

The sources of the plant leaf extracts was *G. sepium* (Jacq.), *T. grandis* Linn, *O. gratissimum* Linn. *A. occidentales* Linn. and *Carica papaya* Linn. Fresh leaves of *G. sepium, T. grandis, O. gratissimum, A. occidentales* and *C. papaya* were washed with tap water and surface sterilized by soaking them for 60 s in 1% sodium hypochlorite (NaOCl) and later rinsed with sterile distilled water. The leaves mere separately crushed with mortar and pestle in distilled water (w/v 25 g/100 ml) and filtered through muslim cloth (Pandey et al., 1982). The crude extract of each plant leaf was then stored in the laboratory at ambient temperature 28 ± 2°C.

The poisoned techniques described by Nene and Thaphiyal (1979) and Tewari and Nayak (1991) were adopted to study the effect of plant leaf extracts on the radial growth of *L. theobromae* and *F. pallidoroseum.* Each plant extract was evaluated at varying concentration: 2.5, 5.0, 7.5 and 10% concentrations (v/v). 2 ml of each of the extract was added to 15 ml sterilized cooled potato dextrose agar (PDA) in 9 cm Petri dishes. Mycelial discs of 8 mm diameter were cut from the periphery of 5-day-old actively growing cultures of *L. theobromae* and *F. pallidoroseum* using sterile cork borer. Each disc was placed in the centre of Petri dishes containing the treated medium. Three replications were maintained for each concentration. Plates without plant extracts were set up to serve as negative controls. The inoculated plates were incubated at 25°C. Radial growth of the colony in each plate was recorded on the third, fifth and seventh day after inoculation by measuring the diameter of the colony along two perpendicular axes. The average of two measurements was taken as the colony diameter (Raghu and Mohanan, 1997). The percent inhibition of *L. theobromae* and *F. pallidoroseum* was calculated by the equation given by Raghu and Mohanan (1997):

$$I = \frac{C-T}{C} \times 100$$

Where, I = Inhibition of fungal growth; C = growth in control, and T = growth in treatment.

Evaluation of storage duration of leaf extracts and inhibitory effect on mycelial growth of isolates

The plant leaf extracts were stored in round-bottom flask at ambient temperature (28±2°C). Each plant extract was stored in two set of flasks, one set stored for 15 days and another set for 30 days at room temperature. After the end of storage period, the extracts were tested for their effect on the mycelia growth of *L. theobromae* and *F. pallidoroseum.*

Three replications were maintained for each concentration 2.5, 5.0, 7.5 and 10.0%. The plates without leaf extracts served as control. The inoculated plates were incubated at 25°C. Radial growth of the colony in each plate was recorded on the third, fifth and seventh day after inoculation along two perpendicular axes.

The average of two measurements was taken as the colony diameter. The percentage inhibition of mycelia growth of *L. theobromae* and *F. pallidoroseum* was calculated as described subsequently.

RESULTS AND DISCUSSION

All the plant leaf extracts evaluated inhibited the mycelial growth of the fungi which proves the antifungal property of the leaf extracts even at lower concentration of 2.5%. For instance, the mycelial growth inhibition of *L. theobromae* at 2.5% for each of the leaf extract ranged between 17.8 to 43.5% after three days, 9.4 to 28.3% after five days and 17.6 to 35.8% after seven days (Table 1). The percentage inhibition of *L. theobromae* in the presence of each of the fresh leaf extract is given in Table 1. Similarly the percent inhibition of *F. pallidoroseum* in the presence of each of the fresh leaf extract at different concentrations is presented in Table 2. Among the leaf extracts of five plants screened, extract of *O. gratissimum* and *A. occidentales* were very effective in inhibiting the growth of *L. theobromae* and *F. pallidoroseum,* respectively. *O. gratissimum* extract even at five percent concentration caused 33.8% mycelial growth inhibition of *L. theobromae* and at ten percent concentration gave 59.3% mycelia growth inhibition in *L. theobromae* after five days (Table 1). This study revealed that antifungal compounds were present in the five leaf extracts screened since they were able to suppress the growth of the microorganisms tested.

The performance of *A. occidentales* was better than that of *O. gratissimum* in the mycelial inhibition of *F. pallidoroseum.* Whereas *O. gratissimum* gave the highest percent mycelial inhibition in *L. theobromae* culture, the mycelia of *F. pallidoroseum* were more sensitive to the extract of *A. occidentales.* For example, at five percent concentration, *A. occidentales* extract caused 34.4% mycelial growth inhibition in *F. pallidoroseum* compared to 24.0% mycelial inhibition obtained in *O. gratissimum* extract after five days (Table 2). There was no significant difference (P = 0.05) between the leaf extract of *T. grandis* and *O. gratissimum* in the inhibition of *F. pallidoroseum* in culture (Table 2).

The extract of *C. papaya* exhibited less antifungal activity than either *O. gratissimum* or *A. occidentales.* For instance, *C. papaya* extract at 5 percent concentration gave 19.7% mycelial growth inhibition of *L. theobromae* and 12% inhibition of *F. pallidoroseum* after five days. Amadioha (1998) also reported that leaf extracts of *C. papaya* was effective in inhibiting the growth of powdery mildew fungus (*Erysiphe cichoracerarum*) *in vitro.* When *C. papaya* extract was tested at 10% concentration, there was no significant difference (P = 0.05) between its performance and *A. occidentales* in the mycelial inhibition of *L. theobromae* (Table 1). However, there was a significant difference between the performance of *C. papaya* extract and *A. occidentales* in the mycelial

Table 1. Percentage inhibition of mycelial growth of *L. theobromae* at different concentrations of leaf extract.

Plant leaf extracts	3rd day Concentration (%)				5th day Concentration (%)				7th day Concentration (%)			
	2.5	5.0	7.5	10.0	2.5	5.0	7.5	10.0	2.5	5.0	7.5	10.0
G. sepium	43.5	48.9	52.0	57.1	25.1	28.1	33.9	36.9	24.1	40.7	44.4	44.4
T. grandis	39.4	43.1	43.5	45.1	21.7	30.6	30.6	31.4	43.5	20.2	36.4	42.8
A. occidentals	17.8	20.3	21.3	44.5	9.4	10.2	16.4	21.4	17.6	19.3	24.9	32.2
O. gratissimum	35.4	51.9	51.7	62.1	29.3	33.8	36.9	59.3	35.8	35.8	44.8	53.3
C. papaya	28.9	46.6	46.6	48.9	9.7	19.7	25.3	33.1	24.1	32.66	33.0	33.7
LSD (0.05)	8.2				6.2				6.6			

Data are means of 3 replicates.

Table 2. Percentage inhibition of mycelial growth of *F. pallidoroseum* at different concentrations of leaf extract.

Plant leaf extracts	3rd day Concentration (%)				5th day Concentration (%)				7th day Concentration (%)			
	2.5	5.0	7.5	10.0	2.5	5.0	7.5	10.0	2.5	5.0	7.5	10.0
G. sepium	12.6	16.9	23.5	28.9	22.8	23.2	27.2	32.4	25.7	32.4	38.2	39.8
T. grandis	10.9	27.9	28.9	31.1	8.0	15.2	33.2	34.8	12.5	22.3	33.7	38.5
A. occidentals	2.7	10.4	23.5	27.3	28.8	34.4	44.0	46.8	22.5	28.9	44.6	49.6
O. gratissimum	20.8	31.7	36.6	41.5	7.2	24.0	30.8	41.2	18.6	23.16	23.1	38.2
C. papaya	4.4	13.7	23.5	26.2	3.0	12.0	28	38	11.1	19.1	30.5	37.2
LSD(0.05)	4.3				7.3				4.3			

Data are means of 3 replicates.

inhibition of *F. pallidoroseum* (Table 2). Further study on the chemical composition of the leaf extracts is recommended.

There was no significant difference (P = 0.05) between the leaf extracts of *G. sepium* and *T. grandis* in mycelial inhibition of *L. theobromae* after seven days (Table 1). However, *A. occidentales* even at 2.5% concentration performed significantly better in the inhibition of growth of *F. pallidoroseum* compared to *T. grandis, G. sepium* or *O. gratissimum,* respectively (Table 2). This trend was observed at five percent and ten percent concentrations. The results presented in Tables 1 and 2 established the sensitivity of *L. theobromae* or *F. pallidoroseum* to the plant leaf extracts None of the extract was found to exhibit stimulatory effect on the mycelial growth of either *L. theobromae* or *F. pallidoseum. O. gratissimum* at ten percent concentration exhibited more than 50% inhibition of mycelial growth of *L. theobromae* after five days (Table 1). However, Shafique et al. (2007) reported that the toxicity of the extracts against a particular fungal species varied with the test plant species. This study also reported different results for each of the leaf extract. Chemical analysis of the leaf extracts will elucidate the differences in the performance of the leaf extracts.

Similarly, *A. occidentales* extracts at ten percent concentration caused 46.8% inhibition of mycelial growth of *F. pallidoroseum* after five days. The percentage

mycelial growth inhibition of *F. pallidoroseum* at 2.5 percent for each of the leaf extract ranged between 2.7 and 20.8% after three days, 7.2 and 28.8% after five days and 11.1 to 25.7% after seven days (Table 2). *T. grandis* at ten percent concentration also caused 43.5% inhibition of mycelial growth in *L. theobromae* after five days.

All the leaf extracts of the five plants after fifteen days of storage at ambient temperature of 28 ± 2°C inhibited the mycelial growth of *L. theobromae* (Table 3) and *F. pallidoroseum* (Table 4). The same trend was observed when the leaf extracts were stored for 30 thirty days (Tables 5 and 6). The antifungal activity of the extracts did not decrease with the period of storage at either fifteen or thirty days. However, the antifungal activity of *A. occidentales* against *L. theobromae* decreased with the period of storage at either fifteen days (Table 3) or thirty days (Table 5). Thus if extract must be stored, refrigeration of the leaf extract is recommended to maintain their efficacy during storage. Similarly, antifungal activity of the extract of *C. papaya* against mycelial of *F. pallidoroseum* decreased after fifteen days (Table 4) and thirty days (Table 6) of storage.

The fresh extract of *C. papaya* at ten percent concentration caused 37.2% mycelial inhibition of *F. pallidoroseum* (Table 2), when stored for either fifteen days or thirty days, it caused 20.5% (Table 4) and15.9% (Table 6) mycelial inhibition, respectively.

Table 3. Percentage inhibition of mycelial growth of *L. theobromae* at different concentrations of leaf extract stored for 15 days.

Plant extracts leaf	3rd day Concentration (%)				5th day Concentration (%)				7th day Concentration (%)			
	2.5	5.0	7.5	10.0	2.5	5.0	7.5	10.0	2.5	5.0	7.5	10.0
G. sepium	45.2	47.3	52.5	54.2	25.5	28.3	35.3	37.4	23.3	39.7	43.8	44.9
T. grandis	37.5	40.2	42.1	45.8	22.1	31.3	31.9	44.6	18.8	35.6	41.7	44.4
A. occidentales	18.8	18.8	20.4	43.1	11.0	14.3	18.1	38.6	16.9	18.8	24.4	30.7
O. gratissimum	52.7	65.2	69.5	71.8	25.9	34.1	40.7	56.0	30.7	34.8	43.0	54.8
C. papaya	29.2	31.5	41.7	44.4	4.9	18.8	23.1	29.7	16.5	29.0	29.6	34.8
LSD (0.05)	7.1				6.2				5.8			

Data are means of 3 replicates.

Table 4. Percentage inhibition of mycelial growth of *F. pallidoroseum* at different concentration of leaf extract stored for 15 days.

Plant extracts leaf	3rd day Concentration (%)				5th day Concentration (%)				7th day Concentration (%)			
	2.5	5.0	7.5	10.0	2.5	5.0	7.5	10.0	2.5	5.0	7.5	10.0
G. sepium	15.3	24.7	26.8	36.3	14	23.2	35.6	40	27.7	35.1	38.5	41.0
T. grandis	13.1	30.0	33.7	35.8	12.4	17.2	30.8	37.2	16.4	24.9	33.3	38.5
A. occidentales	3.2	12.1	23.2	28.9	12.8	13.2	31.6	33.2	22.3	30.7	46.9	50.8
O. gratissimum	14.2	32.6	40.0	41.6	10.2	20.0	24	29.2	17.7	30.3	37.9	39
C. papaya	2.1	2.6	6.8	12.1	2.8	9.2	13.2	21.2	2.5	9.5	14.6	20.5
LSD (0.05)	6.4				5.5				4.9			

Data are means of 3 replicates.

Table 5. Percentage inhibition of mycelial growth of *L. theobromae* at different concentrations of leaf extracts stored for 30 days.

Plant extracts leaf	3rd day Concentration (%)				5th day Concentration (%)				7th day Concentration (%)			
	2.5	5.0	7.5	10.0	2.5	5.0	7.5	10.0	2.5	5.0	7.5	10.0
G. sepium	43.6	44.7	50.6	56.4	25.1	32.5	34.1	36.7	27.8	38.9	43.8	45.1
T. grandis	36.2	37.2	39.4	45.3	20.2	30.5	32.3	40.5	21.9	37.4	42.6	44.8
A. occidentales	14.5	18.5	18.5	41.9	11.1	14.8	20.0	38.2	16.7	22.2	30.0	31.9
O. gratissimum	29.8	51.1	53.8	69.8	24.6	32.8	40.7	58.5	29.2	35.3	43.3	54.4
C. papaya	9.2	29.1	44.7	48.3	18.9	22.9	34.4	39.8	23.1	29.76	31.4	34.4
LSD (0.05)	11.5				6.8				6.2			

Data are means of 3 replicates.

Table 6. Percentage inhibition of mycelial growth of *F. pallidoroseum* at different concentrations of leaf extract stored for 30 days.

Plant extracts leaf	3rd day Concentration (%)				5th day Concentration (%)				7th day Concentration (%)			
	2.5	5.0	7.5	10.0	2.5	5.0	7.5	10.0	2.5	5.0	7.5	10.0
G. sepium	17.1	24.5	27.5	40.9	17..9	28.8	41.6	43.2	25.1	32.9	37.3	42.6
T. grandis	15.5	29.0	31.1	37.8	11.7	21.0	27.2	40.8	15.1	20.9	29.5	38.6
A. occidentales	5.2	20.7	23.8	31.6	13.2	15.6	31.1	34.6	19.8	29.5	43.3	49.6
O. gratissimum	8.3	31.6	40.4	43.0	6.6	9.3	19.5	29.9	13.1	25.1	36.6	37.4
C. papaya	5.2	6.7	8.3	15.5	5.4	7.8	14.4	22.2	3.4	6.8	9.7	15.9
LSD (0.05)	7.6				4.8				6.4			

Data are means of 3 replicates.

Generally, with the exception of *C. papaya* leaf extract, there was no significant difference (P = 0.05) between the fresh leaf extract and the stored extracts in the inhibition of the mycelial growth of either *L. theobromae* or *F. pallidoroseum*.

The leaf extract of *O. gratissimum* demonstrated strong inhibitory effect even after thirty days of storage (Table 5). Similarly, *T. grandis* and *G. sepium* leaf extracts retained their antifungal properties against *L. theobromae* and *F. pallidoroseum* when stored for either fifteen or thirty days. It is evident from the results presented in Tables 5 and 6 that the concentration of each of the tested extract against *L. theobromae* was maintained during the period of storage.

Conclusion

The use of plant leaf extracts of *G. sepium* (Jacq.) Linn, *T. grandis* Linn. *O. gratissimum* Linn. *A. occidentales* Linn. and *C. papaya* Linn. for the control of kolanut storage disease would be seen as a practical solution to the problem encountered by kola farmers and traders during storage of nuts. Also it would be seen as a positive response to public concern about the adverse effects of the use of pesticides on human health and on the environment.

REFERENCES

Agbeniyi SO (2004). Post Harvest incidence and control of fungi associated with kolanuts (*Cola nitida and Cola acuminata*). Ph.D Thesis, University of Agriculture, Abeokuta, Nigeria.

Alkhail AA (2005). Antifungal activity of some extracts against some plant pathogenic fungi. Pak. J. Biol. Sci. 8(3):413-417.

Amadioha AC (1998). Control of Powdery mildew of pepper (*Capsicum annum* L) by leaf extracts *Carica papaya*. J. Herbs, Spices. Med. Plants 6:41-47.

Fu G, Huang SL, Wei JG, Yuan GQ, Ren JG, Yan WH, Cen ZL (2007) First record of *Jatropha podagrica* gummosis caused by *Botryodiplodia theobromae* in China. Australasian Plant Disease Notes 2:75-76. doi: 10.1071/DN07030.

Nickalls RWD (1986). W.F. Daniell (1817 - 1865) and the discovery that Cola –nuts contain caffeine. Pharmaceutical. J. 236:401-402.

Nene YL, Thapliyal PN (1979). Fungicides in Plant Disease control. Oxford and IBH Publishing Co. New Delhi, Bombay, Calcutta. P. 425.

Oludemokun AA (1982). Processing, storage and utilization of kolanuts. Trop. Sci. 24(2):111-117.

Pandey DK, Chandra H, Tripathi NN (1982). Volatile fungitoxic activity of some higher plants with special reference to that of *Callistemon lanceolatus*. Phytopathology 105:175-182.

Raghu PA, Mohanan C (1997). Fungitoxicity of Certain Plant Extracts Against *Phytophthora palmivora* (Bult.) Bult., The causal organism of black pod disease of cocoa. In : 11th International Cocoa Research Conference, Bahia, Brazil. P. 400.

Shafique S, Javaid A, Bajiwa R, Shafique S (2007). Effect of aqueous leaf extracts of allelopathic trees on germination of seed-borne mycoflora of wheat. Pak. J. Bot. 39(7):2619-2624.

Tewari SN, Nayaki M (1991). Leaf extract activity against fungal rice pathogens. Trop. Agric. (Trinidad) 68(4):373-375.

7

Effect of preharvest application of calcium chloride (CaCl$_2$), Gibberlic acid (GA3) and Napthelenic acetic acid (NAA) on storage of Plum (*Prunus salicina* L.) cv. Santa Rosa under ambient storage conditions

S. N. Kirmani[1], G. M. Wani[1], M. S. Wani[1], M. Y. Ghani[2], M. Abid[3], S. Muzamil[3], Hadin Raja[1] and A. R. Malik[1]

[1]Division of fruit Sciences, Sher-E-Kashmir University of Agricultural Sciences and Technology of Kashmir, Shalimar-191121 Srinagar Kashmir-India.
[2]Department of Plant Pathology, Sher-E-Kashmir University of Agricultural Sciences and Technology of Kashmir, Shalimar-191121 Srinagar Kashmir-India.
[3]Central Institute of Temperate Horticulture, Old Air Field Rangreth Srinagar-190007, Jammu and Kashmir-India.

The present investigation was carried out in the experimental field of Division of Fruit Science, Sher-e-Kashmir University of Agricultural Sciences and Technology of Kashmir, Shalimar, Srinagar during the year 2010 to 2011 with a view to study the various physical changes that occur during storage and to prolong the shelf life of plum under ambient storage conditions by preharvest application of various chemicals. Fruit size, weight and firmness recorded continuous decrease with the advancement of storage period. However, 0.5% calcium chloride (CaCl$_2$) proved to be more efficacious in minimizing these losses. Maximum increase in fruit size and weight at the time of harvest was recorded with the preharvest application of 60 ppm NAA. Physiological loss in weight (PLW) and spoilage followed continuously increasing trend with the advancement of storage period. Among the various preharvest treatments, 0.5% CaCl$_2$ applied 20 and 10 days before the expected date of harvest proved to be the most effective treatment in retaining the fruit quality during the entire storage period. Such fruits exhibited minimum loss in weight, maximum retention in firmness and minimum spoilage on each sampling date. In general, overall acceptability of fruits decreased with the passage of storage time. However, fruits treated with CaCl$_2$ were rated as most acceptable and it was followed by Gibberlic acid (GA$_3$) treatment at the end of storage period under ambient conditions.

Key words: Plum, quality, preharvest, calcium chloride (CaCl$_2$), Gibberlic acid (GA$_3$), napthelenic acetic acid (NAA), storage.

INTRODUCTION

Japanese Plum (*Prunus salicina* Lindl.) is one of the most important temperate zone stone fruit. It ranks next to peaches in economic importance (Westwood, 1993).

Being a delicious juicy fruit, it is used both as fresh and in preserved form. Plum is prized both for its exquisite fresh flavour, aroma, and attractiveness and in fruit preservation industry. Besides having medicinal properties, it is a fairly good source of citric acid, sugars and Vitamin A (Ulrich, 1974).Plum is grown from subtropical plains to the temperate high hills. European plum thrives best at 1300 to 2000 m above mean sea level and require about 1000 to 1200 chilling hours (below 7.2°C) during winter to break rest period, whereas Japanese plum require 700 to 1000 chilling hours (below 7.2°C) which is met in mid hill areas located at an elevation of 1000 to 1600 m above mean sea level. Plum is important fruit of North Indian hills comprising Himachal Pradesh, Jammu and Kashmir, hilly areas of Uttar Pradesh and Assam besides being grown in Nilgiris between 1300 to 1600 m above sea level. 'Santa Rosa', a leading commercial cultivar of Japanese plum, known for its fair quality and characteristic flavour is widely grown in Kashmir valley.

Plum is a highly perishable fruit and cannot be stored for longer periods or transported over longer distances under ambient conditions. The post harvest losses of fruits during transportation and marketing are very high, particularly as slight bruises, hardly noticeable on freshly harvested crops, cause the fruits to rot during transportation under hot and humid conditions. Therefore, it is desirable to have a preharvest treatment, which would retard the deterioration in quality during transportation and storage.

Plum fruit is highly delicate and perishable and demands immediate disposal and utilization. After harvesting, biochemical changes in fruits are continuous which lead to fruit softening and spoilage. If these changes are reduced, the storage life of fresh fruits can be effectively increased and spoilage can be reduced. In recent years, plant growth regulators such as auxins like Napthelenic acetic acid (NAA), gibberellins like gibberellic acid (GA₃) and calcium chloride (CaCl₂) have been extensively used for improving the quality, delaying deterioration in storage and thereby increasing the shelf life of various fruits. In view of these perspectives, an attempt has been made to find out the suitable preharvest treatment which could enhance the storage life and improve the quality of plum fruit under ambient storage conditions out of preharvest application of spray of CaCl₂, GA₃ and NAA.

Low fruit calcium levels have been associated with reduced postharvest life and physiological disorders (Wills et al., 1998). For example, Asrey and Jain (2000) found that 0.05% calcium chloride proved to be best in respect of prolonging shelf life (9 days) and acceptability owing to their better appearance, when fully ripe fruits of strawberry cv.

Chandler were treated with different concentrations of calcium nitrate (0.5, 1.0, and 2.0%), calcium chloride (0.05, 0.10, and 0.20%) and ascorbic acid (0.01, 0.02 and 0.05%) at 10°C for five minutes (Asrey and Jain, 2000). Proebsting and Mills (1966) observed that early Italian prune sprayed with 10 ppm Gibberellic acid were firmer

at harvest. Scott and Wills (1977) treated apple fruits with calcium chloride and observed retention of firmness during storage at ambient temperature.

Simnani (1995) observed that fruit firmness in peach was significantly affected by various concentrations of calcium application; the fruit firmness decreased gradually with the prolongation of storage period and was minimum with calcium treatment compared to control. Pawel (2001) found that "Dabrowicka prune" fruit sprayed with calcium were firmer and more resistant to infection after harvest than control fruits.

MATERIALS AND METHODS

The present investigation carried out in the experimental field/laboratory of Division of Fruit Science, Sher-e-Kashmir University of Agricultural Sciences and Technology of Kashmir, Shalimar, Srinagar situated at an altitude of 1390 m above MSL and between 34° 75' North latitude and 74° 50' East longitude, during the year 2010 to 2011. The experiment was conducted on 24 years old trees of plum cv. 'Santa Rosa' of uniform size and vigour which received uniform cultural operations. At the time of final bloom, uniform trees with uniform crop load were selected for experimental work. Treatments and replications were randomly assigned with a single plot size. The experiment consisted of 10 treatments, replicated thrice with a single tree size in a Randomized Block Design. Application of chemicals as spray solutions on plum fruits, CaCl₂ (Calcium Chloride, Hi Media) (0.1, 0.3 and 0.5%), GA₃ (Gibberlic acid-C₁₉H₂₂O₆-Hi Media, Central Drug House-New Delhi) (20, 40 and 60 ppm) and NAA (1-Napthalenic acetic acid-C₁₂H₁₀O₂-Hi Media, Central Drug House-New Delhi) (20, 40 and 60 ppm) was done twice, 20 and 10 days before harvest. After that the harvested fruits were stored under ambient conditions in the laboratory [At an ambient temperature (26 ± 2°C to 15 ± 2°C) and relative humidity (60 to 70%) during investigating storage period from 8 to 22 July, 2010] for studies on post harvest shelf life for a period of 15 days. The trees were sprayed twice at ten days interval, the first spray being carried out on 19th June (70 DAFB) and second spray on 29th June, 2010 (80 DAFB). At the time of first spray, all the twenty seven trees were sprayed and the control trees were left un-sprayed. The second spray was repeated in the similar way as the first application. The fruits of each treatment were harvested at optimum maturity (When 3/4th of colour of fruit changed to Red colour) (8th July, 2010) (90 DAFB) and immersed in running water to remove field heat and then air dried in shade. The uniformly matured fruits were selected (around 60 fruits from each treatment, 20 / Replication) and packed in standard wooden boxes of standard size for recording the fruit weight (g), fruit length (cm), fruit diameter (cm) and fruit volume (cm³) during storage period and physical parameters such as fruit firmness (kg/cm²), physiological loss in weight (%) and spoilage (%).

The length and diameter of 15 randomly selected fruits from each treatment was measured with the help of digital Vernier calliper (Aerospace) and the average expressed in cm. The weight of 15 randomly selected fruits from each treatment in each replication was taken on a top pan balance (Shimadzu- TX323L- Unibloc) and the average weight per fruit was expressed in grams (g). Volume of the fruit was measured by water displacement method using two litre measuring cylinder. A measuring cylinder was filled with water up to certain graduation and selected fifteen fruits, whose weight was recorded, were fully immersed in it. The difference between final and initial volume of water represented the total volume of fruits and the average fruit volume was expressed in cubic centimetre (cm³) per fruit. Fruit firmness was determined by a pressure tester (penetrometer) (Toshiba-India-mod FT-011). The two readings were taken at shoulder of the fruit at sides and the average reading was expressed in kg/cm^2. For calculating physiological loss in weight (%), at random 15 fruits from each treatment were weighed, labelled and kept separate from other fruits at harvest. Periodical weight of labelled fruits was recorded after every 5 days and subsequent loss was worked out. For obtaining organoleptic rating, fruit samples taken at random from each treatment were put before a panel of four judges (trained panel) for organoleptic evaluation. Organoleptic scoring was done Least acceptable = 1, Less acceptable = 2, Acceptable = 3, and Highly acceptable = 4 on the basis of taste, firmness, crispness, colour, sweetness, etc. The spoilage percentage of each treatment and replication was calculated at the fixed intervals of storage at ambient temperature by the following formulae:

$$\text{Spoilage percentage} = \frac{\text{No. of spoiled fruits}}{\text{Total No. of fruits}} \times 100$$

Statistical analysis

The data generated from the present investigations were put to statistical analysis by using R-software. Treatment means were separated and compared using least significant differences (LSD) at P less or equal to 0.05 as per the procedures described by Cochran and Cox (1963).

RESULTS AND DISCUSSION

Fruit size in terms of fruit length and width decreased with the advancement of storage period. The effect of various treatments on the fruit length and fruit width during storage and also the effects of the interactions between storage interval and treatments were found to be non-significant (Table 1). The results have been found to be in conformity with those of Srivastava et al. (1972) who observed that NAA in the range of 10 to 50 ppm did not show any significant effect on fruit size in apricot cv. Kaisha. The fruits treated with 90 ppm NAA (T$_9$) had the highest mean volume (42.83 cm^3) after the storage period which was significantly superior to control which recorded lowest volume (39.51 cm^3). However, the rest of the treatments were on par with the control. The decline in fruit volume during storage intervals during the first 5 days of storage was found to be non-significant whereas afterwards a significant decrease in fruit volume was recorded during rest of storage intervals. Fruit weight was increased significantly by the application of all the treatments. The maximum increase at harvest was observed in response to 60 ppm NAA (43.20 g) (Table 1). Similar increase in fruit weight and volume have also been observed by Srivastava et al. (1973) in peach cv. Alexandra; Khokhar et al. (2004) in strawberry cv. Chandler, upon treatment of fruits with NAA. However, at the end of 15 days of storage 0.1% CaCl$_2$ (T$_1$) retained the maximum weight (37.52 g) followed by 0.3% CaCl$_2$ (T$_2$) (36.63 g) and 0.5% CaCl$_2$ (T$_3$) (36.60 g). Least mean fruit weight was recorded in control (T$_{10}$) (35.13 g). The

decline in the fruit weight during storage was significant while the interactions between treatments and storage interval were found to be non-significant. Fruit growth is caused by cell division followed by cell enlargement. The application of NAA at the preharvest stage might have raised the auxin level in fruits which ultimately might have helped in the improvement of cell size and consequently fruit size, as a direct correlation between the auxin content and fruit growth, in several plants has been reported by Krishnamoorthy (1981). Both the weight and volume of fruits decreased significantly with the increase in storage period.

However, treated fruits maintained higher values of fruit volume and weight as compared to control. The decrease in both weight and volume during storage period may be due to the shrinking of transpiration resulting in retention of better sized fruits during storage. At the end of storage, themaximum weight and volume was observed with CaCl$_2$ 0.1% (T$_1$). The data shows a steady decrease in firmness commensurate with advance in the storage period (Table 2). The most firm fruits at harvest were obtained from trees receiving preharvest application of CaCl$_2$ 0.5% (T$_3$) (4.67 kg/cm^2) and were found to be on par with T$_2$ (4.60 kg/cm^2) and T$_1$ (4.58 kg/cm^2). These fruits also recorded the highest firmness values throughout the 15 days of storage period. The treatments T$_5$ (4.58 kg/cm^2), T$_6$ (4.57 kg/cm^2) and T$_4$ (4.56 kg/cm^2) were also significant as compared to the remaining treatments as well as the controls (Table 2). On an average, mean maximum fruit firmness was recorded in fruits treated with 0.5% CaCl$_2$ (T$_3$) (3.09 kg/cm^2). The control fruits on the other hand recorded the lowest average firmness (2.61 kg/cm^2) after the end of stipulated storage period. Interactions between treatments and storage intervals were found to be non-significant. The fruits treated with CaCl$_2$ maintained higher firmness as compared to GA$_3$ and control, at all storage intervals. 0.5% CaCl$_2$ (T$_3$) treated fruits demonstrated the best effect on maintaining fruit firmness and registered maximum mean fruit firmness (3.09 kg/cm^2) while the control fruits recorded the lowest mean fruit firmness (2.61 kg/cm^2) (Table 2).

Softening of fruits is caused either by breakdown of insoluble protopectin into soluble pectin or by hydrolysis of starch (Matto et al., 1975) or by cellular disintegration leading to increased membrane permeability (Oogaki et al., 1990). The loss of pectic substances in the middle lamellae of the cell wall is perhaps the key step in ripening process that leads to the loss of cell integrity or firmness (Solomes and Latics, 1973). Fruit firmness is one of the most crucial factors in determining the post harvest quality and physiology of fruits. With a decrease in fruit firmness, the tissue rigidity decreases, firstly as a result of hydrolysis of intercellular pectins and secondly by cell turgor pressure decreases due to an increase in permeability of cell membrane to water in the later stages of internal breakdown. The decrease in both the components of fruit firmness appears to contribute to

Table 1. Effect of pre-harvest sprays of various chemicals on fruit length (cm), fruit width (cm), fruit weight (g) and fruit volume (cm³) during ambient storage in plum cv. Santa Rosa (*Prunus salicina* L.).

Treatments (T)	Fruit length (cm)					Fruit width (cm)					Fruit weight (g)					Fruit volume (cm³)				
	Storage intervals in days (I)					Storage intervals in days (I)					Storage intervals in days (I)					Storage intervals in days (I)				
	0	5	10	15	Mean	0	5	10	15	Mean	0	5	10	15	Mean	0	5	10	15	Mean
T$_1$ CaCl$_2$ 0.1%	4.47	4.44	4.41	4.15	4.37	4.28	4.27	4.22	4.15	4.23	43.88	43.22	41.05	37.52	41.42a	42.87	42.40	41.14	38.26	41.17d
T$_2$ CaCl$_2$ 0.3%	4.33	4.30	4.28	4.13	4.26	4.27	4.24	4.21	4.15	4.22	42.44	41.87	39.90	36.63	40.21b	41.32	41.01	40.00	37.14	39.87g
T$_3$ CaCl$_2$ 0.5%	4.35	4.33	4.30	4.12	4.28	4.24	4.23	4.19	4.16	4.20	42.11	41.62	39.87	36.60	40.05b	41.00	40.75	39.77	37.35	39.72h
T$_4$ GA$_3$ 20 ppm	4.49	4.48	4.33	4.18	4.37	4.27	4.25	4.20	4.14	4.21	43.55	42.76	40.00	35.97	40.57b	42.85	42.30	40.00	37.44	40.65f
T$_5$ GA$_3$ 40 ppm	4.45	4.43	4.33	4.16	4.34	4.30	4.27	4.22	4.17	4.24	44.05	43.32	40.72	36.05	41.03b	43.05	42.47	40.60	36.77	40.72e
T$_6$ GA$_3$ 60 ppm	4.51	4.50	4.31	4.16	4.37	4.33	4.29	4.22	4.18	4.25	45.15	44.36	40.48	36.35	41.58a	44.30	43.66	40.14	37.36	41.44c
T$_7$ NAA 20 ppm	4.52	4.50	4.36	4.17	4.38	4.38	4.31	4.23	4.15	4.27	46.80	45.75	40.85	35.94	42.33a	45.87	44.25	40.75	36.59	41.86b
T$_8$ NAA 40 ppm	4.52	4.49	4.35	4.16	4.38	4.40	4.33	4.24	4.16	4.28	47.00	45.59	40.58	35.70	42.22a	46.31	44.03	41.05	36.45	41.96b
T$_9$ NAA 60 ppm	4.54	4.48	4.30	4.18	4.37	4.41	4.33	4.24	4.18	4.29	49.20	46.71	41.34	35.55	43.20a	47.70	44.90	41.56	37.17	42.83a
T$_{10}$ Control	4.34	4.29	4.25	4.17	4.26	4.22	4.21	4.18	4.15	4.19	38.77	38.00	34.27	29.47	35.13c	41.00	40.70	39.38	36.97	39.51i
Mean	4.44	4.42	4.32	4.15		4.31	4.27	4.21	4.16		44.30a	43.3a	41.05b	35.5c		43.6a	42.6b	40.4c	37.1d	
Lsd (P≤0.05)	Treatment (T): NS					Treatment (T): NS					Treatment (T): 2.14					Treatment (T): 0.11				
	Intervals (I): NS					Intervals (I): NS					Intervals (I): 1.35					Intervals (I): 0.07				
	T × I: NS					T × I: NS					T × I: NS					T × I: NS				

Lowercase letters indicate statistical differences amongst the means.

tissue softening (Pollard, 1974).

The desired effect of calcium on maintaining fruit firmness may be due to the calcium binding to free carboxyl groups of polygalacturonate polymer, stabilizing and strengthening the cell wall (Rees, 1975). Calcium binding may strengthen tissue and make it more resistant to hydrolytic enzyme activity as reported in tomato (Wills and Rigney, 1979) where Ca inhibits the polygalacturonase activity in cell walls (Buescher and Hobson, 1982).

Effectiveness of GA$_3$ in maintaining fruit firmness may be due to the reason that might reduce various physiological activities related with softening of fruits (Rees, 1975). A similar reduction in the firmness loss following the pre-harvest application of CaCl$_2$ has been reported by Siddiqui and Bangerth (1995) in apples during storage; Simnani (1995) in peach. There was a continuous increase in physiological loss in weight (PLW) under all the treatments as the storage period progressed. There was a progressive and significant increase in PLW of fruits with an increase in storage duration for both treated and untreated fruits. However, the increase in PLW of calcium chloride treated fruits (T$_3$ - 4.53%, T$_2$ - 4.76% and T$_1$ - 5.04%) was relatively slower and consequently these fruits exhibited significantly lower overall losses as compared to other treatments and control (T$_{10}$ - 9.66%). The treatment consisting of 0.5% CaCl$_2$ (T$_3$) proved to be the most effective in reducing PLW (4.53%) and it was found statistically at par with T$_2$ and T$_1$.

It was followed by T$_5$ (5.80%), T$_4$(6.26%) and T$_6$(6.87%), respectively. However, control fruits (T$_{10}$) exhibited highest PLW on each sampling date thereby recording the highest mean PLW (9.66%) which was significant in comparison to

other treatments. Interactions between treatments and storage intervals were also found to be significant (Table 2). The results have been found to be in conformity with those of Gupta et al. (1984) who also reported that PLW during 9 days of storage in peach fruit was 31% with preharvest application of 1% CaCl$_2$ as compared to 36.90% under control. Preharvest spray of calcium chloride has also been reported to be effective in reducing PLW during storage of apple (Baneh et al., 2003).

Fresh fruits and vegetables can be regarded as water infancy and expensive packages, some of which may be lost during storage or marketing. This water loss leads to loss of weight and thus, is a direct loss in marketing. Fruits, in general, possess considerable resistance to moisture loss, as their water vapour pressure is lower than that of water at the same temperature because of

Table 2. Effect of preharvest sprays of various chemicals on fruit firmness (kg/cm^2), PLW(%), Organoleptic rating and spoilage(%) during ambient storage in plum cv. Santa Rosa (*Prunus salicina* L.) (lowercase letters are used to indicate statistical differences amongst the means).

Treatments (T)	Fruit firmness (kg/cm^2) Storage intervals in days (I)					Physiological loss in weight (%) Storage intervals in days (I)					Organoleptic rating Storage intervals in days (I)					Spoilage (%) Storage intervals in days (I)				
	0	5	10	15	Mean	0	5	10	15	Mean	0	5	10	15	Mean	0	5	10	15	Mean
T_1 CaCl$_2$ 0.1%	4.58	4.00	2.30	1.00	2.97[b]		1.50 (1.22)	5.03 (2.22)	8.60 (2.92)	5.04 (2.12)[a]	3.16	3.15	3.06	2.41	2.94[b]		4.94 (2.22)	23.75 (4.87)	48.98 (6.99)	25.89 (5.09)[c]
T_2 CaCl$_2$ 0.3%	4.60	4.08	2.40	1.08	3.05[a]		1.35 (1.16)	4.71 (2.15)	8.22 (2.86)	4.76 (2.06)[a]	3.21	3.20	3.11	2.46	2.99[a]		2.72 (1.65)	20.95 (4.58)	46.21 (6.79)	23.29 (4.82)[b]
T_3 CaCl$_2$ 0.5%	4.67	4.11	2.49	1.09	3.09[a]		1.12 (1.06)	4.30 (2.04)	8.18 (2.85)	4.53 (1.98)[a]	3.24	3.23	3.14	2.56	3.05[a]		2.72 (1.65)	18.68 (4.32)	44.94 (6.70)	22.11 (4.70)[a]
T_4 GA$_3$ 20 ppm	4.56	3.88	2.18	0.85	2.87[b]		1.85 (1.36)	6.60 (2.54)	10.33 (3.21)	6.26 (2.37)[b]	3.18	3.17	3.09	1.90	2.83[c]		4.94 (2.22)	26.69 (5.17)	52.11 (7.22)	27.91 (5.28)[d]
T_5 GA$_3$ 40 ppm	4.58	3.99	2.24	0.89	2.92[b]		1.65 (1.28)	6.00 (2.42)	9.74 (3.11)	5.80 (2.27)[b]	3.23	3.22	3.11	1.95	2.87[c]		7.16 (2.67)	26.69 (5.17)	48.88 (6.99)	27.58 (5.25)[d]
T_6 GA$_3$ 60 ppm	4.57	3.94	2.20	0.86	2.89[b]		1.80 (1.34)	8.60 (2.92)	10.20 (3.18)	6.87 (2.48)[c]	3.25	3.22	3.15	2.00	2.90[b]		4.94 (2.22)	21.43 (4.62)	47.56 (6.89)	24.64 (4.96)[b]
T_7 NAA 20 ppm	4.44	3.61	1.89	0.66	2.65[c]		2.00 (1.41)	9.80 (3.10)	12.00 (3.45)	7.93 (2.65)[d]	3.13	3.11	3.06	1.80	2.77[d]		2.72 (1.65)	27.77 (5.27)	53.62 (7.32)	28.03 (5.29)[d]
T_8 NAA 40 ppm	4.41	3.60	1.89	0.64	2.63[c]		2.25 (1.50)	10.70 (3.23)	13.00 (3.60)	8.65 (2.78)[d]	3.16	3.12	3.08	1.70	2.76[d]		4.94 (2.22)	28.40 (5.33)	52.11 (7.22)	28.48 (5.34)[d]
T_9 NAA 60 ppm	4.41	3.59	1.87	0.60	2.62[c]		3.00 (1.73)	11.00 (3.30)	14.00 (3.73)	9.33 (2.92)[e]	3.25	3.15	3.08	1.66	2.77[d]		7.16 (2.67)	31.45 (5.61)	55.67 (7.46)	31.42 (5.60)[e]
T_{10} Control	4.40	3.56	1.88	0.60	2.61[c]		3.50 (1.87)	11.50 (3.38)	14.01 (3.74)	9.66 (2.99)[e]	3.13	3.05	3.05	1.60	2.71[e]		7.16 (2.67)	33.80 (5.81)	61.21 (7.82)	34.06 (5.84)[f]
Mean	4.52a	3.84b	2.13c	0.83d			2.00 (1.39)[a]	7.82 (2.73)[b]	10.83 (3.27)[c]		3.19[a]	3.15[b]	3.1[c]	1.9[d]			2.00 (1.39)[a]	5.14 (2.27)[b]	25.96 (5.10)[c]	51.13 (7.15)[g]
Lsd (P≤0.05)	Treatment (T) 0.11	Intervals (I) 0.07	T×I NS			Treatment (T) 0.16	Intervals (I) 0.10	T×I 0.09			Treatment (T) 0.06	Intervals (I) 0.03	T×I 0.08			Treatment (T) 0.11	Intervals (I) 0.06	T×I 0.18		

Data in parentheses is square root transformation of original data.

dissolved substances, mostly sugars. The entire weight loss is not due to water loss alone, for respiration may also account for a part of it. The average score for overall acceptability (organoleptic rating) at harvest was maximum (3.25) in response to 60 ppm GA$_3$ (T$_6$) and 60 ppm NAA (T$_9$) and these treatments were followed by T$_3$(3.24) and T$_5$(3.23). The data also indicates that that the score for overall acceptability decreased under all treatments during the entire 15 days of storage. The decrease in score was fastest in the control fruits (T$_{10}$) which therefore exhibited the lowest average score of only 2.71. However, 0.5% $CaCl_2$ (T_3) treatment resulted in maximum overall acceptability rating of fruit (3.05) during storage. Interactions between treatments and storage intervals were found to be significant. The extent of spoilage at an average was found to be lowest (22.11%) in fruits that had received a preharvest treatment of 0.5% $CaCl_2$ (T_3) and it was significant than all the other treatments. It was followed by treatments with concentrations of 0.3% $CaCl_2$ (23.29%) and 0.1% $CaCl_2$(25.89%) and 60 ppm GA_3 (24.64%) then by T_5(27.58%) and T_4(27.91) and NAA treatments with their effects being proportional to their concentrations applied. These treatments also caused significant reductions in spoilage as compared to control (T_{10}) (34.06%) where maximum spoilage was observed on all sampling days. Interactions between treatments and storage intervals were found to be significant (Table 2).

Calcium is known to act as an anti-senescent agent as it provides cellular disintegration by maintaining protein and nucleic acid synthesis (Faust and Klein, 1973). It is also reported to be effective in decreasing the respiration rates of several commodities (Faust, 1978). During the present study also calcium chloride treatments have been observed to be most effective in reducing PLW of fruits during storage whereas control fruits exhibited maximum loss. The increased weight loss in untreated fruits could be due to increased storage breakdown, which is associated with higher rate of respiration as compared to calcium treated fruits.

Conclusion

The objectives of the investigation were to study the effect of preharvest chemical treatment viz. $CaCl_2$, GA_3 and NAA on physical attributes of plum as well as on the storage life of plum fruits under ambient conditions. The results obtained during the course of investigation showed that maximum increase in fruit size, weight and volume were recorded with preharvest application of 60 ppm NAA. The fruit size and volume followed a declining trend commensurating with advancement in storage period. 0.5% $CaCl_2$ treatments proved to be more efficacious in minimizing the loss. The firmness of fruits showed a decline during storage, the decrease being minimum in 0.5% $CaCl_2$. There was an increase in physiological loss in weight of fruits during storage. However, preharvest application of 0.5% $CaCl_2$ proved to be efficacious in minimizing weight loss during storage. Preharvest application of 0.5% $CaCl_2$ resulted in better retention of sensory quality attributes during storage as a result of which fruits from these treatments were most acceptable at all storage intervals. Spoilage of the fruits was found to be substantially lower in fruits that were given $CaCl_2$ treatments. GA_3 treatments also resulted in lower spoilage compared to the control fruits.

From the studies, it may be concluded that storage life of plum fruits could be prolonged with the preharvest application of calcium chloride ($CaCl_2$). Preharvest application of $CaCl_2$ at 0.5% proved most beneficial in enhancement of quality in terms of improving fruit firmness stimulating organoleptic taste as well as prolonged shelf-life under ambient storage conditions. Hence, it represents the best preharvest treatment for getting better quality 'Santa Rosa' plum for better remuneration to the orchardist.

REFERENCES

Asrey R, Jain RK (2000). Effect of certain post harvest treatments on shelf life of strawberry cv. Chandler. Acta Hort., 696:2.

Baneh HD, Hassani A, Majidi A, Zomorodi S, Hassani G, Malakooti MJ (2003). Effect of calcium chloride concentration and number of foliar application on the texture and storage quality of apple (Red Delicious) in Orumia region. Agric. Sci Tabriz 12(4) 47-54.

Buescher RW, Hobson GE (1982). Role of calcium and chelating agents in regulating the degradation of tomato fruit tissue by polygalacharonase. J. Food. Biochem., 6:78-084.

Cochran GC, Cox GM, (1963). Experimental Designs. Asia Publishing Home, Bombay. p. 611.

Faust M (1978). The role of calcium in the respiratory mechanism of apples colloques. International du Centre National de la Racherche Scientifique, 238: 87-92.

Faust M, Klein JD (1973). Levels and sites of metabolically active calcium in apple fruit. J. Am. Soc. Hort. Sci., 99: 93-94.

Gupta OP, Singh BP, Singh SP, Chauhan KS (1984). Effect of calcium compounds as preharvest spray on the shelf life of peach cv. Sharbati. Punjab Hort. J., 24(1/4):105-10.

Khokhar UU, Prashad J, Sharma MK (2004). Influence of growth regulators on growth, yield and quality of strawberry cv. Chandler. Haryana J. Hort. Sci., 33(3-4): 186-188.

Krishnamoorthy HN (1981). Plant growth substances including applications in agriculture. Tata McGraw Hill Publishing Co. Ltd., New Delhi, pp. 3-87.

Matto AK, Murata T, Pantastico EB, Chactin K, Ogata K, Phan CT (1975). Chemical changes during ripening and senescence. In : Postharvest Physiology, Handling and Utilisation of Subtropical Fruits and Vegetables (Ed. E.B. Pantastico). AVI Publishing Co. Inc. Westport, Connecticut, pp. 103-127.

Oogaki C, Wang HG, Gemma H (1990). Physiological and biochemical characteristics and keeping qualities of temperate fruits during chilled storage. Acta Hort. 279:541-558.

Pawel W (2001). 'Dabrowicka Prune' fruit quality as influenced by calcium spraying. J. Plant Nutr., 24:1229-1241.

Pollard JE (1974). Pectinolytic enzyme activity and changes in water potential components association with internal breakdown in McIntosh apples. J. Am. Soc. Hort.Sci., 100: 642-649.

Proebsting EL, Missls HH (1966). Effect of GA3 and other regulators on quality of Early Italian prunes (Prunus domestica L.). Proceedings of the Am. Soc.Hort.Sci.,89:135-139.

Rees DA (1975). Steriochemistry and binding behaviour of carbohydrate chains. In Biochemistry of Carbohydrates . MTP I (Ed. W.J. Whelan), Butterworths, London, Unv. Park Press, Baltimore, Inst. Rev. Sci. Biochem. Ser., p. 5.

Scott KJ, Wills RBH (1977). Vacuum infiltration of calcium chloride; A method for reducing bitter pit and senescence of apples during storage of ambient temperature. HortScience, 12:71-72.

Siddiqui S, Bangerth F (1995). Effect of preharvest application of calcium on flesh firmness and cell wall composition of apples, influence of fruit size. J. Hort.Sci., 70: 263-69.

Simnani SSA (1995). Effect of calcium nitrate and hydrocooling on the storage life of Shan-I-Punjab peach. M. Sc. Thesis, Punjab University, Ludhiana.

Solomes T, Latics GG (1973). Cellular organisation and fruit ripening.

Nature 245: 390-391.

Srivasta RP, Mishra RS, Bana DS, Verma VK (1973). Effect of PGRs on fruit drop, fruit size, maturity and quality of peach. Progre. Hort., **5**(2):11-21.

Srivastava RP, Misra RS, Bana DS (1972). Effect of PGRs on fruit drop, fruit size, maturity and quality of apricot var. Kaisha. Progres. Hort., **3**(1):51-60.

Ulrich R (1974). Organic acids. In : Biochemsitry of Fruit and their Products (Ed. A.C. Hulme).Academic Press, London and New York, 1:189-218.

Wills RBH, Rigney CJ (1979). Effect of calcium on activity of mitchondria and pectin enzymes isolated from tomato fruits. J. Food Biochem., **3**:103-110.

Wills R, McGlasson B, Graham D, Joyce D (1998). Postharvest: An Introduction to the physiology and handling of fruit, vegetables and ornamentals. UNSW Press, Sydney, p. 262.

Westwood MN (1993). Temperate Zone Pomology. W.H. Freeman and Company, San Francisco, California, USA, p. 223.

8

Crambe seeds quality during storage in several conditions

Lílian Moreira Costa[1], Osvaldo Resende[2], Douglas Nascimento Gonçalves[3] and Anderson Dinis Rigo[3]

[1]Federal Institute of Education, Science and Technology of Goiás (Instituto Federal de Educação, Ciência e Tecnologia Goiano – IF Goiano) – Rio Verde Câmpus, GO, Rodovia Sul Goiana, Km 01 - Zona Rural - CEP: 75901-97, Brazil.
[2]Board of Undergraduate Studies, IF Goiano – Rio Verde Câmpus, GO, Rodovia Sul Goiana, Km 01 - Zona Rural - CEP: 75901-970, Brazil,
[3]PIBIC/CNPq scholar, IF Goiano – Rio Verde Câmpus, GO, Brazil.

Crambe abyssinica seeds are spherical and surrounded by a structure called the pericarp. The basic function of the pericarp is to protect the grains against abrasion and shocks, to function as a barrier against microorganisms and to allow the seeds to be stored for long periods of time. The aim of this study was to evaluate the *C. abyssinica* seed quality stored under different environmental conditions without the pericarp. Crambe seed with 6.5% w.b. moisture content were used. Measurements for the electrical conductivity, water uptake, germination percentage and index of germination velocity (IGV) were performed at the beginning of the experiment (zero months) and every two months for a period of a year. The seed were stored under three environmental conditions: Room temperature (26±3°C, 55±12% relative humidity [RH]), a cold room (5±1°C, 79±5% RH) or a climate-controlled chamber (18±1°C, 53±7% RH). The climate-controlled chamber maintained the best quality in the crambe seed, with better germination percentage and IGV than the other conditions. The storage conditions promoted decrease in the crambe seed quality. It was possibly visualize that there was a loss of dry matter during storage, especially lipids adhered in Kraft paper bags.

Key words: *Crambe abyssinica*, seeds quality, storage environment.

INTRODUCTION

Currently, businesses, as well as state and federal agencies, have prioritized the search for alternative raw materials for biodiesel production and are continuously evaluating the effect of their attributes, such as oil content, yield, production system and crop cycle. *Crambe abyssinica* is a winter crop that has great potential to become a raw material for biodiesel in addition to being a good candidate for crop rotation.

Crambe (*C. abyssinica*) is a member of the Brassicaceae family that is native to Mediterranean regions

and is considered a potential crop for the production of biodiesel due to its cultivation characteristics and high oil content. Studies conducted with crambe grown in Mato Grosso do Sul - Brazil (FMS Brilliant variety) have shown a 40% oil content in the seed (Souza et al., 2009).

Crambe shows important characteristics, such as a low production cost, a short growing cycle, tolerance to drought and low temperatures; it is one of winter crops, but can be grown in other times and it can be planted at later times when there is too much risk to other crops in the Midwest region of Brazil (Pitol et al., 2010).

The fruits have been reported to show low bulk density, approximately 328 kg m^{-3} (Reuber et al., 2001), which is the major problem when setting up a production chain due to the high transportation and storage costs. Peeling (pericarp removal) of the crambe fruit could significantly reduce the operational costs of the crop. However, there is no adequate information on the seed quality of crambe when it is stored without the pericarp.

C. abyssinica seeds are spherical and surrounded by an integument structure called the pericarp (Ruas et al., 2010). The pericarp, which remains attached to the seeds after harvest, representing 25 to 30% of the total weight of the fruit, has high lignin content (40%) and is approximately 41% cellulose (Gastaldi et al., 1998).

Several studies have been conducted to study the influence of storage conditions on seed quality of crops, including annual ryegrass (Eichelberger et al., 2003), annatto (Corlett et al., 2007), arnica (Melo et al., 2007), sorghum-sudan (Toledo et al., 2007), *Coffea arabica* (Vieira et al., 2007) papaya (Berbert et al., 2008), cotton (Queiroga et al., 2009), castor (Fanan et al., 2009) and soybean (Forti et al., 2010).

The main purpose of storage, which actually begins before the harvest when the seeds reach their physiological maturity and continues until the time of sowing, is to maintain seed quality and minimize deterioration, as the quality of the seed is determined during development and cannot later be improved, even under ideal storage conditions (Baudet, 2003). Seed deterioration is irreversible and cannot be stopped, but it is possible to reduce the rate of deterioration through proper handling and efficient environmental storage conditions (Baudet, 2003).

Vieira et al. (2004) have suggested that the determination of electrical conductivity of seeds was a sensitive test for the evaluation of vigor because in the deterioration process, one of the first events is the loss of membrane integrity. Seeds with low vigor tend to show disorganization of the cellular membrane structures, allowing for the increased leaching of solutes, such as sugars, amino acids, organic acids, proteins, phenolic substances and inorganic ions, including K^+, Ca^{2+}, Mg^{2+} and Na^{2+} (Vanzolini and Nakagawa, 2005; Dias et al., 2006). Therefore, the objective in this study was to evaluate the crambe seed quality when stored under three different environmental conditions without the pericarp.

MATERIALS AND METHODS

Crambe (*Crambe abyssinica*) seeds (cultivar FMS Brilhante) produced at the Fundação MS were used in this study. The fruit was harvested in August 2008 using a harvesting machine adapted for the process, and the pericarp was mechanically removed. The experiment was conducted at the Laboratory of Postharvest of Plant Products and Laboratory of Seeds at the Instituto Federal de Educação Ciência e Tecnologia Goiano - Câmpus Rio Verde, GO,

Brazil (IF Goiano-Câmpus Rio Verde). The seeds were conditioned in three replicates in Kraft paper bags, unifoliate, with an initial seed moisture content of 6.5% (wet base, w.b.) that was determined gravimetrically using an incubator at 105±3°C for 24 h (Brazil, 2009). Approximately 0.4 kg of seed was used in the paper bag, which were kept under three conditions: room temperature (26±3°C, 55±12% relative humidity [RH]), a cold room (5±1°C, 79±5% RH) and a climate-controlled chamber (18±1°C; 53±7% RH). During storage, the RH and temperature were recorded by a digital datalogger.

The seeds were stored from August 11th, 2009, to August 11th, 2010, and the samples were evaluated at 0, 2, 4, 6, 8, 10 and 12 months, in three repetitions, for water absorption, bulk density, electrical conductivity, germination and the index of germination velocity (IGV).

To determine the water uptake, the samples were subjected to hydration in distilled water for a period of 12 h. Absorption was performed in a chamber with a controlled temperature at 25 ± 2°C and plastic cups (100 - ml capacity) containing 75 ml of distilled water with 15 g of seeds (a 5:1 mass ratio). The samples were gently agitated so that all of the seeds were completely submerged. After hydration, the samples were removed from the cups and placed on a filter paper to blot for two minutes and then weighed to 0.01 g. The moisture content after absorption was obtained by the following equation:

$$U^* = \frac{M_e - M_s}{M_s}$$

(1)

where, U^* = moisture content of the product (decimal dry base, d.b.), M_e = mass after water absorption (kg) and M_s = mass of the dry product (kg).

The bulk density (ρ_{ap}), expressed in kg m^{-3}, was determined with an electronic hectoliter scale with a 0.01 g resolution using a 90 ml container. The electrical conductivity (EC) of the crambe seeds was measured without the pericarp using the method described by Vieira and Krzyzanowski (1999). Fifty seeds were used for four replicates of each treatment and weighed accurately to two decimal places (0.01 g). The samples were soaked in plastic cups (100 ml capacity) containing 75 ml of water and kept in a BOD chamber with a controlled temperature at 25 ± 2°C for 24 h. Solutions containing the seeds were lightly agitated for uniformity of the leaching and immediately measured using a portable digital conductivity meter (model CD-850 "INSTRUTHERM"). The results were divided by the mass of 50 seeds and expressed in µS cm^{-1} g^{-1}.

The germination test was conducted with four subsamples of 30 seeds from each treatment. The fruits were packed in Gerbox boxes on blotting paper moistened with distilled water, which was equivalent to 2.5 times the dry substrate mass, to achieve adequate moisture and uniformity of the test. The samples were kept in a Mangelsdorf germinator set at a constant temperature of 25 ± 2°C. The evaluations were performed every two days from the second day after sowing until 32 days were completed according to the criteria established in the Rules for Seed Analysis (Brazil, 2009). The average germination percentage was calculated, and the IGV was calculated as follows: IGV = $n_1.d_1^{-1} + n_2.d_2^{-1} + n_3.d_3^{-1}... n.d_n^{-1}$; where n_1 is the number of seeds germinated on the first day of counting; n_2 is the number of seeds germinated on the second day of counting; n_3 is the number of seeds germinated on the third day of counting; n_n is the number of seeds germinated on the nth day of counting; d_1 is the first day; d_2 is the second day; d_3 is the third day; and d_n is the nth day (Maguire, 1962).

The experiment was designed according to a subdivided plot scheme, with the three storage conditions (room temperature, cold room and climate-controlled chamber) plots and the evaluation months as the subplots. The averages were compared by Tukey's test at 5% significance.

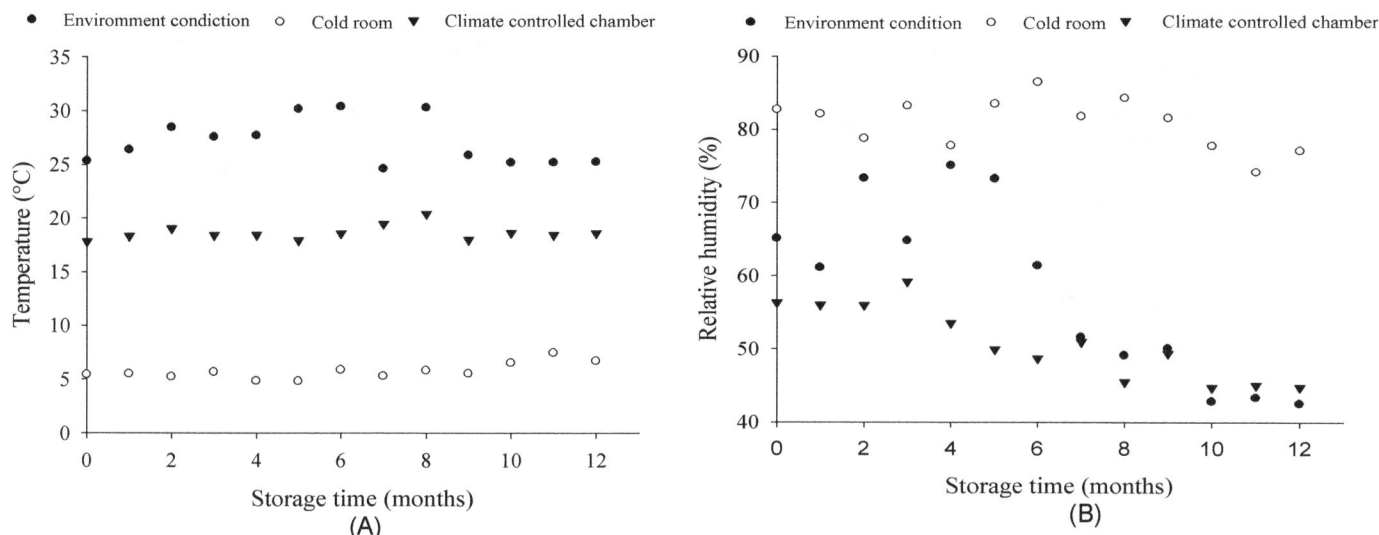

Figure 1. (A) Average temperature during storage in the three conditions: Room temperature (26±3°C), a cold room (5±1°C) and a climate-controlled chamber (18±1°C). (B) Average relative humidity during storage: room temperature (55±12%), a cold room (79±5%) or a climate-controlled chamber (53±7%).

RESULTS AND DISCUSSION

Figure 1 shows the average monthly values of the temperature and the relative humidity of the air in the three storage chambers for crambe seeds without the pericarp. The cold room provided the highest relative humidity due to the low temperature of the environment, and showed the lowest changes over time. The environment condition showed the greatest changes in the temperature and relative humidity, which can be attributed to the changes in thermal and moisture regimes associated with the change of seasons. The climate-controlled chamber showed the lowest relative humidity due to the cooling system, which removed water vapor from the chamber. A summary of the analysis of variance for the variables analyzed during the storage of crambe seeds without the pericarp under three environmental conditions is shown in Table 1.

Table 2 shows the moisture content values of the crambe seeds without the pericarp. Seeds stored under room temperature and in the cold room showed large variations in their moisture content during storage. The changes in the air conditions caused constant alterations in the moisture content of the seeds stored in bags permeable to water vapor.

It was observed at the end of the storage period that there was a decrease in the moisture content of the seeds stored in the cold room, due to the moderately low temperature and relative humidity under this condition (Figure 1), which caused the seeds to attain equilibrium with the environmental conditions in question. This result was similar to those observed by Catunda et al. (2003) for the storage of passion fruit seeds under three different conditions for 10 months.

Table 3 shows the variations in the bulk density of the crambe seeds during their storage under the three different environmental conditions. Although the values were influenced by the time and storage conditions, they did not, however, exhibit a defined behavior. The values ranged from 513.63 to 573.79 kg m^{-3}. Therefore, in this research, verified that seeds without pericarp presented the bulk density more than Reuber et al. (2001) for crambe fruit approximately 328 kg m^{-3} (with pericarp). Thus, it appears that the seed without pericarp present decreasing the storage and transportation costs.

According Pitol et al. (2010), one of the alternatives to solve the transportation problem of seeds is the peel of the fruit of crambe (removal of the pericarp) before shipping, so the product has bulk density peeled around 740 kg m^{-3}.

Water uptake by the seeds was not different under the three storage conditions (Table 4); however, over storage time, significant differences did arise for this parameter. Costa et al. (2012a) when storing fruits of crambe (crambe with pericarp) under the same conditions and storage period of this study, which concluded at the end of 12 months of storage, the water absorption is not different from baseline for the three conditions storage.

Water uptake and distribution in the seeds, which are regulated by the cellular water potential, occur as much by capillary action as by diffusion in the high-to-low water potential direction. According to Ullmann et al. (2010), water absorption is a good parameter for the evaluation of mechanical damage, as their values are linked to damage caused in the integument and seed structure. It is possible that no physical damage occurred with the removal of the pericarp, which remained attached to the crambe seed after harvest, and this could have influenced the values of water absorption during storage.

Regarding electrical conductivity, there was an increase

Table 1. Moisture content, bulk density, water absorption, electrical conductivity, germination rate and index of germination velocity during storage of crambe seeds without the pericarp under different environmental conditions for 12 months.

Variables analyzed	Source of variation	Mean squared	CV (%)
Moisture content	Environment	34.92**	9.30
	Months	6.91**	9.59
	Environment x Months	4.84**	
Bulk density	Environment	388.32*	1.27
	Months	1393.69**	2.04
	Environment x Months	419.13**	
Water absorption	Environment	0.00033^{NS}	4.69
	Months	0.011*	3.18
	Environment x Months	0.0010^{NS}	
Electrical conductivity	Environment	60153.47**	6.04
	Months	34343.84**	4.94
	Environment x Months	4464.97**	
Germination	Environment	1891.74**	12.17
	Months	1885.80**	16.28
	Environment x Months	247.60**	
IGV	Environment	53.75**	15.68
	Months	34.99**	21.09
	Environment x Months	4.71*	

**Significant at 1% by F test. *Significant at 5% by F test. ^{NS}Not significant.

Table 2. Moisture content in crambe seeds without the pericarp (% w.b.) subjected to storage under room temperature, cold room or climate-controlled conditions for 12 months.

Environments	Storage period (months)						
	0	2	4	6	8	10	12
Room temperature	6.54^{aAB}	5.11^{aBC}	7.04^{aA}	6.32^{bAB}	6.01^{bABC}	4.59^{bC}D	3.12^{bD}
Cold room	6.54^{aBC}	5.36^{aC}	7.25^{aB}	8.93^{aA}	8.74^{aA}	10.08^{aA}	6.55^{aBC}
Climate -controlled chamber	6.54^{aA}	5.80^{aA}	5.69^{aA}	5.12^{cAB}	5.86^{bA}	4.20^{bB}	3.83^{bB}

Means followed by a same lower case letter in the columns and uppercase in lines do not differ by Tukey's test at 5% probability.

Table 3. Bulk density (kg.m^{-3}) of crambe seeds without the pericarp stored at room temperature, in a cold room or in a climate-controlled chamber for 12 months.

Environments	Storage period (months)						
	0	2	4	6	8	10	12
Room temperature	513.63^{aB}	547.46^{bA}	530.63^{bAB}	545.58^{aA}	534.52^{aAB}	549.59^{aA}	540.73^{abAB}
Cold room	513.63^{aD}	573.79^{aA}	560.99^{aAB}	532.78^{aCD}	537.53^{aBCD}	531.46^{aCD}	557.63^{aABC}
Climate -controlled chamber	513.63^{aB}	540.95^{bAB}	544.58^{abA}	548.70^{aA}	542.40^{aA}	535.02^{aAB}	525.70^{bAB}

Means followed by a same lowercase letter in the columns and uppercase in lines do not differ by Tukey's test at 5% probability.

in the amount of electrolytes released by the seeds during storage. The EC tended to increase more under room temperature and in the climate-controlled chamber, confirming the influence of time and storage conditions in the amount of solutes leached into the solution. In general, seeds stored in the refrigerated chamber showed lower electrical conductivity values than those kept under the other storage conditions, indicating that refrigeration of the seeds results in the least amount of electrolyte leakage (Table 5). These results agree with those obtained by Pontes et al. (2006) for *Caesalpinia peltophoroides* seeds stored for a period of 240 days at 5

Table 4. Water absorption of crambe seeds without a pericarp (decimal db) stored at room temperature, in a cold room or in a climate-controlled chamber for 12 months.

Environments	Storage period (months)						
	0	2	4	6	8	10	12
Room temperature	1.09	1.05	1.10	1.06	1.01	1.13	1.10
Cold room	1.09	1.05	1.11	1.05	0.99	1.13	1.09
Climate -controlled chamber	1.09	1.03	1.10	1.10	1.05	1.09	1.06
Averages	1.09^{AB}	1.04^{C}	1.10^{A}	1.05^{BC}	1.04^{C}	1.11^{A}	1.08^{ABC}

Means followed by a lowercase letter in the columns and uppercase in lines do not differ by Tukey's test at 5% probability.

Table 5. Electrical conductivity of crambe seeds without the pericarp ($\mu S\ cm^{-1}\ g^{-1}$) stored at room temperature, in a cold room or in a climate-controlled chamber 12 months.

Environments	Storage period (months)						
	0	2	4	6	8	10	12
Room temperature	329.57^{aE}	433.69^{aCD}	382.65^{aD}	465.45^{aC}	541.33^{aB}	589.49^{aAB}	600.36^{aA}
Cold room	329.57^{aC}	365.75^{bBC}	325.46^{bC}	386.56^{bB}	439.22^{bA}	406.15^{bAB}	414.45^{bAB}
Climate -Controlled chamber	329.57^{aD}	367.41^{bBCD}	350.22^{abCD}	392.28^{bABC}	413.82^{bAB}	443.18^{bA}	427.58^{bA}

Means followed by a lowercase letter in the columns and uppercase in lines do not differ by Tukey's test at 5% probability.

Table 6. Germination and index of germination velocity (IGV) of crambe seeds without the pericarp stored at stored at room temperature, in a cold room or in a climate-controlled chamber for 12 months.

Environments	Germination (%)						
	Storage period (months)						
	0	2	4	6	8	10	12
Room temperature	72.87^{aA}	66.67^{aAB}	65.00^{aAB}	49.44^{bBC}	36.39^{bCD}	35.00^{bCD}	22.50^{bD}
Cold room	72.87^{aA}	74.72^{aA}	75.83^{aA}	46.11^{bBC}	27.50^{bC}	55.83^{aAB}	45.00^{aBC}
Climate -controlled chamber	72.87^{aAB}	72.78^{aAB}	82.22^{aA}	68.89^{aAB}	60.83^{aAB}	68.89^{aAB}	55.56^{aB}
	IGV						
Room temperature	8.53^{aA}	7.73^{aAB}	8.33^{aA}	5.3^{bABC}	4.02^{bBC}	3.50^{bC}	2.46^{bC}
Cold room	8.53^{aAB}	9.70^{aA}	9.77^{aA}	5.14^{bC}	2.75^{bC}	6.03^{bABC}	5.60^{aBC}
Climate -controlled chamber	8.53^{aA}	9.94^{aA}	10.83^{aA}	8.6^{aA}	7.52^{aA}	9.33^{aA}	7.22^{aA}

Means followed by a lowercase letter in the columns and uppercase in lines do not differ by Tukey's test at 5% probability.

or 20°C with 70 or 62% relative humidity, respectively. In addition, Borba Filho and Perez (2009) have stored the seeds of white (*Tabebuia roseoalba*) and purple (*T. serratifolia*) ipê for 300 days at laboratory room temperature (21 to 31°C, 40 to 78% RH), in a refrigerated chamber (4 to 6°C, 38 to 43% RH) and in a cooled chamber (14 to 20°C, 74 to 82% RH). They reported that the highest values of electrolytes were leached from the seeds kept in the laboratory environment, which became apparent after only 60 days of storage.

Table 6 shows the values for the germination percentage and index of germination velocity (IGV) of the crambe seeds stored for 12 months in all three environments. For each month of storage, the germination potential was higher in the seeds stored in the climate-controlled chamber. Crambe seeds had a higher percentage

of germination at month zero, the beginning of the storage period, under all storage conditions. As this may indicate an absence of dormancy, it is important to note that these seeds started the storage period 12 months after their harvest. The dormancy mechanism is common in the seeds of several species after harvesting (Brazil, 2009). A study on crambe seeds dormancy would be necessary to consider the possibility that the germination potential increased with the time interval following harvest. Costa et al. (2012b) and Faria et al. (2012) have studied the viability of crambe seeds subjected to different drying conditions and moisture content and have observed a low germination percentage. In addition, Oliva (2010) found than low germination after drying crambe seeds and storing them for 8 months in unifoliate paper bags.

During storage, the percentage of germination increased in cold room and Climate controlled chamber until the fourth month, from which time, the values decreased and fluctuated (Table 6). Crambe seeds are protected by the pericarp when the fruits are left intact after harvest. It is possible that the process of removal pericarp affected the seed quality because after four months, the seeds stored in Kraft paper bags showed oily patches. It was possibly visualize that there was a loss of dry matter during storage, especially lipids adhered in Kraft paper bags, which are one of the major reserve substances found in crambe seeds.

The seeds rich in lipids have limited longevity due to their specific chemical composition. For example, sunflower seed storage demands special attention due to high oil content, otherwise processes may occur that lead to loss of germination ability and seed viability (Christensen, 1971)

Different longevity of seed storage as well as storage conditions exerts significant influence on seed germination (Nkang and Umoh, 1997). The results of Sharma (1977) clearly pointed out to declining trends in total oil content and seed germination during storage of oilseed species. Seed aging during storage is an inevitable phenomenon, but the degree and speed of decline in seed quality depend strongly, beside storage conditions, on plant species stored and initial seed quality (Elias and Copeland, 1994) as well as on seed genetic traits (Malenčić et al., 2003).

It was possibly visualize that there was also a higher incidence of fungal infection in the seeds stored in the cold room. This fact is due to the higher values of water activity in the seeds under this condition (higher than 0.75, Figure 1), showing high levels of water throughout the whole storage period.

The index of germination velocity (IGV) was also higher in the seeds stored in the climate-controlled chamber, as observed for the germination percentage. This index increased significantly until the fourth month of storage, from which time the values decreased and fluctuated (Table 6).

Santos and Paula (2005) have shown that "branquilo" seeds exhibited a decrease in the percentage and rate of germination when stored for five months in paper bags, indicating that, in addition to the storage environment, a reduction in germination may also be associated with the packaging used.

Conclusions

Based on the results presented, we conclude that: The climate-controlled chamber retained the best quality of crambe seeds lacking the pericarp and provided a higher percentage of germination and higher IGV values than for the other two environments, and the storage conditions promoted decrease in the crambe seed quality. It was possibly visualize that there was a loss of dry matter during storage, especially lipids adhered in Kraft paper bags.

ACKNOWLEDGEMENTS

The authors extend thanks to CNPq for their financial support, which was indispensable to the execution of this study, and for granting the Master's scholarships to the first author. We also thank CAPES for their financial support for this work.

REFERENCES

Baudet L (2003). Seed storage. In: Peske ST, Rosenthal MD and Rota GRM seeds: fundamentals of science and technology. Pelotas. Editora and UFPel graphical University. pp. 414.

Berbert PA, Carlesso VO, Silva RF, Araújo EF, Thiebaut JTL, Oliveira MTR (2008). Physiological quality of papaya seeds as affected by drying and storage. Braz. J. Seeds 30(1):40-48,

Borba Filho AB Perez SCJGA (2009). Seed storage of White-ipe and purple-ipe in different packaging and environments. Braz. J. Seeds 31(1):259-269.

Brazil Ministry of Agriculture (2009). Livestock and Supply. Rules for Seed Testing. Ministry of Agriculture, Livestock and Supply. Agriculture Defense Department. Brasília, DF: Map/ACS. pp. 395.

Catunda PHA, Vieira HD, Silva RF, Possession SCP (2003). Influence of moisture content, packaging and storage conditions on seed quality of passion fruit. Braz. J. Seeds 25(1):65-71.

Christensen CM (1971). Evaluating conditions and storability of sunflower seeds. J. Stored Prod. Res. 7:163-169.

Corlett FMF, Barros ACSC, Villela FA (2007). Physiological quality of annatto seeds stored in different environments and packaging. Braz. J. Seeds 29(2):148-158.

Costa LM, Resende O, Gonçalves DN, Rodrigues E, Sousa KA, Sales JF, Donadon JR (2012b). The influence of drying on the physiological quality of crambe fruits. Acta Scientiarum. Agron. Maringá. 34(2):213-218..

Costa LM, Resende O, Gonçalves DN, Sousa KA (2012a). Qualidade dos frutos de crambe durante o armazenamento. Revista Brasileira de Sementes 34(2):239-301.

Dias DCFS Bhering MC Tokuhisa D Hilst PC (2006). Electrical conductivity test for evaluation of vigor in onion seeds. Braz. J. Seeds Pelotas 28(1):154-162.

Eichelberger L Maia MS Peske ST Moraes DM (2003). Drying delay effect on physiological quality of stored annual ryegrass seeds. Braz. Agric. Res. 38(5):643-650.

Elias SG, Copeland LO (1994). The effect of storage conditions on canola (Brassica napus L.) seed quality. J. Seed Technol. 18:21-22.

Fanan S Medina PF Camargo MBP Ramos NP (2009). Influence of harvest and storage on physiological quality of seeds of castor. Braz. J. Seeds 31(1):150-159.

Faria RQ, Teixeira IR, Devilla IA, Ascheri DPR, Resende O (2012). Cinética de secagem de sementes de crambe, Revista Brasileira de Engenharia Agrícola e Ambiental 16(5):573–583.

Forti VA, Cicero SM, Pinto TLF (2010). Assessment of damage progression by "moisture" and reduced vigor in soybean seeds, cultivar TMG113-RR during storage, using x-ray images and physiological tests. Braz. J. Seeds 32(3):123-133.

Gastaldi G, Capretti G, Focher B, Cosentino C (1998). Characterization and properties of cellulose isolated from the Crambe abyssinica hull. Ind. Crops Products 8(3):205-218.

Maguire JD (1962). Speed of germination-aid in selection and evaluation for seedlig emergence and vigor. Crop Sci. 2 (1):176-177.

Malenčić Đ, Popović M, Miladinović J (2003). Stress tolerance parameters in different genotypes of soybean. Biol. Plantarum 46:141-143.

Melo PRB, Oliveira JÁ, Pinto JEBP, Castro EM, Vieira AR, Evangelista

JRE (2007). Germination of arnica (Lychnophora pinaster Mart.) Stored in different conditions, Agrotechnol. Sci. 31(1):75-82.

Nkang A, Umoh EO (1997). Six month storability of five soybean cultivars as influenced by stage of harvest, storage temperature and relative humidity. Seed Sci. Technol. 25(1):93-99.

Oliva ACE (2010). Quality of crambe seed subjected to drying methods and storage periods. Thesis (MS) - Universidade Estadual Paulista, School of Agricultural Sciences, Botucatu. pp. 78.

Pitol C, Broch DL, Roscoe R (2010). Technology and Production: Crambe 2010. Maracaju: MS Foundation. pp. 60.

Pontes CA, Corte VB, Borges EEL, Silva AG, Borges RCG (2006). Influencia da temperatura de armazenamento na qualidade das sementes de *Caesalpinia peltophoroides* Benth. (sibipiruna). Revista Árvore. 30(1):43-48.

Queiroga VP, Castro LBQ, Gomes JP, Silva AL, Alves NMC, Araujo DR (2009). Physiological quality of cotton seeds stored as a function of different treatments and cultivars. J.Agro-ind. Products 11(1):43-54.78(6):661-664.

Ruas RAA, Nascimento GB, Bergamo EP, Daur Jr RH, Arruda RG (2010). Embebição e germinação de sementes de crambe (*Crambe abyssinica*). Pesquisa Agropecuária Trop. 40(1):61–65.

Reuber MA, Johnson LA, Watkins LR (2001). Crambe seed dehulling for improved oil extraction and meal quality. J. Am. Oil Chemist's Soc.

Santos SRG, Paula CR (2005). Electrical conductivity test to evaluate the physiological quality of seeds of Sebastiania commersoniana (Bail) Smith and Dows - Euphorbiaceae. Braz. J. Seeds 27(2):136-145.

Sharma KD (1977). Biochemical changes in stored oil seeds. Indian J. Agric. Res. 11(3):137-141.

Souza RB, Vianna ACA, Soares CM, Ida El Oliveira LCS, Favaro SP (2009). Chemical characterization of seeds and jatropha cakes turnip fodder and crambe. Braz. Agric. Res. 44(10):1328-1335.

Toledo MZ, Cavariani C, Nakagawa J, Alves E (2007). Effects of storage environment on seed quality of sorghum-sudan. Braz. J. Seeds 29(2):44-52.

Ullmann R, Resende O, Sales JF, Chaves TH (2010). Quality of jatropha seeds subjected to artificial drying. Agronomic Sci. Mag. 41 (3): 442-447.

Vanzolini S, Nakagawa J (2005). Electrical conductivity test in peanut seeds. Braz. J. Seeds 27(2):151-158.

Vieira AR, Oliveira JA, Guimaraes RM, Pereira CE, Carvalho FE (2007). Storage of coffee seeds. environments and methods of drying. Braz. J. Seeds 29(1):76-82.

Vieira RD, Krzyzanowski FC (1999). Electrical conductivity test. In: Krzyzanowskl F.C. Vieira RD and France Neto JB (Ed.). Seed vigor: concepts and tests. London: Abrates. pp. 1-26.

Vieira RD, Scappi Neto A, Bittencourt SRM, Panobianco M (2004) .Electrical conductivity of the seed soaking solution and soybean seedling emergence. Scientia Agricola 61 (2):164-168.

9

Gamma irradiation can control the number of psychotrophic bacteria in Agaricus bisporus during storage

Meire C. N. Andrade[1], João P. F. Jesus[1], Fabrício R. Vieira[1], Sthefany R. Viana[1], Marta H. F. Spoto[2] and Marli T. A. Minhoni[1]

[1]Universidade Estadual Paulista, UNESP, Faculdade de Ciências Agronômicas, FCA, Departamento de Produção Vegetal/Defesa Fitossanitária, Módulo de Cogumelos. Rua José Barbosa de Barros, 1780 - Fazenda Lageado. Caixa Postal 237, 18610-307. Botucatu, SP, Brasil.
[2]Universidade de São Paulo, USP, Escola Superior de Agricultura, ESALQ, Departamento de Agroindústria, Alimentos e Nutrição. Av. Pádua Dias, 11, Caixa Postal 9, 13418-900. Piracicaba, SP, Brasil.

We evaluated the effect of gamma irradiation doses (0, 125, 250, and 500 Gy) in control of psychrotrophic bacteria in different strains of *Agaricus bisporus* (ABI-07/06, ABI-05/03, and PB-1) during storage, cultivated in composts based on oat straw (*Avena sativa*) and *Brachiaria* spp. The experimental design was completely randomized in a factorial scheme $4 \times 2 \times 3$ (irradiation doses $\times$ composts $\times$ strains), with 24 treatments, each consisting of 2 replicates, totaling 48 experimental units (samples of mushrooms). The mushrooms collected from all culture conditions were packaged in plastic polypropylene with 200 g each and subjected to Cobalt-60 irradiator, type Gammacell 220, and dose rate 0.740 kGy h^{-1}, according to the treatments. Subsequently, the control (nonirradiated) and other treatments were maintained at 4 ± 1°C and 90% relative humidity (RH) in a climatic chamber to perform the microbiological analysis of mushrooms on the 1st and 14th day of storage. According to the results, it was found that the highest mean colony psychotrophic count, after 14 days of storage, was observed in strain ABI-07/06 [1.30×10^{8} g^{-1} most probable number (MPN)] in nonirradiated mushrooms, coming from *Brachiaria* grass-based compost, and this same strain under the same storage conditions, coming from the same type of compost that underwent a dose of 500 Gy, obtained a significant reduction in mean colonies of psychrotrophic bacteria (2.25×10^{4} g^{-1} MPN). Thus, the irradiation doses tested favored reducing the number of colonies of psychrotrophic bacteria, regardless of the type of compound and strain of *A. bisporus*.

Key words: Champignon, shelf life, postharvest, mushrooms.

INTRODUCTION

Edible mushrooms are a food of high nutritional quality. However, the useful life of the mushrooms as well as their nutritional value vary depending on the species, strain, processing postharvest, developmental stage of mushroom, and the type of substrate used (Andrade et al., 2008; Bononi et al., 1995; Minhoni et al., 2005). As mushrooms are highly perishable, they tend to lose the quality immediately after harvest. The short shelf life (1 to 3 days at room temperature) is a disadvantage for the distribution and marketing of fresh product. In this period, changes occur in the mushroom, such as darkening, opening of the pileus, stem elongation, increase in the diameter of the hat, weight loss, and change in texture due to high respiration rate and lack of physical protection to prevent water loss or microbial attack (Akram and Kwon, 2010; Sommer et al., 2010; Singh et al., 2010).

Food irradiation is one of the best and most satisfactory techniques for food preservation. However, the irradiation dose required to control microorganisms on food depends on various factors such as the strength of each particular kind and degree of contamination of the food. For edible mushrooms, irradiation commonly is performed in many countries with low doses (1 to 3 kGy), aiming to reduce the number of spoilage microorganisms and prolonging its life and sensory qualities (Fernandes et al., 2012; Farkas, 2006).

In Brazil, it is still common practice to irradiate mushrooms as a method for postharvest storage, and for *Agaricus bisporus*, the production is largely marketed in the preservative solution, for instance, calcium chloride and glucanalactone (Kuyper et al., 1993; Rodrigo et al., 1999). However, it is known that such solutions change the original taste of fresh mushrooms and their physicochemical and sensory properties. Recently, Moda (2008) evaluated the shelf life of *Pleurotus sajor-caju* irradiated with 125, 250, 500, and 750 Gy and found that a dose of 750 Gy was the most suitable and its estimated shelf life of five days, being this time superior for nonirradiated mushrooms. However, in Brazil, no scientific reports were published of the use of gamma radiation to *A. bisporus*, being necessary to check the feasibility of applying this technique in growing conditions prevailing in the country. Thus, we evaluated the effect of irradiation doses (0, 125, 250, and 500 Gy) in control of psychrotrophic bacteria in strains of *A. bisporus* (ABI-07/06, ABI-05/03, and PB-1) grown in two kinds of composts (brachiaria and oats) during 14 days of storage at 4 ± 1°C and 90% RH.

MATERIALS AND METHODS

The production of mushrooms was developed on the premises Mushroom Module, Department of Plant Protection, Faculty of Agronomic Sciences (FCA/UNESP), Botucatu/SP, Brazil.

Experimental design and treatments

The experimental design was a factorial 4 × 3 × 2 (irradiation doses × strains × composts), with 24 treatments, each consisting of 2 replicates, totaling 48 experimental units (samples of mushrooms).

Data were subjected to analysis of variance, and means were compared by Tukey test, using the SISVAR 4.2 statistical program, developed by Department of Mathematical Sciences from UFLA (Federal University of Lavras), MG, Brazil.

Strains of *Agaricus bisporus*

We used pure culture (primary matrix) strains ABI-07/06, ABI-05/03, and PB-1; the strain ABI-07/06 originated from Piedade/SP, the strain ABI-05/03 originated from Cabreúva/SP, and the strain PB-1 from the company Brasmicel (Suzano-SP), and that, according to the same log files, was acquired by Paul Stamets (Washington, USA) in 2000. These strains are maintained in culture stock, the base culture medium (CA compost agar) submersed in mineral oil sterilized and maintained in biological oxygen demand (BOD) adjusted to 8°C, the Bank Matrix Module mushrooms, located at the Department of Production Plant of the Faculty of Agricultural Sciences, UNESP, Botucatu/SP, Brazil.

Composting, growing, and harvesting

We formulated two types of composts based on brachiaria (*Brachiaria* spp.) and oat (*Avena sativa*) straws, with an initial C/N ratio of about 25/1, for the cultivation of *A. bisporus* strains. All stages of composting, growing, and harvesting followed procedures used by Andrade et al. (2008).

Postharvest

The mushrooms collected from all culture conditions proposed were packaged in plastic polypropylene boxes each containing 200 g. Transport of samples was done in cool boxes to the Center for Nuclear Energy in Agriculture (CENA), University of São Paulo (USP), Brazil, where they were irradiated with 125, 250, and 500 Gy in Cobalt-60 irradiator, type Gammacell with 220 kGy dose rate 0.740 h^{-1}. The control (nonirradiated) and other treatments were maintained at 4 ± 1°C and 90% of moisture in an incubator for the realization of physicochemical analyzes on the 1st and 14th day of storage.

Sample preparation and dilutions

Analyses were performed on samples of 25 g of fresh mushrooms, weighed aseptically and placed in sterile elemeyers with 225 ml of peptone water (0.1%) sterile, constituting 10^{-1} dilution after stirring for 2 min in a peristaltic homogenizer. One milliter of 10^{-1} dilution was pipetted in 9 ml of sterile peptone water (0.1%) and from this dilution 10^{-2}, 10^{-3} to 10^{-6} dilutions were tested.

Psychrotrophic count

We used for the enumeration of psychrotrophic bacteria the dilutions 10^{-3} to 10^{-6} in duplicates. We added 0.1 ml of dilution in Petri dishes containing culture medium plate count agar (PCA) previously prepared and sterilized, these being subsequently inverted and incubated at 7°C for 10 days. After this period, we counted the number of colonies by MPN g^{-1} food (most probable number). This same methodology procedure and evaluation was repeated after 14 days of storage of the mushroom samples. Thus, microbiological evaluations were made for two periods (on the 1st and 14th day of storage), totaling 384 Petri dishes, with 192 in each period.

Table 1. F values from analysis of variance of the number of psychrotrophic microorganisms colonies present in mushrooms of *Agaricus bisporus* strains ABI-05/03, ABI-07/06, and PB-1, newly irradiated (initial), and stored at 4 ± 1°C for 14 days (final), grown on two kinds of composts based on oat straw and *Brachiaria* and submitted to irradiation doses 0, 125, 250, and 500 Gy.

Variation cause	Psychrotrophic (initial)	Psychrotrophic (final)
Strain (S)	236.75**	90.66**
Compost (C)	7.84**	29.25**
Dose irradiation (D)	1450.94**	261.72**
S × C	10.59**	10.58**
S × D	173.98**	91.43**
C × D	12.90**	25.62**
S × C × D	42.29**	11.20**
CV (%)	14.7	37.4

** Significant at 1%.

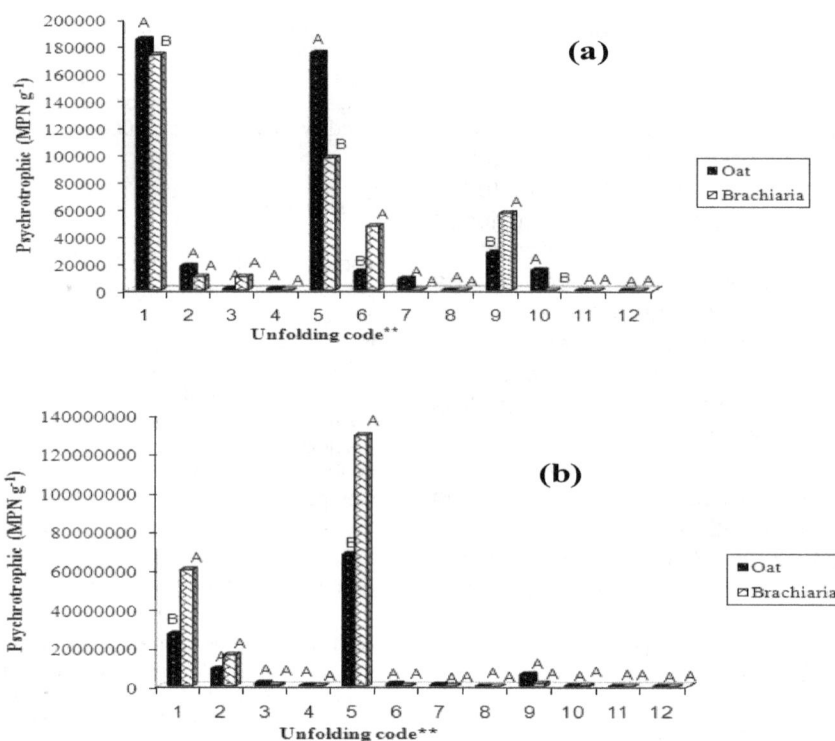

Figure 1. Comparison of mean number of psychrotrophic microorganisms colonies in newly irradiated mushrooms (a) and stored at 4 ± 1°C for 14 days (b), in function of the type of compost. Means followed by the same letter within each evaluation period and unfolding encoding are not statistically different from each other (Tukey, 5%).**Agaricus bisporus* Strain - Dose Irradiation:1- ABI 05/03 - 0, 2- ABI 05/03 - 125, 3- ABI 05/03 - 250, 4- ABI 05/03 - 500, 5- ABI 07/06 - 0, 6- ABI 07/06 - 125, 7- ABI 07/06 - 250, 8- ABI 07/06 - 500, 9- PB-1 - 0, 10- PB-1 - 125, 11- PB-1 - 250, 12- PB-1 - 500.

RESULTS AND DISCUSSION

F values from analysis of variance of the number of psychrotrophic bacterial colonies present in *A. bisporus* are shown in Table 1. For all strains of *A. bisporus* tested at doses of 250 and 500 Gy, there was no significant difference in the number of psychrotrophic bacterial colonies from mushrooms grown on composts based on oats and *Brachiaria* grass, both in newly irradiated and in stored mushrooms (Figure 1). Moreover, in nonirradiated

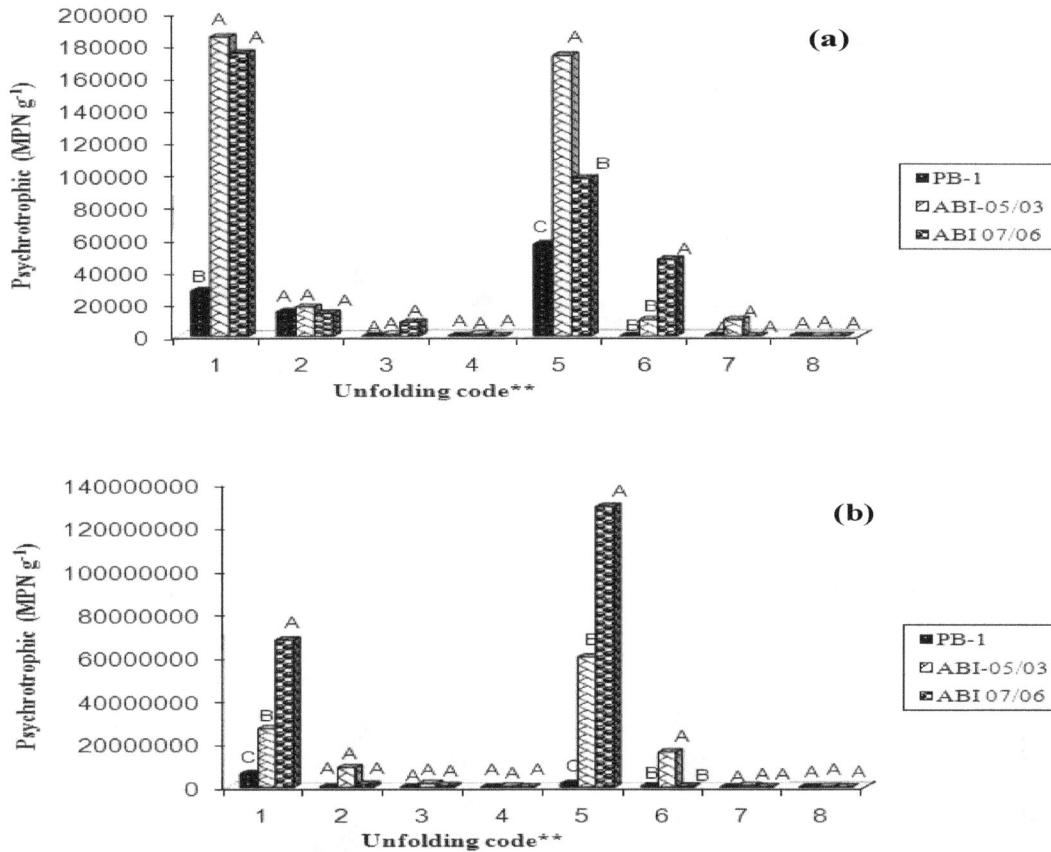

Figure 2. Comparison of mean number of psychrotrophic microorganisms colonies in newly irradiated mushrooms (a) and stored at 4 ± 1°C for 14 days (b), in function of the strain of Agaricus bisporus. Means followed by the same letter within each evaluation period and unfolding encoding are not statistically different from each other (Tukey, 5%).**Compost - Dose Irradiation: 1- Oat - 0, 2- Oat - 125, 3- Oat - 250, 4- Oat - 500, 5- Brachiaria − 0, 6- Brachiaria - 125, 7- Brachiaria - 250, 8- Brachiaria - 500.

(control) mushrooms, the number of psychrotrophic bacteria was influenced by the compost type. These results indicate that irradiation at doses of 250 and 500 Gy was effective in eliminating the effect of the compost on the number of psychrotrophic bacteria and indicated a significant reduction of microorganisms compared to control in both periods.

The highest average number of psychrotrophic colonies, after 14 days of storage, was observed in the ABI-07/06 strain (1.30×10^8 MPN g^{-1}) in nonirradiated mushrooms, coming from Brachiaria compost (Figure 1). This same strain under the same storage conditions, coming from the same compost type but after a dose of 500 Gy, showed a significant reduction in mean number of psychrotrophic bacterial colonies (2.25×10^4 MPN g^{-1}). These results differ from those obtained by Moda (2008), which evaluated the increased shelf life mushroom P. sajor-caju with application of gamma radiation and also found that the average psychrotrophic bacteria at a dose of 500 Gy (4.5×10^8 MPN g^{-1}) in mushrooms P. sajor-caju stored for 10 days was not lower than the average obtained on samples of nonirradiated mushrooms (2.5

$\times 10^7$ MPN g^{-1}).

It is known that the type of mushroom and irradiation dose applied directly influence the results of increasing the mushrooms' shelf life. About this, Rivera et al. (2011) evaluated the effect of gamma irradiation on the microbial population of Tuber melanosporum; during 35 days of storage at 4°C, they found that the dose of 1500 Gy did not increase the useful life of mushrooms and maintenance of the quality of truffles. Moreover, Jiang et al. (2010), evaluating the effect of integrated application of gamma irradiation (1000, 1500 and 2000 Gy) and modified atmosphere on the microbiological properties of shiitake mushroom (Lentinula edodes), found that the dose of 1000 Gy was the most efficient in maintaining the firmness level of mushrooms.

Comparing the mean number of psychrotrophic bacterial colonies of strains of A. bisporus, it was found that there were significant differences in the nonirradiated treatments in two periods (Figure 2). Furthermore, in samples of mushrooms subjected to doses of 250 and 500 Gy, the values for the strains of A. bisporus were similar to each other as was the reduction in the number

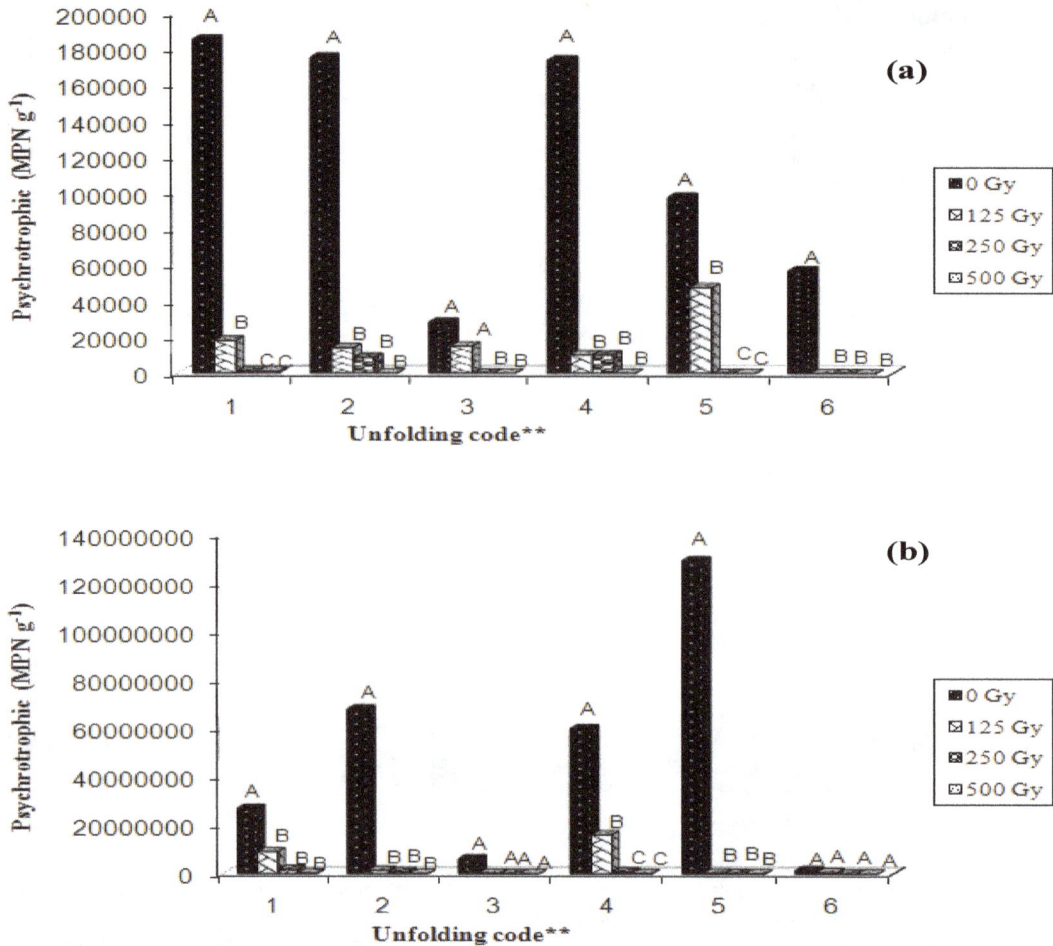

Figure 3. Comparison of mean number of psychrotrophic microorganisms colonies in newly irradiated mushrooms (a) and stored at 4 ± 1°C for 14 days (b), in function of irradiation dose. Means followed by the same letter within each evaluation period and unfolding encoding are not statistically different from each other (Tukey, 5%). **Compost –*Agaricus bisporus* Strain: 1- Oat – ABI-05/03, 2- Oat – ABI-07/06, 3- Oat – PB-1, 4- Braquiaria - ABI-05/03, 5- Braquiaria – ABI-07/06, 6- Braquiaria – PB-1.

of psychrotrophic bacteria (compared to control) in both irradiated and fresh mushrooms stored, resulting in an increase in the mushrooms' shelf life. Beaulieu et al. (2002) also reported the benefit of preserving irradiation on the mushroom *A. bisporus* reporting that the useful life was extended to 4 days with a dose of 4500 Gy of irradiation. Also, Lescano (1994) reports that the dose of 3000 Gy, combined with Poly-Vinyl Chloride (PVC) film packaging and storage of 10 ± 2°C, increased the shelf life of mushrooms *A. bisporus*, providing a white color retention, and growth and the opening of the pileus, acceptable for commercialization for 11 days and for consumption up to 16 days of storage.

Regarding the mean colonies of psychrotrophic bacterial colonies on mushrooms when submitted to irradiation, it was found that all doses tested favored reducing these colonies compared to control (Figure 3). For ABI-05/03 strain grown on oat-based compost, the average psychrotrophic colony count in the nonirradiated

treatment was 1.85×10^5 MPN g^{-1}. Already at a dose of 500 Gy, this number was reduced to 1.00×10^3 MPN g^{-1}. Similar results were obtained by Moda (2008) who also found a reduction of the number of psychrotrophic bacteria in *P. sajor-caju* between 1 day before and 1 day after irradiation (500 Gy) of 6.2×10^6 MPN g^{-1} for 1.3×10^4 MPN g^{-1}, respectively.

Conclusion

The irradiation doses tested favored the reducing number of psychrotrophic bacteria colonies, regardless of composts and *A. bisporus* strains.

ACKNOWLEDGMENT

To the coordination of improvement of personnel of

superior level, by the conception of the master scholarship to the first author.

REFERENCES

Akram K, Kwon JH (2010). Food irradiation for mushrooms: A review. J. Korean Soc. Appl. Biol. Chem. 53:257-265.

Andrade MCN, Zied DC, Minhoni MTA, Kopytowsky FJ (2008). Yield of four *Agaricus bisporus* strains in three compost formulations and chemical composition analyses of the mushrooms. Braz. J. Microbiol. 39:593-598.

Beaulieu M, D'Aprano G, Lacroix M (2002). Effect of dose rate of gamma irradiation on biochemical quality and browning of mushrooms *Agaricus bisporus*. Radiat. Phys. Chem. 63:311-315.

Bononi VL, Capelari M, Trufem SFB (1995). Cultivo de cogumelos comestíveis. Ícone, São Paulo, P. 206.

Farkas J (2006). Irradiation for better foods. Trends Food Sci. Technol. 17:148-152.

Fernandes A, Antonio AL, Oliveira MBPP, Martins A (2012). Effect of gamma and electron beam irradiation on the physico-chemical and nutritional properties of mushrooms: A review. Food Chem. 135:641-650.

Jiang T, Luo S, Cheng Q, Shen L, Ying T (2010). Effect of integrated application of gamma irradiation and modified atmosphere packaging on physicochemical and microbiological properties of shiitake mushroom (*Lentinula edodes*). Food Chem. 122:761-767.

Kuyper L, Weinert IAG, Mc Gill AEJ (1993). The effect of modified atmosphere packaging and addition of calcium hypochlorite on the atmosphere composition, colour and microbial quality of mushrooms. Food Sci. Technol. 26:14-20.

Lescano G (1994). Extension of mushroom (*Agaricus bisporus*) shelf life by gamma irradiation. Postharvest Biol. Technol. 4:255-260.

Minhoni MTA, Kopytowski Filho J, Andrade MCN (2005). Cultivo de *Agaricus blazei* Murrill ss. Heinemann. Fundação de Estudos e Pesquisas Agrícolas e Florestais, Botucatu, P. 141.

Moda EM (2008). Aumento da vida útil de cogumelos *Pleurotus sajor-caju in natura* com aplicação de radiação gama. Esalq/USP, Piracicaba, P. 105.

Rivera CS, Venturini ME, Marco P, Oria R, Blanco D (2011). Effects of electron-beam and gamma treatments on the microbial populations, respiratory activity and sensory characteristics of Tuber melanosporum thuffes packaged und modified atmospheres. Food Microbiol. 28:1252-1260.

Rodrigo M, Calvo C, Sanchez T, Rodrigo C, Martinez A (1999). Quality of canned mushrooms acidified with glucana-lactone. Int. J. Food Sci. Technol. 34:161-166.

Singh P, Langowski HC, Wanib AA, Saengerlaub S (2010). Recent advances in extending the shelf life of fresh *Agaricus* mushrooms: A review. J. Sci. Food Agric. 90:1393-1402.

Sommer I, Schwartz H, Solar S, Sontag G (2010). Effect of gamma-irradiation on flavour 50-nucleotides, tyrosine, and phenylalanine in mushrooms (*Agaricus bisporus*). Food Chem. 123:171-174.

10

A comparative analysis of barn and platform as storage structures for yam tuber in Ibadan, Nigeria

Y. Mijinyawa and John O. Alaba

Department of Agricultural and Environmental Engineering, Faculty of Technology, University of Ibadan, Oyo State, Nigeria.

An experimental study was undertaken to evaluate and compare the performances of a local barn and a platform, as storage structures for yam tubers (*Dioscorea rotundata poir.*). The criteria used for evaluation and comparison were the degree of weight loss during storage, tuber sprouting and rotting of yam tubers during 17 weeks storage duration between March and June 2008. Measurements of temperatures and relative humidity in the storage environment were taken thrice daily during the period. Weight loss in each tuber was measured weekly while sprouts were removed from tubers fortnightly. Results show that, the average temperature and relative humidity on the platform were 30.4°C and 57.3% respectively while for the barn, they were 26.5°C and 55.5%, respectively. The average weight loss in tubers in the barn during the duration was 32.8% while for tubers on the platform, it was 30.3%. Yam tubers on the platform recorded 5.4% sprouting while those in the barn had 4.9% sprouting. Palm leaves cover for yam tubers on the platform protected the tubers from excessive heat and moisture loss. Rotting was observed in 10% of the tubers stored in the barn but was completely absent from those stored on the platform.

Key words: Barn, platform, sprouting, weight loss, yam tuber.

INTRODUCTION

Yam belongs to the genus *Dioscorea* which has over 600 species but only about 6 of which are cultivated for human consumption while a few non-edible ones are cultivated for industrial raw materials. Yam plays a prominent role in a variety of human food diets and livestock feed in many of the areas where it is cultivated (Lancaster and Coursey, 1984; Opara, 1999; IITA, 2008). Yam has socio-economic and cultural values in many parts of the world, these being manifested in the celebration of traditional ceremonies to usher in the new yam season (Opara, 1999). In some parts of Nigeria, it is customary for the parents of a bride to offer her yams for planting as a resource to assist them in raising the family. The meals offered to gods and ancestors by some traditionalists consist principally of mashed yam. A well-built and well stocked yam barn is one of the major factors through which a man gains prestige in some communities (Lancaster and Coursey, 1984; Opara, 1999).

Although, yam can be cultivated in many environments with a temperature of between 25 and 30°C, an annual rainfall of about 1,000 to 1,500 mm distributed evenly over the vegetative period of 5 to 6 months and deep, fertile, friable, and well-drained soils, majority of the world's yams are cultivated in Sub-Saharan Africa. The 2005 world yam production records show that, of the annual production of 48.7 000 000 tonnes from about 47 countries of the tropical and sub-tropical regions, 97% was produced in Sub-Saharan Africa with Nigeria alone accounting for about 70%. Ghana is the world's largest exporter of yam with an annual export of about 12,000 tonnes (IITA, 2008).

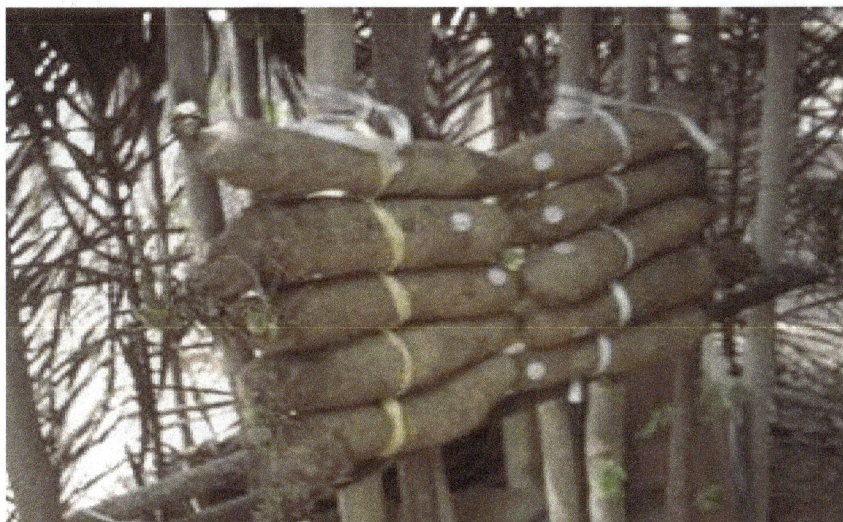

Plate 1. An interior section of the yam barn.

Most edible yams species reach maturity in 8 to 11 months after planting. As a seasonal crop, harvested yam tubers are stored to meet the demand during the off-season period. Adequate aeration, reduction of temperature, protection from direct sunlight and flood, and regular inspection of produce are the basic requirements for successful and long term storage of yam tubers (Wilson, 1980; Lancaster and Coursey, 1984; Orhevba and Osunde, 2006). Ventilation prevents moisture condensation on the tuber surface and assists in removing the heat of respiration. Low temperature is necessary to reduce losses from respiration, sprouting and rotting; while regular inspection is important to remove sprouts, rotted tubers, and to monitor the presence of rodents and pests. Dormancy in stored yam tubers, is the period after harvest during which sprouting is inhibited. It is influenced by the yam species, temperature and relative humidity of the storage environment. At lower temperatures, the rate of respiration is reduced, the formation of germ is delayed and the onset of sprouting can be prolonged leading to longer storage periods (Orraca-Tetteh, 1978; Knoth, 1993; Shiwachi et al., 2002).

The structures used for the storage of yam tubers are numerous. Some of the storage structures include trench or clamp silos, underground pits, barns of various designs, shelves in specially constructed or improvised sheds, raised huts, and assorted platforms. The popularity of these structures varies from one region to another, and the choice made depends on the volume to be stored and what the farmer can afford.

Oyo State lies between latitude 7°03' and 9° 23' N and between longitude 2°47' and 4°35'E, with an annual rainfall of about 1,300 mm distributed over the period of 8 months. Yam is one of the most popular crops cultivated by the farmers, which after harvest is stored either as chips or tubers. The barn and platform are among the most popular structures used for the storage of the tubers. The work reported here was undertaken to evaluate and compare the performances of these two structures for the storage of the yam tuber.

MATERIALS AND METHODS

Experimental site

A hilltop under a tree shade within the premises of the Department of Agricultural and Environmental Engineering, University of Ibadan, was selected to simulate the traditional practice whereby farmers prefer elevated sites under shades with adequate drainage and unobstructed ventilation for the storage of yam tubers.

Storage structures

The storage structures used for this study were a barn and platform. The barn was constructed of main vertical poles from *Gliricidia sepium* species. Holes of 30 cm depth and spaced 30 cm apart were dug along the perimeter of the 2.7 m by 1.7 m barn for the support of the main vertical members. Poles of average diameter of 5 cm were used as main vertical members. The main members were held together by smaller diameter poles placed horizontally and tied using cordage. The exterior of the barn was covered with palm leaves tied to the poles. The palm leaves were also used as roofing material for the barn which was 1.9 m above the floor. A small entrance was created for entry into the barn for the purpose of inspection. The yam tubers were placed horizontally along the height of the main members, one above the other and tied using cordage (Plate 1).

For the construction of the platform, 4 holes to a depth of 30 cm were dug at the corners of a 0.9 m × 0.9 m square. Four 1.2 m high Y-fork columns of diameter 8 cm were inserted in the holes and compacted. Metal sheets were cut, shaped and wound round the columns to serve as rodent guards. Split *Bambusa vulgaris* were laid horizontally on top of the columns to form the platform. The yam tubers were then heaped on the platform and covered with

Plate 2. A platform with yam in storage.

palm leaves (Plate 2).

Sourcing of yam tubers

The yam tubers used for this study were white yam (*Dioscorea rotundata poir.*) which is the most commonly cultivated species among the farmers in the area of study. The tubers used were harvested from a farm at Igboho, a popular yam growing community in Oyo state. The authors undertook the harvesting in order to ensure that there was no mix up of species. The harvesting was carefully done to ensure that no mechanical damages were inflicted on the tubers. The tubers were cleaned by trimming off roots attachment

Experimentation

The experimentation consisted of monitoring the temperature and relative humidity of the storage environments, the weight losses in the tubers and the rate of sprouting in the tubers.

Temperatures and relative humidity were monitored using wet and dry bulb thermometers and hygrometers. Three readings; at 8.00 am; 12.00 noon, and 6.00 pm were taken daily and averaged on weekly basis.

The tubers were labeled for ease of identification and weighed before the commencement of the experiment. The weights of the tubers were taken every week and the difference between subsequent weights represented the weekly loss. Weights were measured with Digital and SK 2000 weighing balance sensitive to 1

g and 2 kg capacity. The sprouts were removed fortnightly and weighed.

During the weekly weighing periods, the tubers were also observed for physical defects such as rotting and insect attack. The experiment was carried out for a period of 17 weeks between March and June 2008.

RESULTS AND DISCUSSION

Storage environment

The platform and barn temperatures are presented in Figure 1. While the temperatures in the platform varied from 23.1 to 32°C with an average value of 30.4°C, those within the barn varied from 23.0 to 31.7°C with an average value of 26.5°C. The temperatures within the barn were generally lower than those recorded in the platform for all periods throughout the experimentation period.

The relative humidity ranged from 44 to 78.3% with an average value of 55.5% for the barn while for the platform, the range was from 47 to 80% with an average of 57.3%. Although, there were a few overlaps, the relative humidity within the barn was generally lower than for the platform (Figure 2).

The variation in the environmental conditions within the

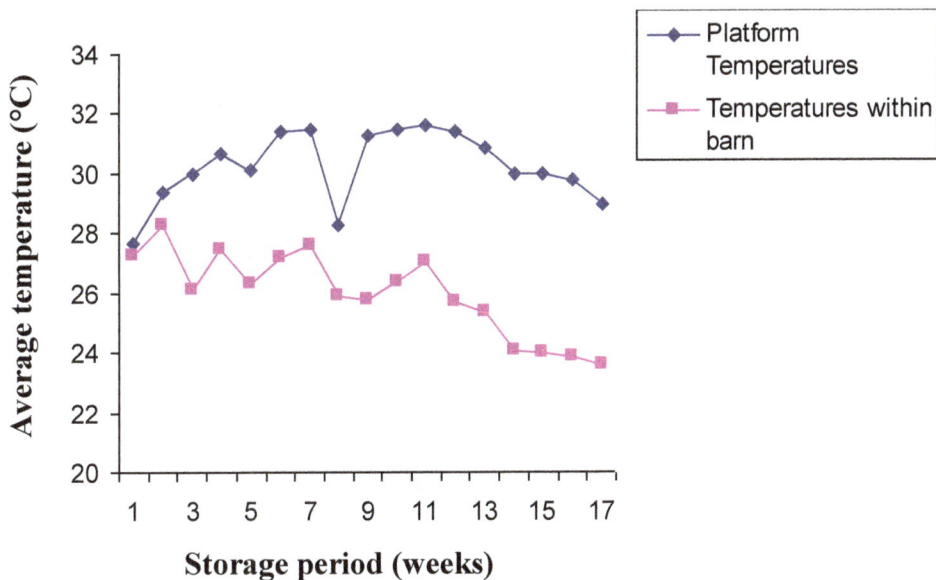

Figure 1. Temperature fluctuations with storage period.

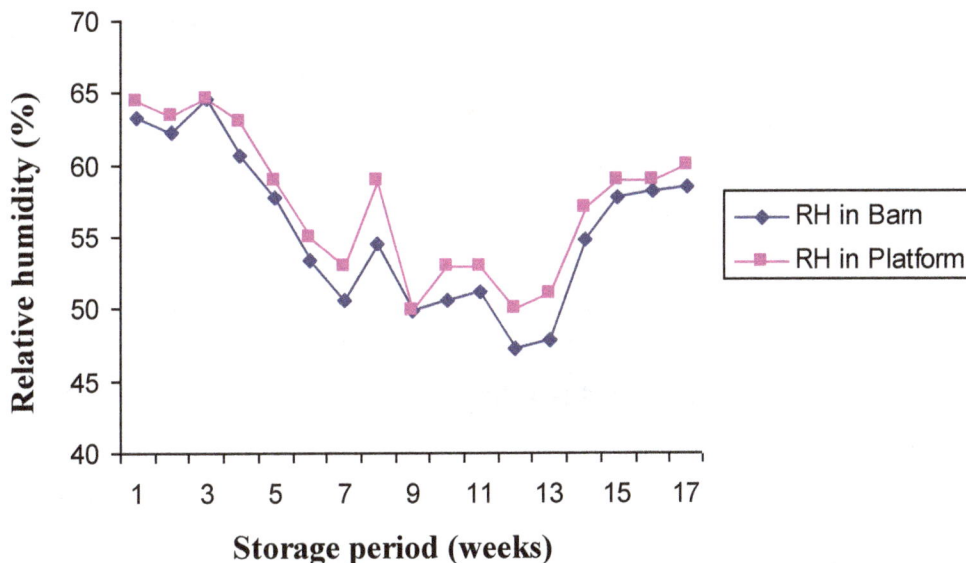

Figure 2. Relative humidity versus storage period.

two structures is attributed to the mulching effect of the palm leaves used on the platform. There is more ventilation within the barn which favours the release of heat and moisture from the enclosure unlike the platform in which the palm leaves acted as a barrier and inhibited the escape of heat and moisture.

Weight loss

The weekly rate and cumulative weight losses observed in the stored yam tubers are presented in Figures 3 and 4 while the summary of the statistical analysis is presented in Table 1. The values of the weekly weight losses were higher in the barn than the platform except for the 2nd, 4th and 11th weeks when the rate was observed to be higher on the platform than the barn. Although there was no significant difference ($p > 0.05$) between the rates of weight losses in the barn and the platform, the mean value for the barn (1.93) was higher than the platform (1.78). Weight loss in stored yam tubers is attributed to three factors. These are moisture loss through

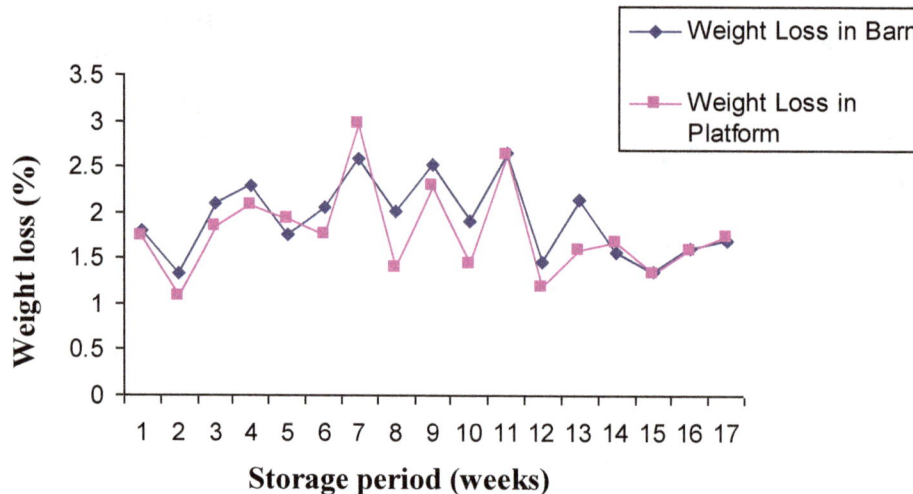

Figure 3. Weekly weight losses.

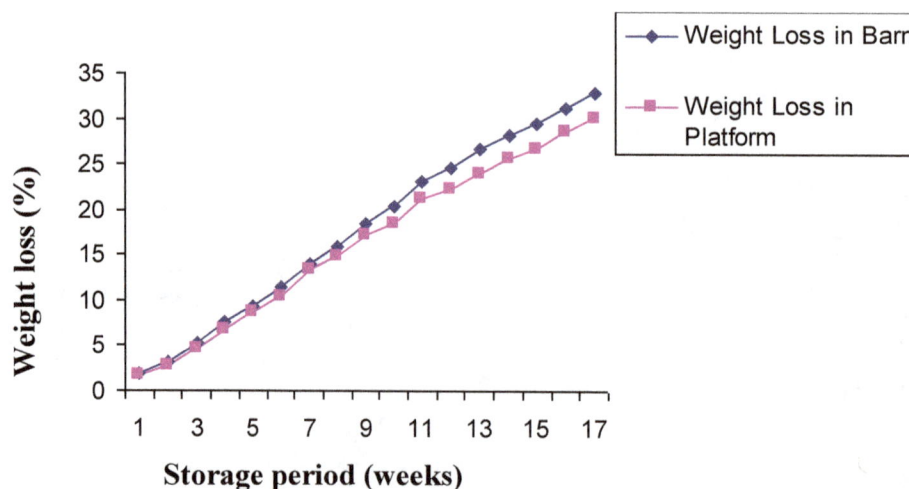

Figure 4. Cummulative weight loss.

Table 1. T-test for weight loss in tubers stored in barn and platform.

Group	N	$\bar{x}$	SD	df	t	Sig. t
Barn	17	1.93	0.42			
Platform	17	1.78	0.49	32	0.96*	0.86

*Not significant at $p > 0.05$.

transpiration, respiration and sprouting which exhaust the food stored in the yam. Among the three factors, moisture loss is reported to contribute the highest percentage to the weight loss even though such loss may not be in terms of the edible portion of the tuber (Orraca-Tetteh, 1978; Wilson, 1980).

The observed difference in weight loss between the yam tubers stored in the two structures can be attributed to the amount of moisture loss. The palm leaves used for covering acted as mulch which restricted the rate of moisture loss from the tubers unlike the tubers in the barn which were not well protected.

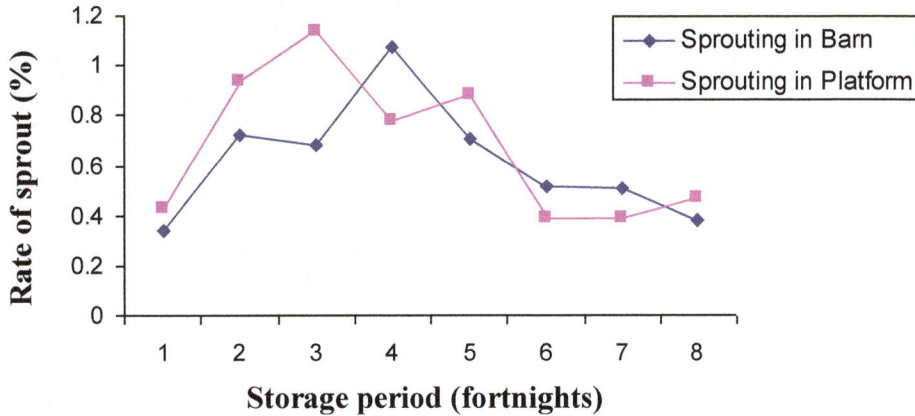

Figure 5. Rate of sprouting.

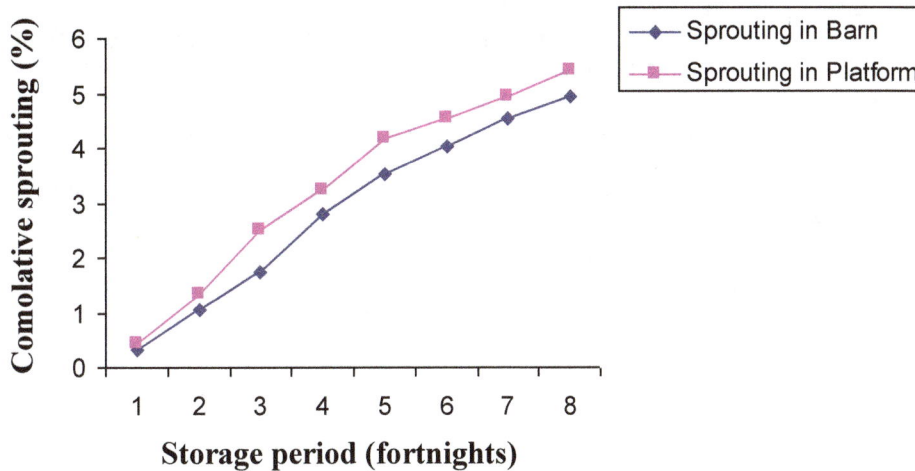

Figure 6. Cumulative sprouting.

Table 2. T-test for sprouting in tubers stored in barn and platform.

Group	N	$\bar{x}$	SD	df	t	Sig. t
Barn	8	0.62	0.23	14	0.46*	0.21
Platform	8	0.68	0.29			

*Not sig. at p > 0.05.

Sprouting

The weekly and cumulative data of sprouts in the yam tubers stored in the two structures are presented in Figure 5 and 6 while the summary of the statistical analysis is presented in Table 2. Although, there was no significant difference (p > 0.05) between the sprouting in barn and the platform, the mean value for the barn (0.62) was lower than the platform (0.68). Sprouting is promoted by humid environment and high temperatures. The higher relative humidity and temperatures within the yam tubers stored on the platform are the major factors responsible for the higher sprouting in the platform than the barn.

Physical observation

Rotting of yam tubers was one of the physical parameters

considered in this study. At the 5th week of storage, rotting was initiated in three tubers within the barn and by the end of the 14th week, a tuber had completely rotted. There was however no incidence of rotting among the yam tubers stored on the platform throughout the period of the experiment.

CONCLUSIONS AND RECOMMENDATIONS

Yam tubers were stored in barn and platform over a period of 17 weeks. While the barn maintained an environment of 26.5°C and 55.5% relative humidity, the platform environment was 30.4°C and 57.3% relative humidity. The yam tubers stored on the platform were found to have sprouted more (5.42%) than those in the barn (4.93%) but the overall weight loss in the yam tubers was more in the barn (32.8%) than on the platform (30.3%). The mulching effect of the palm leaves cover for the yam tubers stored on the platform was considered a possible factor. There was no rotting observed among the tubers stored on the platform as against the barn where rotting was observed in about 10% of the yam tubers. The platform is able to reduce weight loss which is to the advantage of the farmer as yams are priced on weight basis and it may therefore be preferred. Further work which should involve longer storage periods and determination of the qualities of the stored yam tubers is recommended.

ACKNOWLEDGEMENTS

The authors are grateful to Messrs B. Aluogho and A. N. Kronakumo for their assistance in the collection of materials and construction of the storage structures.

REFERENCES

International Institute for Tropical Agriculture (IITA) (2008). Yam http://www.iita.org/cms/details/yam_project_details.aspx?zoneid=63&articlei d=268. Accessed May 19, 2008.

Knoth J (1993). Traditional Storage of Yams and Cassava and Its Improvement. http://www.fao.org/inpho/content/documents//vlibrary/gtzhtml/x0066e/X0066 E00.htm#Contents. Accessed May 19, 2008.

Lancaster PA, DG Coursey (1984). Traditional Postharvest Technology of Perishable Tropical Staples. FAO Agricultural Services Bulletin No. 59. Http://Www.Fao.Org/Docrep/X5045e/X5045e00.Htm#Contents. Accessed May 19, 2008.

Opara LU (1999). Yam storage. In CIGR Handbook of Agricultural Engineering Volume IV: Agro Processing. The American Society of Agricultural Engineers, St. Joseph, MI. pp. 182-214.

Orhevba BA, Osunde ZD (2006). Effects of Storage Condition and CIPC Treatments on Sprouting and Weight Loss of Stored Yam Tubers. Proceedings of the 28th Annual Conference of the Nigerian Institution of Agricultural Engineers. 28:352-360.

Orraca-Tetteh R (1978). A Note on Post-Harvest Physiology and Storage of Nigerian Crops. Food Nutr. Bull. 1:1. http://www.unu.edu/Unupress/food/8F011e/8F011E00.htm#Contents. Accessed May 19, 2008.

Shiwachi H, Ayankanmi T, Asiedu R, Onjo M (2002). Induction Of Sprouting In Dormant Yam (*Dioscorea* Spp.) Tubers With Inhibitors of Gibberellins.http://journals.cambridge.org/action/displayAbstract?fromPage=online&aid=1 47025.

Wilson JE (1980). Careful Storage of Yams. Commonwealth Secretariat, London. P. 8. http://www.ctahr.hawaii.edu/adap2/Publications/Ireta_pubs/yam_storage.pdf. Accessed May 20, 2008.

11

Diallel analysis for pod yield and its components traits in vegetable Indian bean (*Dolichous lablab* L.)

Patil Atul B., Desai D. T., Patil Sandip A. and Ghodke Umesh R.

ACHF Farm, Navsari Agriculture University, Navsari-396 450, Gujarat, India.

The nature and magnitude of genetic variance for yield and its component traits were studied in Indian bean using diallel analysis. The estimates of general combining ability (GCA) variance were much higher than specific combining ability (SCA) variance except days to 50% flowering, number of pod per cluster and fiber content; this indicated the importance of both additive as well as non-additive gene effects are involved in the expression of these characters. Genotypes NIB-69 and NIB-54 were identified as good general combiner for pod yield per plant. The cross combination viz., NIB-57 x NIB-69, NIB-69 x NIB-80, NIB-32 x NIB-54, NIB-41 x NIB-69 and NIB-23 x NIB-54 were the most promising crosses for improvement of pod yield. In the light of present study, the use of good general combining parents in the hybridization programme, selection of the desirable segregants from the segregating generations by adopting progeny selection method for exploiting additive genetic variance would lead to rapid improvement in this crop.

Key words: Genetic variance, diallel analysis, Indian bean.

INTRODUCTION

Indian bean (*Lablab purpureus* L.) is an important pulse crop of Gujarat. There are two cultivated types of Indian bean *viz.,* typicus and lignosus (Shivashankar et al., 1971). Typicus is a garden type and is cultivated for its soft and edible pods. Lignosus is known as field bean and mainly cultivated for dry seed as pulse and is more popularly recognized as 'Wal', 'Wal–papdi' or 'Valor' in Gujarat state. The green pods are used for vegetable purpose whereas; ripe and dried seeds are consumed as split pulse. The seeds can sometimes be soaked in water overnight and when germination initiates, they can be sun-dried and stored for future use. The fodder has good palatability and the cattles are nourished well. It can also be used as nitrogen fixing pulse crop. The fresh/immature pods contain 4.5% proteins and 10% carbohydrates.

For the development of elite strain, the identification of genetically superior parent is an important prerequisite. Combining ability studies reveal the nature of gene action and lead to identification of parents with high general combining ability effect and the cross combination with high specific combining ability effects. This in turn helps in choosing the parents to be included hybridization or population breeding programme. Among the different biometrical methods employed to study combining is the one proposed by Griffing's (1956).

MATERIALS AND METHODS

Studies were conducted by using 8 genotype being maintained at the Regional Horticulture Research Station, Navsari Agricultural

Table 1. Analysis of variance for combining ability in Indian bean.

S/N	Characters	Mean squares			
		GCA	SCA	Error	GCA/SCA
1	Days to 50% flowering	18.79	43.23**	9.69	0.45
2	Plant height (cm)	71.53**	64.01**	11.99	1.11
3	Number of branches per plant	0.25**	0.54**	0.05	1.46
4	Number of pods per cluster	3.57	0.58**	0.10	6.1
5	Number of pods per plant	230.19**	124.32**	6.19	1.85
6	Pod length (cm)	0.50**	0.32**	0.04	1.55
7	Seeds per pod	1.07**	0.45**	0.05	2.37
8	Pod yield per plant (g)	1357.91**	1547.64**	21.83	0.87
9	Average pod weight (g)	5.89**	5.41**	0.69	1.08
10	Protein content (%)	4.14**	1.39**	0.16	2.97
11	Fiber content (%)	0.16	0.05	0.04	3.2

*,**Significance at 5 and 1% level respectively.

University, Navsari. Twenty eight F_1's were developed half diallel mating design and evaluated along with parents in randomized block design with three replications during 2009-10. Plant to Plant and row to row distance was maintained as 20 and 60 cm respectively. All the cultural practices were followed to raise the normal crop. Data were recorded on five randomly selected plants in each treatment for eleven characters viz. Days to 50% flowering, plant height, number of branches per plant, number of pod per cluster, number of pods per plant, pod length, seeds per pod, green pod yield per plant, average pod weight, protein content and fiber content. General and specific combining ability effects were estimated according to method describe by Griffing's (1956).

RESULTS AND DISCUSSION

The analysis of variance for combining ability presented in Table 1 revealed that mean squares due to general combining ability were highly significant for all the characters except days to 50% flowering, number of pods per cluster and fiber content. Similarly variance for specific combining ability was highly significant for all the characters except fiber content. The analysis of variance for combining ability revealed that variances due to general combining ability (GCA) and specific combining ability (SCA) were highly significant for plant height, number of branches per plant, number of pods per plant, pod length, seed per pod, pod yield per plant, average pod weight and protein content indicating the importance of both additive as well as non-additive genetic components of variation in the expression of these economic characters. However, the additive variance was considerably higher than non-additive variance for all the attributes except green pod yield per plant indicating predominance of additive variance in controlling expression of these characters. The present findings also supported the results of Kabir and Sen (1990), Basu et al. (2002) and Kannan et al. (2003).

The estimates of general combining ability effects of parental lines (Table 2) revealed that the NIB-54 showed significant positive GCA effects for green pod yield per plant, days to 50% flowering, plant height, number of branches per plant, number of pod per cluster, number of pods per plant, pod length, seeds per pod, average pod weight and protein content, which indicates the best general combiner among the parents. Among the parents, NIB-69 was also found to be the good general combiner for all the characters under study except days to 50% flowering, pod length, protein content and fibre content. Whereas, NIB-80 for number of pod per plant, average pod weight and protein content. Similar result also recorded by earlier workers, Kannan et al. (2003) and Gavali et al. (2011) for number pods per plant, pod yield per plant, Basu et al. (2002) and Virja et al. (2006) for average pod weight, Sawant et al.(2006) for pod length.

The SCA effects for hybrids pertaining to different characters are given in Table 3. In the present study, best cross combinations involved good x good, good x poor and even poor x poor SCA effects. The top cross combinations NIB-57 x NIB-69 (poor x Good), NIB-69 x NIB-80 (Good x poor) and NIB-32 x NIB-54 (Poor x Good) had exhibited highest significant specific combining ability effects for green pod yield per plant, plant height, branches per plant, pod per plant and average pod weight, having at least one parent is good general combiner. Which indicating that hybrids having one parent with high GCA effect are expected to produce segregants of fixable nature in segregating generations though simple pedigree method. The hybrids *viz.,* NIB-57 x NIB- 80, NIB-23 x NIB- 54 and NIB-41 x NIB-69 showed highest significant positive SCA effects for days to 50% flowering, indicates earliness for vegetable purpose. For protein content NIB-41 x NIB-54, NIB-69 x NIB- 80 and NIB-41 x NIB-69 and the cross NIB-69 x NIB- 80 for fibre content appeared highest SCA effects in desired direction. Similar results also obtained by Singh et al. (1980), Bagade et al. (2002), Valu et al. (1999) and

Table 2. Estimation of general combining ability (GCA) effects of parents for various characters in Indian bean.

Parent	DFF	PH	NBP	NPC	NPPP	PL	SPP	PYPP	APW	PC	FC
NIB 23	1.234	0.116	-0.182*	-0.111	-5.767**	-0.008	-0.378**	-11.609**	-0.342	0.562**	0.202**
NIB 32	1.262	1.571	0.036	0.072	-3.758**	-0.063	-0.254**	-4.649**	0.237	-0.011	-0.073
NIB 41	0.669	2.376*	-0.133	-0.281**	-3.380**	0.329**	0.316**	-7.565**	-0.065	-0.809**	0.128*
NIB 54	-1.894*	-2.630*	0.165*	0.394**	3.682**	0.238**	0.142*	15.226**	0.634*	0.888**	-0.085
NIB 56	1.476	1.484	-0.117	0.278**	4.077**	-0.063	-0.083	-3.221*	-1.055**	-0.166	-0.185**
NIB 57	-0.094	1.305	-0.060	0.649**	-4.041**	-0.122	0.493**	-4.622**	0.612*	-0.731**	0.095
NIB 69	-1.641	-5.444**	0.284**	0.293**	7.204**	0.088	0.156*	21.224**	1.004**	-0.389**	-0.070
NIB 80	-1.044	1.224	0.007	-1.292**	1.983**	-0.399**	0.392**	-4.785**	-1.025**	0.654**	-0.011
S.E (g$_i$) ±	0.92122	1.02452	0.06911	0.09517	0.73627	0.06155	0.06688	1.38227	0.24619	0.12184	0.05989
S.E (g$_i$-g$_j$) ±	1.39276	1.54893	0.10449	0.14389	1.11313	0.09306	0.10112	2.08979	0.37221	0.18421	0.09055

DFF = Days to 50% flowering; PH = plant height (cm); NBPP = number of branches per plant; NPC = number of pod per cluster; NPPP = number of pods per plant; PL = pod length (cm); SPP = seeds per pod; PYPP = pod yield per plant (g); APW = average pod weight (g); PC = protein content (%); FC = fiber content (%). *,**Significance at 5 and 1% level respectively.

Table 3. Estimation of specific combining ability (specific combining ability (SCA) effect of hybrids for various characters in Indian bean.

S/N	Crosses	Days to 50% flowering	Plant height (cm)	No. of primary branches/ plant	No. of pods per cluster	No. of pods per plant	Pod length (cm)	Seeds per pod	Pod yield per plant (g)	Average pod weight (g)	Protein content (%)	Fiber content (%)
1	NIB-23 x NIB-32	1.313	0.539	0.215	0.014	2.921	0.274	0.704**	7.399	1.082	-0.084	0.251
2	NIB-23 x NIB-41	4.875	4.594	0.311	-0.302	-8.357**	-0.634**	0.370	14.006**	3.081**	0.623	-0.140
3	NIB-23 x NIB-54	-13.272**	-13.740**	1.056**	1.013**	18.291**	0.680**	0.810**	58.244**	2.491**	0.400	0.207
4	NIB-23 x NIB-56	-0.901	3.304	0.318	0.169	5.623*	-0.822**	0.375	2.911	-0.097	0.950*	-0.233
5	NIB-23 x NIB-57	3.668	1.185	0.151	0.418	4.184	0.207	0.249	8.292	0.703	0.686	0.167
6	NIB-23 x NIB-69	0.215	6.490*	-0.363	0.114	-11.574**	0.067	-1.234**	-44.863**	-2.372**	0.023	-0.128
7	NIB-23 x NIB-80	0.718	1.026	-0.933**	-0.222	-9.763**	0.364	0.154	-19.355**	-0.276	-0.350	0.233
8	NIB-32 x NIB-41	1.847	2.070	-0.973**	0.064	-11.722**	-0.549**	0.182	-4.985	-1.071	0.697	-0.232
9	NIB-32 x NIB-54	-3.443	-14.417**	1.022**	0.859**	19.236**	0.745**	1.062**	64.284**	3.166**	1.120**	-0.258
10	NIB-32 x NIB-56	1.707	1.335	0.130	0.455	3.724	0.283	0.280	-2.049	-2.225**	0.027	0.161
11	NIB-32 x NIB-57	0.770	4.898	0.103	0.204	-1.295	0.112	0.095	-7.658	-0.388	0.520	-0.128
12	NIB-32x NIB-69	1.897	4.969	0.079	-0.310	-0.929	-0.301	-2.259**	-19.194**	-1.754*	-0.303	0.137
13	NIB-32x NIB-80	-1.080	3.109	0.036	-0.489	3.271	-0.337	-0.760**	-1.375	0.062	0.324	0.328
14	NIB-41x NIB-54	6.003*	3.574	-1.062**	0.403	2.078	0.660**	-0.254	-28.570**	-3.556**	2.247**	0.101
15	NIB-41x NIB-56	-4.647	2.847	-0.661**	0.499	3.260	0.271	0.011	3.627	0.120	0.191	-0.090
16	NIB-41x NIB-57	1.383	-0.647	0.572*	0.608*	-0.793	0.380*	-0.425*	-7.332	0.083	-0.063	0.251
17	NIB-41x NIB-69	-12.594**	-5.666	1.111**	1.364**	18.929**	0.541**	0.702**	63.383**	3.092**	1.591**	0.006

Table 3. Contd.

18	NIB-41x NIB-80	-1.687	2.77	0.565**	-1.052**	4.557**	-0.669**	0.390	-7.188	0.991	-0.709	0.167
19	NIB-54 x NIB-56	-11.547**	-13.160**	0.888**	0.724*	6.844**	0.629**	0.294	42.886**	3.567**	0.674	0.014
20	NIB-54 x NIB-57	0.542	9.879**	-1.116**	-0.177	-13.475**	-0.942**	-0.291	-46.983**	-2.290**	0.680	-0.305
21	NIB-54 x NIB-69	7.463**	10.100**	0.080	-0.641*	-9.059**	-0.169	-0.305	-29.709**	-0.318	-0.063	-0.050
22	NIB-54 x NIB-80	8.076**	5.503	-0.033	-0.657**	-2.495	-0.652**	-0.676**	0.830	-1.422	-1.206**	0.101
23	NIB-56 x NIB-57	3.726	4.478	-0884**	-0.611**	3.630	0.392*	-0.017	-1.966	-1.104	0.144	-0.286
24	NIB-56 x NIB-69	4.973	5.626	-0.099	-0.515	-8.108**	-0.388*	-0.140	-20.692**	0.401	-0.249	0.069
25	NIB-56 x NIB-80	3.496	1.459	0.438*	0.039	-3.054	-0.051	-0.622**	-1.523	0.580	0.738*	0.110
26	NIB-57 x NIB-69	-3.457	-18.585**	1.525**	1.454**	24.080**	0.281	1.444**	102.849**	4.924**	-1.293**	0.120
27	NIB-57 x NIB-80	-13.611**	-1.502	0.272	0.308	5.060*	0.088	0.289	1.748	-1.217	1.394**	-0.289
28	NIB-69 x NIB-80	-10.857**	-16.537**	0.047	0.954**	15.889**	1.345**	1.469**	70.992**	3.682**	1.991**	-0.644**
	S.E (sij) ±	2.82394	3.14059	0.21186	0.29174	2.25697	0.18868	0.20502	4.23724	0.75468	0.37350	0.18359
	S.E (sij-sik) ±	4.17828	4.64678	0.31347	0.43166	3.33939	0.27917	0.30335	6.26938	1.11662	0.55263	0.27164
	S.E (sij-skl) ±	3.93932	4.38103	0.29554	0.40697	3.14841	0.26321	0.28600	5.91082	1.05276	0.52103	0.25611

*, **, Significance at 5 and 1% level respectively.

Gavali et al. (2011).

Therefore, the present investigation revealed that the parents NIB-54 and NIB-69 were good general combiners for pod yield per plant and they can be use for future breeding programme for vegetable purpose. Among specific combinations NIB-57 x NIB- 80, NIB-23 x NIB- 54 and NIB-41 x NIB-69 were identified as most promising hybrids for pod yield per plant for vegetable lablab bean.

REFERENCES

Bagade AB, Patel DU, Singh B, Desai NC (2002). Heterosis & combining ability for yield and yield components in Indian bean. (*Dolichos lablab* L.). Indian J. Pulses Res. 15 (1): 6-48.

Basu MR, Muthaiah AR, Ashok S (2006). Combining ability studies in Pigeonpea. Crop Res. 31(3):396-398.

Gavali, ST, Khandelwal V, Chauhan DA, Lodam V, A (2011). Heterosis and combining ability in Indian bean. J. Food Legumes 24(2):145-147.

Griffing (1956). Concept of general and specific combining ability in relation to diallel crossing system. Aust. J. Biol. Sci. 9:463-493.

Kabir J, Sen S (1990). Diallel analysis for combining ability in Dolichos bean (Lablab niger Medik. and *Dolichos uniflorus* Lam.). Trop. Agric. 67(2):123-126.

Kannan B, Paramasivam KS, Paramasivam K (2003). Combining ability analysis for yield and its component in lablab. Legume Res. 26(3): 188-191.

Sawant SS, Bendale VW, behave SG, Jadhav BB (2006). Studies on combining ability for yield components and yield realization in lablab bean. J. Maharashtra. Agric. Univ. 31(1):77-81.

Shivashankar G, Srirangasayi I, Kempanna C, Viswanatha SR (1971). Day- Neutral varieties of *Dolichos lablab* L. Mysore J. Agric. Sci. 5:216-218.

Singh SP, Singh HN, Gupta KK (1980). Combining ability and inheritance studies through diallel in Indian bean. (*Dolichos lablab*). Indian J. Hort. 37(4):392-396.

Valu MG, Pandya HM, Dhaduk HL Vaddoria MA (1999). Gene action in Indian bean. G.A.U Res. J. 25(1):32-34.

Virja, NR, Bhatiya, VJ, Poshiya VK (2006). Heterosis and combining ability in Indian Bean (*Lablab purpureus* L. Sweet). Agric. Sci. Digest 26(1):6-10.

12

Impact of nitrogen fertilization on the yield and content of protein fractions in spring triticale grain

K. Wojtkowiak[1], A. Stępień[2], M. Tańska[3], I. Konopka[3] and S. Konopka[4]

[1]Department of Fundamentals of Safety, University of Warmia and Mazury in Olsztyn, Poland.
[2]Department of Agriculture Systems, University of Warmia and Mazury in Olsztyn, Poland.
[3]Department of Food Plant Chemistry and Processing, University of Warmia and Mazury in Olsztyn, Poland.
[4]Department of Working Machine and Methodology of Research, University of Warmia and Mazury in Olsztyn, Poland.

Spring triticale cv. Andrus was cultivated in 2010 and 2011 with the application of different combinations of nitrogen fertilization with a total dose of 80 or 120 kg ha^{-1}. The nitrogen fertilizer was applied into soil or both into soil and on leaves (foliar application) with and without microelements. The average grain yield was 6.21 t ha^{-1} with a range from 5.89 to 6.64 t ha^{-1} depending on the variant of fertilization. The average protein content at the dose of 80 and 120 kg N ha^{-1} was 14.43 and 14.74 g per 100 g of grain, respectively. A higher dose of nitrogen resulted in an increase of albumins with globulins and ω and α/β prolamins in grain. This indicates that, even though some variants of fertilization with nitrogen favored the accumulation of protein in the grain, it was mainly the content of monomeric proteins that increased. The increase in their mass, due to a significant predominance in triticale grain, is undesired for potential use in baking. It suggests a lack of possibility for improvement of baking properties in triticale grain as a result of tested variants of fertilization.

Key words: Triticale, nitrogen fertilization, glutelins, prolamins.

INTRODUCTION

Triticale is a hybrid cereal plant derived from wheat and rye genomes. In 2011, the sown area of this species (both winter and spring cultivars) in the world amounted to 3853,000 ha (FAO, 2013; http://faostat.fao.org). The advantages of triticale include high grain yield, resistance to biotic and abiotic stresses and valuable grain composition (Zecevic et al., 2010). Despite these features, triticale grain is mainly used as feedstuff material (McGoverin et al., 2011). The low level of use of this species in the baking industry results from low baking properties, which are determined by the content and functional parameters of storage proteins. Although, the average content of protein in triticale grain ranges from 18 to 20% (Gulmezoglu and Aytac, 2010; Zecevic et al.,

2010), the yield of gluten and its parameters are worse than those of wheat grain. For many years, breeders have been allowed to select species for improved triticale gluten quality. Igrejas et al. (1999) have shown that, triticale exhibit great genetic diversity among the group of storage proteins. Triticale proteome is cultivar-dependent and similar to wheat and rye proteomes. However, as with other cereals, it can be modified by the environment. The main agrotechnical factor that influences grain yield is mineral fertilization (Nefir and Tabără, 2011). The availability of nitrogen for plants depends on its form. Urea is the most common form of nitrogen fertilizer, yet the efficacy of nitrogen utilization from urea is conditioned by plant species and the method of fertilization. Results

Table 1. Scheme of field nitrogen fertilization.

Object	Available nitrogen (kg ha^{-1})	Fertilizer type and applying time (rate kg ha^{-1})		
		Before sowing	(BBCH 23-29)	(BBCH 31-32)
K1	80	-	$CO(NH_2)_2$ (40)	$CO(NH_2)_2$ (40)
K2	80	-	$CO(NH_2)_2$ (20) + azofoska[#] (20)	$CO(NH_2)_2$ (40)
K3	80	-	$CO(NH_2)_2$ (40)	$CO(NH_2)_2$ (40)*
K4	80	-	$CO(NH_2)_2$ (40)	$CO(NH_2)_2$ (32) + ekolist[#*] (8)
K5	120	NH_4NO_3 (40)	$CO(NH_2)_2$ (40)	$CO(NH_2)_2$ 40
K6	120	NH_4NO_3 (40)	$CO(NH_2)_2$ (20) + azofoska[#] (20)	$CO(NH_2)_2$ (40)
K7	120	NH_4NO_3 (40)	$CO(NH_2)_2$ (40)	$CO(NH_2)_2$ 40*
K8	120	NH_4NO_3 (40)	$CO(NH_2)_2$ (40)	$CO(NH_2)_2$ (32) + ekolist[#*](8)

$CO(NH_2)_2$ - urea; NH_4NO_3 – ammonium nitrate; [#] multifertilizers, [*]foliar fertilization.

Table 2. Climate conditions during triticale vegetation.

Year	Temperature (°C)						Average
	March	April	May	June	July	August	
2010	2.1	8.1	12.0	16.4	21.1	19.3	13.2
2011	1.6	9.1	13.1	17.1	17.9	17.6	12.7
1961-2005	1.2	6.9	12.8	15.9	17.8	17.7	12.1
	Precipitation (mm)						
2010	36.7	18.2	131.9	84.8	80.4	95.3	74.6
2011	16.3	22.5	51.1	81.7	202.0	82.1	76.0
1961-2005	27.6	35.7	51.9	78.5	75.1	66.1	55.8

of numerous studies have indicated better utilization of nitrogen from foliarly-applied urea than from the soil. It has been shown that, the foliar application of nitrogen in a late phase of vegetation generates an increase in yielding and protein content by approximately 7 and 9%, respectively (Kinaci and Gulmezoglu, 2007). Mut et al. (2005) and Nefir and Tabără (2011) recorded an increase in triticale grain yield together with an increase in the dose of nitrogen, including the applied foliarly and supplemented with multi-component fertilizers. Supplementation of basic fertilization with copper, zinc, manganese and iron is particularly important (Nadim et al., 2012). The use of mineral fertilization combined with microelements increases grain yield and simultaneously improves the nutritional value of cereal grain (Malakouti, 2008).

This paper discusses the impact of nitrogen fertilization applied at doses of 80 and 120 kg ha^{-1} into soil and into both soil and foliarly, with and without multi-component fertilizers as well as the impact of the year of harvesting on the yield of grain and protein and its composition in spring triticale grain.

MATERIALS AND METHODS

Spring triticale cv. Andrus was cultivated in 2010 to 2011 in the

Education and Research Station of University of Warmia and Mazury in Tomaszkowo (53°72 N; 20°42 E), Poland, on typical brown soil with a texture of light loam. The soil was characterised by an acidic reaction, a low content of organic carbon (7.7 g kg^{-1}), a high content of phosphorus (10.9 mg kg^{-1} in d.m and potassium (20.7 mg kg^{-1} in d.m), a medium content of magnesium (5.0 mg kg^{-1} in d.m), an average content of available zinc (14.5 mg kg^{-1} in d.m), manganese (182 mg kg^{-1} in d.m) and iron (1100 mg kg^{-1} in d.m) and a low copper content (2.1 mg kg^{-1} in d.m). The experiment was set in a randomized block model in 3 replications. The doses of 30.2 kg P ha^{-1} as triple superphosphate and 83.1 kg K ha^{-1} as potassium salt were applied on all experimental objects. The fertilization with nitrogen was only a differentiating factor (Table 1).

The average temperature and precipitations during vegetation of triticale in the month of March to August period are presented in Table 2. The temperatures and their monthly distributions did not differ from the multi-annual mean (12.1°C). More diversified were precipitations. The average monthly volume of precipitation in the month of March to August period was 74.6 mm in 2010 and 75.9 mm in 2011 and was higher by approximately 35% than the multi-annual mean.

Protein content and yield

Triticale grain samples were collected at harvest. The grain was then dried to approximately 14% and cleaned from dust and tailings in a laboratory sieve-air separator. Finally, the grain was milled to particles below 300 nm. Ground samples were stored in a refrigerator before the analyses. The content of nitrogen in the milled grain samples was determined with the Kjeldahl method and

Table 3. Impact of fertilization variant, year of harvesting and nitrogen dose on the grain (t ha^{-1}) and protein yield (kg ha^{-1}) and the composition of proteins in grain (g per 100 g of grain).

Fertilization variant		Grain yield	Protein yield	Albumins + globulins	Prolamins	Glutelins	Sum of protein fractions
Average		**6.21**	**786.4**	**2.25**	**8.79**	**3.65**	**14.69**
Year	2010	6.07^a	773.2^a	2.31^a	8.77^a	3.77^a	14.85^a
	2011	6.35^a	799.6^a	2.19^b	8.81^a	3.53^b	14.53^b
Object	K1	5.89^a	754.4^a	1.96^{abc}	8.52^c	3.18^d	13.66^c
	K2	6.14^a	788.5^a	2.29^{abc}	8.73^{bc}	3.62^a	14.64^{bc}
	K3	6.04^a	778.0^a	2.15^c	8.43^c	3.37^a	13.95^c
	K4	6.33^a	817.3^a	2.29^{bc}	8.86^b	3.79^a	14.94^{ab}
	K5	5.95^a	750.3^a	2.41^{bc}	9.21^a	3.80^a	15.42^a
	K6	6.64^a	813.3^a	2.23^{bc}	8.71	3.79^a	14.73^b
	K7	6.35^a	819.1^a	2.39^{bc}	9.20^a	3.84^a	15.43^a
	K8	6.36^a	770.1^a	2.25^{bc}	8.66	3.81^a	14.72^b
Nitrogen dose	80	6.13^a	778.4^a	2.18^a	8.72^a	3.53^a	14.43^b
	120	6.12^a	783.5^a	2.29^a	8.81^a	3.64^a	14.74^a

Means in the same column (separately for year, object and nitrogen dose) followed by different letters are significantly different (α ≤ 0.05).

the content of total protein was then calculated using a 5.7 multiplier. The yield of protein was calculated based on the grain yield and content of protein in the grain.

Protein extraction and analysis

The quantitative and qualitative protein characteristics were determined with the RP-HPLC technique according to Konopka et al. (2007). The content of albumins and globulins, prolamins and glutelins was analyzed. The assays were performed with a Hewlett Packard 1050 series apparatus. Detection of protein fractions was carried out at the 210 nm wavelength and their identification consisted in an analysis of spectra and standard protein retention times. The content of protein is expressed in g per 100 g of grain using the standard curves for bovine serum albumin (BSA) and gliadins and glutenins of wheat cv. Tonacja.

Statistical analysis of data

Data were statistically processed with the ANOVA and a "post-hoc" Duncan test. The comparisons between the average values were preformed separately for the year of cultivation, variant of nitrogen fertilization and the total dose of nitrogen fertilizer. The calculations were performed at level of significance α = 0.05 with STATISTICA v.10 software (StatSoft, Inc.).

RESULTS AND DISCUSSION

Grain yield

The average grain yield was 6.21 t ha^{-1} with the variation ranging from 5.89 to 6.64 t ha^{-1}, depending on the variant of fertilization with nitrogen (Table 3). The recorded differences were, however, statistically insignificant. The yield volume was significantly higher than the average

spring triticale yielding in other EU countries (FAO, 2013; http://faostat.fao.org). According to COBORU (2013 http://coboru.pl), with a transition from average to intensive fertilization, the grain yield for this cultivar may increase from 5.79 to 6.67 t ha^{-1}. However, Piekarczyk et al. (2011) found only a minor increase in wheat grain yield as a result of the application of nitrogen at the dose of 40 to 160 kg ha^{-1}. According to Nefir and Tabără (2011), the increase in the dose of nitrogen from 80 kg ha^{-1} to 160 kg ha^{-1} generates an increase in yielding as high as approximately 12%. These authors have also shown that, mineral fertilization with dose of 160 kg N, 60 kg P, and 60 kg K per ha promoted the growth of production yield by 44% comparing to unfertilized plot. The increase in spring triticale yielding is favored by a combination of nitrogen fertilization with the foliar application of zinc (Knapowski et al., 2009).

Protein grain yield

The protein yield ranged from 704.9 to 861.7 kg ha^{-1} with the total average value of 786.4 kg ha^{-1} (Table 3). Neither the year of cultivation nor any of the fertilization variants generated a statistically significant change in the protein yield. Only a tendency towards an increase in the protein yield in 2011 and under the influence of higher nitrogen dose (120 kg ha^{-1}) was noted. Alaru et al. (2003) showed that, the main factor influencing the content of protein in triticale grain was the cultivar. Weather conditions in the growth period have a lesser impact, and nitrogen fertilization being the least important. According to these authors, fertilization with nitrogen at tillering caused an average increase in the protein content in triticale grain of

Table 4. Impact of fertilization variant, year of harvesting and nitrogen dose on the characteristics of storage proteins in grain (g per 100 g of grain).

Fertilization variant		Prolamins			Glutelins	
		ω	α/β	γ	HMW	LMW
Average		0.54	4.90	3.35	1.26	2.39
Year	2010	0.54[a]	4.86[a]	3.36[a]	1.35[a]	2.42[a]
	2011	0.54[a]	4.93[a]	3.34[a]	1.17[b]	2.36[b]
Object	K1	0.55[b]	4.73[bc]	3.25[a]	1.24[bc]	1.94[d]
	K2	0.64[a]	4.78[bc]	3.30[a]	1.23[bc]	2.38[b]
	K3	0.51[bc]	4.66[c]	3.26[a]	1.20[c]	2.17[c]
	K4	0.55[b]	4.93[b]	3.38[a]	1.29[bc]	2.50[a]
	K5	0.62[ab]	5.14[a]	3.46[a]	1.31[a]	2.48[a]
	K6	0.48[c]	4.92[b]	3.31[a]	1.23[bc]	2.56[a]
	K7	0.51[bc]	5.20[a]	3.49[a]	1.30[ab]	2.53[a]
	K8	0.47[c]	4.80[bc]	3.38[a]	1.26[abc]	2.54[a]
Nitrogen dose	80	0.54[b]	4.85[a]	3.33[a]	1.25[a]	2.28[b]
	120	0.58[a]	4.88[a]	3.35[a]	1.26[a]	2.39[a]

Means in the same column (separately for year, object and nitrogen dose) followed by different letters are significantly different ($\alpha \leq 0.05$).

1.57%. Lestingi et al. (2010) proved that, the optimal dose of nitrogen for maintaining good quality of triticale grain was approximately 50 kg ha[-1].

Protein characteristics

The average protein content in the grain of tested triticale cultivar was 14.69 g per 100 g of grain (Table 3). Of this, prolamins constituted 59.8 and glutelins 24.8%, whilst the amount of albumins and globulins was the lowest, amounting to 15.4% in total. The year of harvesting and the total dose of nitrogen fertilizer impacted the total concentration of protein (higher values were recorded in 2010 and after the application of 120 kg N ha[-1]). Moreover, statistically significant differences were found between individual fertilization variants. The difference between the highest and the lowest total protein content was 1.77 g per 100 g of grain. The comparison between K3 and K7 variants showed that, the application of additional pre-sowing fertilization with nitrogen at the dose of 40 kg ha[-1] generated an increase in the relative protein content in the grain by as much as 10.6%.

Examples of chromatograms depicting the individual protein fractions in the tested triticale grain are presented in Figure 1. Comparison of these data to typical chromatogram of storage proteins in wheat grain (cv. Tonacja) showed that, these species are distinctly different. Analyzed triticale grain contained less prolamins with retention times up to approximately 12 min, and in glutelins composition, it can be a visible balance between the subunits of high (with retention times up to 10 min) and low molecular weight. Under the influence of lower nitrogen dose were detected statistically lower contents of glutelins and albumins + globulins in the grain harvested in 2011 (except for high molecular weight (HMW) glutelins). The large difference in the content of albumins and globulins was stated between K1 and K5 variants, and for prolamins between K3 and K7 (Tables 3 to 4). The application of additional pre-sowing nitrogen fertilization generated an increase of 22.9 and 9.1%, respectively. Within the prolamins, the applied fertilization variants only generated a change in ω and α/β subunits and statistically significant differences were detected mainly between variants with additional pre-sewing nitrogen fertilization: K1 and K5 (ω and α/β) and K3 and K7 (α/β). It was also found that, a higher dose of nitrogen favored the accumulation of ω gliadins and the increase was 7.4%. The variability in the content of glutelins under different variants of fertilization was also high and ranged from 3.18 to 3.84%. Within this group of proteins, the low-molecular-weight (LMW) glutelins showed more pronounced changes (from 1.94 to 2.56%) than HMW subunits (from 1.20 to 1.31%). The amount of both fractions was statistically lower in 2011, whereas the content of LMW glutelins was significantly higher in grain fertilized with nitrogen at 120 kg ha[-1].

Many authors have indicated the potential utility of triticale grain in the baking industry (Amiour et al., 2002; Martinek et al., 2008). The baking value of triticale grain depends on the amount and quality of storage proteins, which is influenced by genetic and environmental factors (Erekul and Köhn, 2006; Burešová et al., 2010; McGoverin et al., 2011). The hallmark feature of the

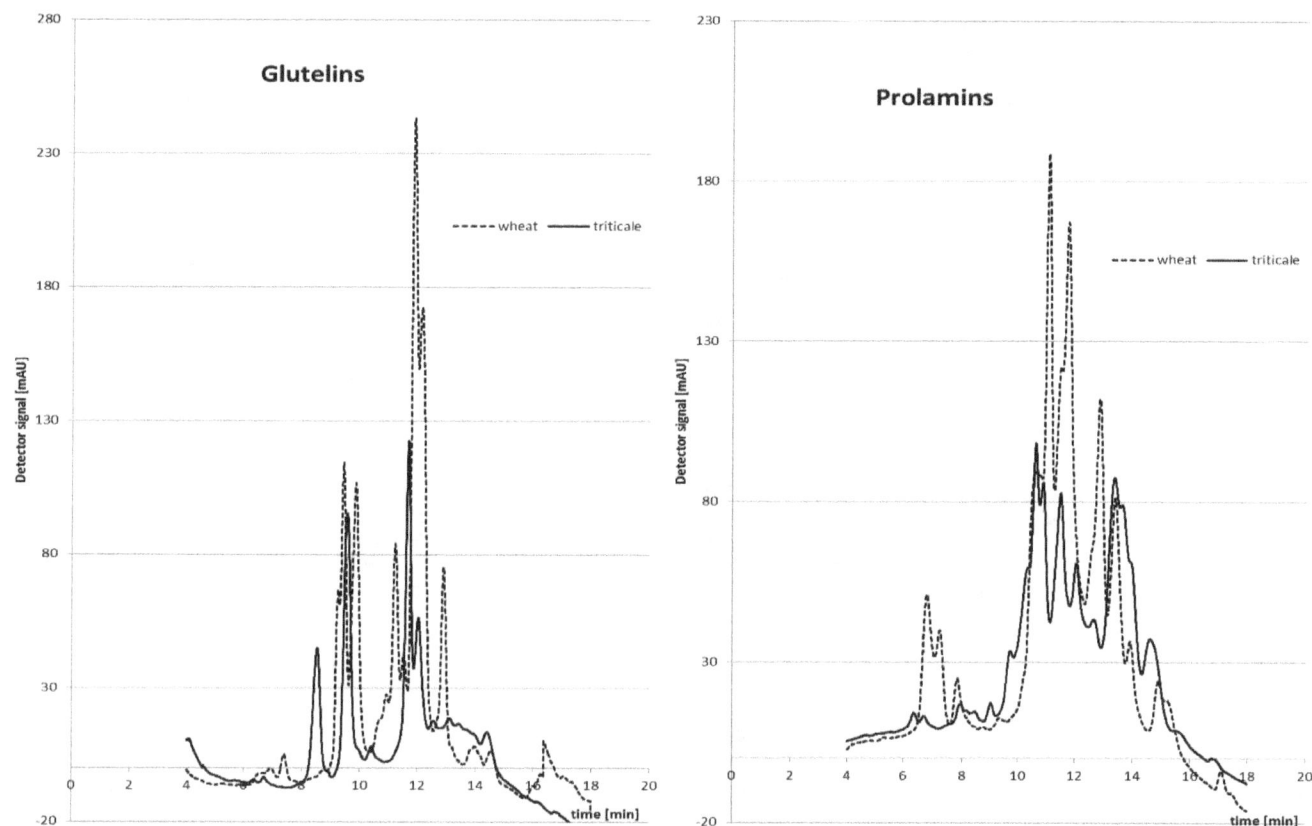

Figure 1. Comparison of storage proteins in triticale cv. Andrus and wheat cv. Tonacja grain.

Andrus cultivar is a high thousand grain weight with a low protein content (COBORU, 2013). Improvements in grain quality parameters can be achieved with adequate agricultural engineering procedures. According to Luo et al (2001), the content of HMW and LMW fractions, although genetically determined, may increase slightly under the influence of nitrogen application in later growth stages.

High temperatures after anthesis and abiotic stress in the early stages of grain filling exert a negative impact on the accumulation of protein (Knezevic et al., 2007), while drought stress has a positive effect (Fernandez-Figares et al., 2000). The present study confirmed the significant impact of the climate on the characteristics of protein. An analysis of meteorological data (Table 2) has shown that, the average temperature during vegetation in 2011 year was slightly lower than in 2010. The biggest difference (3.2°C) was recorded in July. Furthermore, both years were much abundant in rainfall than the multi-annual mean. This indicates that, the climatic conditions in 2011 were less favorable for the accumulation of storage proteins. The more intensive fertilization with nitrogen contributed to an increase in the ω fraction and LMW glutelins. A similar phenomenon was observed by Wieser and Seilmeier (1998) who suggested that, nitrogen fertilization generates a higher increase in the content of

gliadins than of glutelins. This results in an increase of monomeric proteins and a reduction of polymeric proteins. The ratio of prolamins to glutelins in bread wheat grain should approximately be 1:1 (Singh and MacRitchie, 2001; Shewry and Halford, 2002). In the grain of tested triticale cultivar, this value ranged from 2.2 to 2.7:1. Some variants of used fertilization favored the accumulation of prolamins, what can additionally increase viscous properties of protein over its elasticity. This indicates a lack of potential for improvements in the baking properties in triticale grain as a result of proposed fertilization variants.

REFERENCES

Alaru M, Laur U, Jaama E (2003). Influence of nitrogen and weather conditions on the grain quality of winter triticale. Agron. Res. 1:3-10.

Amiour N, Dardevet M, Khelifi D, Bouguennec A, Branlard G (2002). Allelic variation of HMW and LMW glutenin subunits, HMW secalin subunits and 75K gamma-secalins of hexaploid triticale. Euphytica 123:179-186.

Burešová I, Sedláčková I, Faměra O, Lipavský J (2010). Effect of growing conditions on starch and protein content in triticale grain and amylose content in starch. Plant. Soil Environ. 56(3):99-104.

COBORU (2013). www.coboru.pl/dr/charaktodmiany.aspx (accessed 2013).

Erekul O, Köhn W (2006). Effect of weather and soil conditions on yield components and bread-making quality of winter wheat (*Triticum* aestivum L.) and winter triticale (Triticosecale Wittm.) varieties in north-east Germany. J Agron. Crop. Sci. 192:452-464.

FAO (2013). Food and agricultural commodities production. FAOSTAT. http://faostat.fao.org/site/567/DesktopDefault.aspx?PageID=567#anc or (accessed 16.01.2013).

Fernandez-Figares I, Marinetto J, Royo C, Ramos JM, Garcia del Moral LF (2000). Aminoacid composition and protein and carbohydrate accumulation in the grain of triticale grown under terminal water tress simulated by a senescing agent. J. Cereal Sci. 32:249-258.

Gulmezoglu N, Aytac Z (2010). Response of grain and protein yields of triticale varieties at different levels of applied nitrogen fertilizer. Afr. J. Agric. Res. 5(18):2563-2569.

Igrejas G, Guedes-Pinto H, Carnide V, Branlard G (1999). Seed storage protein diversity in triticale varieties commonly grown in Portugal. Plant Breed. 118(4):303-306.

Kinaci E, Gulmezoglu N (2007). Grain yield and yield components of triticale upon application of different foliar fertilizers. Interciencia 32:624-628.

Knapowski T, Ralcewicz M, Barczak B, Kozera W (2009). Effect of nitrogen and zinc fertilizing on bread-making quality of spring triticale cultivated in Notec Valley. Pol. J. Environ. Stud. 18:227-233.

Knezevic D, Paunovic A, Madic M, Dukic N (2007). Genetic analysis of nitrogen accumulation in four wheat cultivars and their hybrids. Cereal. Res. Commun. 35(2):633-636.

Konopka I, Fornal Ł, Dziuba M, Czaplicki S, Nałęcz D (2007). Composition of proteins in wheat grain obtained by sieve classification. J. Sci. Food. Agric. 87(12):2198-2206.

Lestingi A, Bovera F, De Giorgio D, Ventrella D, Tateo A (2010). Effects of tillage and nitrogen fertilization on triticale grain yield, chemical composition and nutritive value. J. Sci. Food. Agric. 90:2440-2446.

Luo C, Griffin WB, Branlard G, McNeil DL (2001). Comparison of low- and high molecular-weight wheat glutenin allele effects on flour quality. Theor. Appl. Genet. 102(6-7):1088-1098.

Malakouti MJ (2008). The effect of micronutrients in ensuring efficient use of macronutrients. Turk. J. Agric. For. 32:215-220.

Martinek P, Vinterová M, Burešová I, Vyhnánek T (2008). Agronomic and quality characteristics of triticale (X Triticosecale Wittmack) with HMW glutenin subunits 5+10. J. Cereal Sci. 47:68-78.

McGoverin CM, Snyders F, Muller N, Botes W, Fox G, Manley M (2011). A review of triticale uses and the effect of growth environment on grain quality. J. Sci. Food. Agric. 91:1155-1165.

Mut Z, Sezer I, Gűlűmser A (2005). Effect of different sowing rates and nitrogen levels on grain yield, yield components and some quality traits of triticale. Asian J. Plant Sci. 4(5):533-539.

Nadim MA, Awan IU, Baloch MS, Khan EA, Naveed K, Khan MA (2012). Response of wheat (Triticum aestivum L.) to different micronutrients and their application methods. J. Anim. Plant. Sci. 22(1):113-119.

Nefir P, Tabără V (2011). Effect on products from variety fertilization and triticale (Triticosecale Wittmack) in the experimental field from răcăşdia caras-severin country. Res. J. Agric. Sci. 43(4):133-137.

Piekarczyk M, Jaskulski D, Gałęzewski L (2011). Effect of nitrogen fertilization on yield and grain technological quality of some winter wheat cultivars grown on light soil. Acta Sci. Pol. Agricultura, 10(2):87-95.

Shewry PR, Halford NG (2002). Cereal seed storage proteins: structures, properties and role in grain utilization. J. Exp. Bot. 53(370):947-958.

Singh H, MacRitchie F (2001). Application of polymer science to properties of gluten – mini review. J. Cereal Sci. 33:231-243.

Wieser H, Seilmeier W (1998). The influence of nitrogen fertilisation on quantities and proportions of different protein types in wheat flour. J. Sci. Food. Agric. 76(1):49-55.

Zecevic V, Knezevic D, Boskovic J, Milenkovic S (2010). Effect of nitrogen and ecological factors on quality of winter triticale cultivars. Genetika 42(3):465-474.

13

Laboratory Evaluation of Cotton (*Gossypium hirsutum*) and Ethiopian Mustard (*Brassica cariata*) Seed Oils as Grain Protectants against Maize Weevil, *Sitophilus zeamais* M. (Coleoptera:Curculionidae)

Fekadu Gemechu[1], Dante R. Santiago[2] and Waktole Sori[1]

[1]Department of Plant Sciences and Horticulture, Jimma University College of Agriculture and Veterinary Medicine, P. O. Box-307, Jimma, Ethiopia.
[2]Department of Environmental Sciences and Technology, Jimma University College of Public Health and Medical Sciences, P. O. Box-378, Jimma, Ethiopia.

Maize is the second most widely grown cereal crop in Ethiopia. In storage, maize is severely destroyed by storage insect pests, mainly maize weevil *(Sitophilus zeamais)*. In an effort to develop a non-synthetic pesticide control approach, a study was conducted to determine the efficacy of two cooking oils, Ethiopian mustard (*Brassica carinata*) and cotton (*Gossypium hirsutum*), to control *S. zeamais* under laboratory conditions. The oils were applied at the rate of 0.2 to 0.5 ml per 250 g of grain and compared with untreated control and malathion super dust as standard check. The study was laid-out in completely randomized design (CRD) with three replications for each treatment. The efficacy of the oils was assessed on the basis of total insect mortality, median lethal time (LT$_{50}$), weevil progeny emergence, seed hole's number, weight loss and germination rate. The results showed that the oils caused 25 to 100% mortality at the different concentrations used. Both oils had LT$_{50}$ of 0.5 day when applied at the concentration of 0.5 ml. At concentration of 0.3 to 0.5 ml, both oils caused zero weevil progeny emergence, minimum seed damage, zero grain weight loss and 89.2 to 95.5% seed germination rate which were similar to those of malathion (Diethyl succinate) and significantly different from those of the untreated control. The tests demonstrated that the two oils are effective stored maize grain protectants and can be used as components of maize weevil integrated pest management option.

Key words: Cooking oils, maize weevil, *Sitophilus zeamais,* mortality, stored grain.

INTRODUCTION

In Ethiopia, maize is the second widely grown cereal crop with annual land coverage of 1.8 million ha and average yield of 2.2 tons/ha (CSA, 2007). The importance of maize for Ethiopia is underscored by its use as food and as raw material for various industries (Meseret, 2011). A major constraint in the storage of maize is the damage caused by the maize weevil, *Sitophilus zeamais* (Abraham,

1991, 1995; Emana and Tsedeke, 1999; Negeri and Adisu, 2001; Sori and Ayana, 2012). The insect bores a hole of 1 mm in diameter into the grain, deposits the eggs and seals the opening with a gelatinous waxy secretion. The post-embryonic immature stages develop in the grain and emerge as adults from the grain feeding on the content that leads to total grain loss (Boxall et al., 2002;

Table 1. Treatments (concentrations) of the chemicals applied on maize for the control of *S. zeamais*.

Oil	Dosage (Treatments)	Local name	Common name
Untreated control	-	-	-
Malathion dust (5%)	0.125 g	-	-
G. hirsutum	0.20 ml 0.30 ml 0.40 ml 0.50 ml	Tit/ Jirbi	Cotton
B. carinata	0.20 ml 0.30 ml 0.40 ml 0.50 ml	Gomenzer	Ethiopian mustard

Rees, 2004).

At present, the use of synthetic pesticides is the only recourse for farmers in protecting their maize stored grains. However, synthetic pesticides have a number of unwanted side-effects such as poisoning among handlers, toxic residues in food and feedstuff, ecological disruption as well as chronic and genetic maladies (Dubey et al., 2007; Kumar et al., 2007). This situation calls for the development of alternative methods of maize grain protection that are safer, health and eco-friendly. One attractive approach is the use of cooking and essential oils as seed protectants. Their modes of action include fumigant effect, topical toxicity, antifeedant or repellent activity, among others (Muhammad, 2009).

In a previous study, Fekadu et al. (2012) tested several botanical powders and two cooking oils, cotton (*Gossypium hirsutum*) and Ethiopian mustard (*Brassica carinata*) seed oils, for their activity against *S. zeamais*. The results indicated that the two oils were effective in killing the weevil and preventing damage to maize grains. This study was conducted to evaluate the efficacy of the two cooking oils at different application rates under laboratory conditions. The goal was the generation of a new technology for safe, low-cost, easy and efficacious control of maize weevil on stored maize grains.

MATERIALS AND METHODS

Description of the study area

The experiment was conducted at room temperature in entomology laboratory of Jimma University College of Agriculture and Veterinary Medicine (JUCAVM), Southwestern Ethiopia from July to December, 2011. JUCAVM is located 354 km Southwest of Addis Ababa (capital city of Ethiopia) at an approximate geographical coordinates of latitude (06°36' N) and longitude (37°12' E) at an altitude of 1710 m above sea level. The mean maximum and minimum temperatures are 26.8 and 11.4°C, respectively and the mean maximum and minimum relative humidity are 91.4 and 39.92%, respectively.

Sitophilus rearing conditions

S. zeamais adults were collected from stored cereal grains from shops in Jimma town market and brought to the laboratory of the College of Agriculture and Veterinary Medicine, Jimma University, Ethiopia. The weevils were reared on maize grains in jars covered with perforated caps for gas exchange at 27°C and 70% RH in an incubator (heating incubator made in Spain called T.P. Selecta s.a). Newly emerged F_1 adults (1:1 male-female ratio) were used as the test insects.

Maize grains

Maize variety BH-660 grains were obtained from Nekempte Cereal Division and Distribution Enterprise. This variety, developed by the National Maize Research Program, Bako, Western Ethiopia, is the most commonly grown hybrid in the country and considered to be susceptible to insect infestation (Abraham and Basedow, 2004). The grains were kept at -6°C for 7 days to kill any infesting insects, cleared of broken kernels and debris and then graded manually according to size. Similar sized grains were selected for the experiment.

Cooking oils and bioassay procedure

Refined Ethiopian mustard (*B. carinata*) and unrefined cotton (*G. hirsutum*) seed oils were obtained respectively from Haranghe Fadis and Addis Mojo oil refineries, Eastern Ethiopia and stored in the laboratory at room temperature until needed.

The oils were separately mixed by handshaking with 2 ml of acetone in such a way that the final dosages were 0.2 to 0.5 ml per 250 g maize grains. Malathion super dust (5%) formulation was used as the positive control at a dosage of 0.125 g/ 250 g maize grain and maize grains treated with acetone alone served as the negative control. Treatments (Table 1) were replicated three times and laid out using completely randomized design (CRD). In all cases, 12 × 16 × 7 cm³ plastic microwave boxes with perforated covers (for aeration) were used as containers (experimental jars). After thorough mixing of the chemicals with the maize grains by shaking for 5 min, the boxes were uncovered to allow the solvent to evaporate for 2 h.

To each box were added 20 adult weevils and the insects, maintained at 27°C and 70% RH in incubator. Mortality of the

Table 2. The Effect of different concentrations of *the chemicals used* on *S. zeamais*.

Treatment	Dosage	Mortality (%)	Median lethal time, LT_{50} (day)	Confidence Interval		Slope
				Lower	Upper	
G. hirsutmn oil	0.2 ml	30	25.8[b]	13.9	94.9	1.43 ± 0.29
	0.3 ml	45	8.4[c]	-	-	0.78 ± 0.54
	0.4 ml	85	0.6[d]	-	-	1.18 ± 0.69
	0.5 ml	100	0.5[d]	0.0	0.92	3.33 ± 1.56
B.carinata oil	0.2 ml	25	50.9[a]	13.9	94.9	1.43 ± 0.23
	0.3 ml	35	15.5[c]	-	-	0.78 ± 0.80
	0.4 ml	80	1.3[d]	0.18	2.2	1.18 ± 0.40
	0.5 ml	90	0.6[d]	0.01	0.23	3.33 ± 0.44
Malathion	0.125 g	90	0.14[d]	-	-	1.01 ± 1.03
Control	-	5	NA	NA	NA	-

-, The confidence intervals are not provided because of the very low LT_{50} values are beyond the computing capacity of the soft ware (USEPA probit analysis program); NA, not applicable.

weevils was recorded daily from day 1 to 5 and at days 10, 15 and 20 (Parugrug and Roxas, 2008). At the end of the observation period, all adult weevils were removed from the boxes and the grain seeds were then returned to the incubator. From days 20 to 40 of adult weevil's introduction to experimental jars, the grains from each box were sieved to determine the number of progenies that emerged. On day 45, samples of the grains were taken for the determination of grain damage holes, weight loss and germination rate.

For grain damage determination, 10 seeds were taken randomly from each replication and examined for exit holes and occurrence of larvae within the seeds, if any. Weight loss was obtained using the formula described by Ileke and Oni (2011) which is given by:

$$Percentage\ Weight\ Loss = \frac{Initial\ Weight - Final\ Weight}{Initial\ Weight} \times 100$$

The percentage seed germination (viability index) was determined from randomly selected 110 grains placed in Petri dishes containing moist filter papers which were incubated at 30°C for 7 to 10 days (Ogendo et al., 2004).

Data analysis

The insect percentage mortality and treatment median lethal time were determined using the US Environmental Protection Agency probit analysis program version 1.5 (Finney, 1971). Mortality rate were corrected using Abbott (1925) formula. Prior to analysis, data on progeny emergency, seed damage, weight loss and germination were square root-transformed to reduce variance heterogeneity (Gomez and Gomez, 1984). One-way analysis of variance and Tukey's honestly significant difference post hoc test at 5% significance level were performed using SAS version 9.2 software.

RESULTS

Fatal effects of oils on *Sitophilus zeamais*

The efficacy of the oils as killing agents on *Sitophilus* is presented in Table 2. Both cotton and mustard seed oils caused appreciable mortality of the weevils at all dosages used. Maximum total mortality (100%) of *S. zeamais* was caused when unrefined cotton oil was applied at 0.5 ml/250 g of maize grain followed by mustard oil at 0.5 ml and malathion, both inducing 90% total mortality. On the other hand, the lowest dosage of cotton and mustard oils (0.2 ml per 250 g of grain) respectively resulted in minimum total mortality among the concentration used, 30 and 25% in 20 days. The killing efficacy of the oils increased with increasing dosage. At a dosage of merely 0.4 ml/ 250 g (1.6 ml/kg), both cotton and mustard seed oils killed respectively 85 and 80% or more of the weevils and maximum insect fatality (90 and 100% mortality) was achieved at a dosage of 0.5 ml/250 g (2 ml/kg) in 1 day. The percentage mortality caused to weevils due to the toxicity of the oils were comparable with that of the positive check, malathion at 0.125 g/250 g (90%) and significantly different from that of the untreated check (5% total mortality). Consequently, the median lethal time for both oils at 0.5 ml/250 g was less than 1 day (0.5 and 0.6 days for both cotton and mustard, respectively), an indicative of the high efficacy of both oils against *Sitophilus*. As the concentration of the oils decreases, the time taken to kill 50% of the test insect (weevils) was longer and vice versa.

Effects of oils on weevil progeny emergence

Both oils inhibited the reproduction of the weevils at all dosages significantly ($P \le 0.05$) (Table 3). Maximum and significantly more number of progenies (8 F_1 progenies) were emerged from the maize grain untreated with the oils. This was followed with the lowest concentration of both cooking oils. At the lowest dosage (0.2 ml/250 g) of

Table 3. Number of progenies of *S. zeamais* emergency from maize grains treated with different concentrations of the chemicals used.

Treatment	Dosage	Progeny number
G. hirsutmn oil	0.2 ml	3 (1.9)[b]*
	0.3 ml	0 (0.7)[d]
	0.4 ml	0 (0.7)[d]
	0.5 ml	0 (0.7)[d]
B. carinata oil	0.2 ml	1 (1.2)[c]
	0.3 ml	0 (0.7)[d]
	0.4 ml	0 (0.7)[d]
	0.5 ml	0 (0.7)[d]
Malathion	0.125 g	0 (0.7)[d]
Control	-	8 (2.9)[a]
P value		0.0001
HSD		0.37
CV(%)		15.3

*The numbers inside parentheses are the transformed data ($\sqrt{x+0.5}$) and means with the same letters within the column are not significantly different ($P < 0.05$)

both oils, insect progeny production was less than half (1 to 3 progenies) of that of the untreated control (8 progenies) and at higher oil dosages, as with the positive check, reproduction was completely inhibited, that is, no progeny emerged. These results are of course related to the relatively high fatal effects of the oils on the introduced weevils.

Effects of oils on weevil damaging activity

Except for cotton oil at a dosage of 0.2 ml/250 g, both oils at all dosages resulted in significantly less ($P < 0.0001$) damaged grains than did the untreated control (Table 4). Maximum and significant number of bored grains (0.54 out of 10 seeds), percentage weight loss (0.8%), and minimum percentage germination (84.8%) were obtained from the untreated control. On the contrary, significantly less number of bored seeds, weight loss and more percentage germinated seeds were obtained from the maximum concentration of the two cooking oils and the positive control. The number of holes of grains treated with lower dosages (0.2 and 0.3 ml/250 g) of the oil were less than half of that of the untreated check (0.2 versus 0.54 holes per 10 grains). At the dosage of 0.5 ml/250 g of the oils, the grains exhibited no holes at all, indicating that the weevils were effectively prevented from laying eggs on the grains. Similarly, weight loss of grains treated with low dosages of oil was half that of the untreated check (0.4 versus 0.8%) and those grains that received the highest dosage showed no loss in weight. The percentage germination data for grains treated with

the oils followed similar trend; the oil-treated grains had germination rates of 88.3 to 95.5% which were significantly higher ($P < 0.0001$) than 84.8% for the untreated control.

DISCUSSION

Fekadu et al. (2012) found that both cotton and Ethiopian mustard seed oils caused, respectively, 100 and 95% mortality of *S. zeamais* with corresponding median lethal time of less than 1 day. In an effort to fine tune the use of these oils at lower dosages of 0.2 to 0.4 ml/250 g of maize grain were tested in addition to the dosage of 0.5 ml/250 g previously tested. The results of the test, using such variables as median lethal time, progeny emergence, seed hole number, weight loss and percentage germination, indicate that the dosage of 0.4 ml/250 g (1.6 ml/kg) is statistically comparable with the slightly higher dosage of 0.5 ml/250 g (2 ml/kg). These levels of treatment correspond respectively to 0.16 and 0.20% (v/w) of the oils. Demissie et al. (2008) obtained 100% mortality of maize weevil with a much higher dosage of 4.0% (w/w) cooking oil. The levels of oil found effective against the weevil in the present study likewise is compared favorably with a dosage of 0.7 ml *Cymbopogon citratus* essential oil per 50 g maize, equivalent to 1.4% (v/w), used by Odeyemi (1993) against *S. zeamais*. Essential oils are highly volatile and do pose fumigant activity leading to stored-product insects mortality (Ahn et al., 1998). These authors tested essential oils obtained from savory, oregano and myrtle and with varying degree of toxicity oils from the three plants have showed insecticidal activity against three species of adult insects namely: *Ephestia kuehniella*, *Plodia interpunctella* and *Acanthoscelides obtectus*.

There was a significant reduction in progeny emergence as a result of the efficacy of the two cooking oils. This might be due to increased adult mortality, ovicidal and/or larvicidal properties of the cooking oils confirming the findings of Selase and Getu (2009) and Bamaiyi et al. (2007). Different concentrations of dry ground leaves of *Chenopodium ambrosoides* resulted in complete (100%) inhibition of oviposition, progeny emergence and mortality of larvae of *Callosobruchus chinensis*, *Callosobruchus maculatus* and *A. obtectus* preventing feeding and damage (Tapondjou et al., 2002). In a separate study, *Chenopodium* leaf powder mixed with maize and sorghum grains at the rates of 2 and 4% w/w caused complete reduction in maize weevil F_1 progeny production (Mekuria, 1995; Dejen, 2002). Plant based products weaken adults weevils leading to laying of fewer eggs ultimately even with reduced hatchability of the eggs to larvae and final metamorphosis to adults. The effectiveness of different non-synthetic chemical products to various storage insect pests of stored products have been reported by several authors (Huang et al., 2000; Tripathi et al., 2000; Mbailao et al., 2006; Negahban et

Table 4. Effect of different concentrations of the chemicals used on maize seed grain percentage weight loss and germination percentage of infested maize by *S. zeamais*.

Treatment	Dosage	Hole number/10 seeds	Weight loss (%)	Germination (%)
G. hirsutum oil	0.2 ml	0.5(1.0)[a*]	0.4(1.2)[b*]	86.5(9.8)[fg*]
	0.3 ml	0.2(0.8)[b]	0.0(0.7)[c]	89.2 (10)[de]
	0.4 ml	0.1(0.8)[c]	0.0(0.7)[c]	92.8 (10.2)[bc]
	0.5 ml	0.0(0.7)[d]	0.0(0.7)[c]	95.5 (10.3)[a]
B. carinata oil	0.2 ml	0.3(0.9)[b]	0.4(1.1)[bc]	88.3 (9.9)[ef]
	0.3 ml	0.2(0.8)[b]	0.4(0.9)[bc]	89.2 (9.97)[de]
	0.4 ml	0.1(0.8)[c]	0.0(0.7)[c]	91.9 (10.1)[cd]
	0.5 ml	0.0(0.7)[d]	0.0(0.7)[c]	93.7 (10.2)[ab]
Malathion	0.125 g	0.0(0.7)[d]	0.0(0.7)[c]	95.5 (10.3)[a]
Control	-	0.54(1.0)[a]	0.8(1.6)[a]	84.8 (9.7)[g]
P value		0.0001	0.0001	0.0001
HSD		0.20	0.37	0.11
CV(%)		6.75	13.9	0.4

*The numbers inside parentheses are the transformed data ($\sqrt{x+0.5}$) and means with the same letters within the columns are not significantly different (P < 0.05)

al., 2007; Oni and Ileke, 2008; Sahaf et al., 2008; Ayvaz et al., 2010; Bachrouch et al., 2010; Sivakumar et al., 2010; Adedire et al., 2011; Ileke and Oni, 2011; Mahmoudvand et al., 2011). In other studies, it was reported that plant based products may act as fumigant, repellent, stomach poison and physical barrier against various insects (Mulungu et al., 2007; Law-Ogbomo and Enobakhare, 2007). Also, Araya (2007) found that fresh *Cymbopogon citratus* essential oil exhibited high mortality (85 to 100%) on mites showing acaricidal activity of the essential oils.

The study of Paranagama et al. (2003) showed that rough rice treated with *C. citratus* essential oil exhibited reduced germination rate in comparison with the control. In contrast, the cotton and mustard seed oils used in the present study did not show any adverse effect on maize seed germination. In agreement with the present study, the investigation of Dejen (2002) showed that powders of *Datura stramonium*, *Jatropha curcas*, *Phytoloca dodecondra* and *Azadrachta indica* applied for the control of *S. zeamais* did not show any significant effect on the germination capacity of stored sorghum grains.

The two cooking oils, *G. hirsutum* and *B. carinata* exhibited toxicity to adult weevils, inhibition of progeny emergency and as a result no damage to the grains throughout the storage period similar to the standard chemical. The toxicity of these cooking oils may be due to their active components responsible for the insecticidal properties against the insect pests including weevils. Oils are known to have toxic effects on insects involving their spiracular system (Cooping and Menn, 2000). Blockage of the spiracles by oils severely limits breathing leading to asphyxiation and death of the insect. The fatty acid composition seemed to be responsible for this acute

toxicity of oils.

Conclusion

The study demonstrated that cotton and Ethiopian mustard seed oils exhibit toxic activity to maize weevil at treatment rates of less than or equal to 0.2% (v/w) when applied on stored maize. These oils are used for cooking and thus are safe for treating maize grains. They pose no danger to humans or animals even when the grains are used for food or feedstuff. It is recommended that studies be conducted to determine the efficacy, technical and economic feasibility of the oils against the maize weevil in pilot scale involving 50 kg of grains and eventually in the scale of tons of grain.

ACKNOWLEDGMENT

The authors thank Jimma University College of Agriculture and Veterinary Medicine for the financial and materials support and provision of laboratory facilities.

REFERENCES

Abbott WS (1925). A method of computing the effectiveness of an insecticide. J. Econ. Entomol. 18:265-267.

Abraham T (1991). The biology, significance and control of the maize weevil, *Sitophilus zeamais* Motsch. (Coleoptera:Curculionidae) on stored maize.. M.Sc. Thesis, Alemaya University of Agriculture, Alemaya, Ethiopia.

Abraham T (1995). Insects and other arthropods recorded from stored maize in western Ethiopia. Afr. Crop Sci. J. 4:339-343.

Abraham T, Basedow T (2004). A survey of insect pest problems and stored product protection in stored maize in Ethiopia in the year 2000. J. Plant Dis. Prot. 111:257-265.

Adedire CO, Obembe OO, Akinkurolele RO, Oduleye O (2011). Responses of *Callosobruchus maculates* (Coleoptera: Chrysomellidae: Bruchinae) to extracts of cashew kernels. J. Plant Dis. Protect. 118(2):75-79.

Ahn YJ, Lee SB, Lee HS, Kim GH (1998). Insecticidal and acaricidal activity of caravacrol and ß-thujaplicine derived from *Thujopsis dolabrata* var. *hondai* sawdust. J. Chem. Ecol. 24(1):81-90.

Araya G (2007). Evaluation of powder and essential oils of some botanical plants for their efficacy against *Zabrotes subfasciatus* (Boheman) (Coleoptera: Bruchidae) on haricot bean (*Phaseolus vulgaris* L.) under laboratory condition in Ethiopia. MSc. Thesis, Addis Ababa University, Addis Ababa, Ethiopia.

Ayvaz A, Sagdic O, Karaborklu S, Ozturk I (2010). Insecticidal activity of essential oils from different plants against three stored-product insects. J. Insect Sci. 10:1-13.

Bachrouch O, Jemaa JMB, Chaieb I, Talou T, Marzouk B, Abderraba M (2010). Insecticidal Activity of *Pistacia lentiscus* Essentail oil on *Tribolium castaneum* as Alternative to Chemical Control in Storage. Tunisia J. Plant Protec. 5(1):63-70.

Bamaiyi LJ, Ndams IS, Toro WA, Odekina S (2007). Laboratory evaluation of mahogany (*Khaya senegalensis* (Desv.) seed oil and seed powder for the control of Callosobruchus maculates (Fab.) (Coleoptera: Bruchidae) on stored cowpea. J. Entomol. 4(3):237-242.

Boxall RA, Brice JR, Taylor SJ, Bancroft RD (2002). Technology and Management of Storage. In: Golob et al. (Eds.) Crop Post-Harvest: Science and Technology, Principles and Practices, Blackwell publisher, London, UK. pp. 141-232.

Cooping LG, Menn JJ (2000). Biopesticides: A review of their action, application and efficacy. Pest Manag. Sci. 56(8):651-676.

CSA (2007). Agricultural sample survey report on area and production of crops (Private Peasant Holding, Meher Season). Statistical Bulletin, No. 388, Addis Ababa, Ethiopia.

Dejen A (2002). Evaluation of some botanicals against maize weevil, *Sitophilus zeamais* motsch. (Coleoptera: Cruculionidae) on stored sorghum under laboratory condition at Sirinka, Pest Manag. J. Ethiopia. 6:73-78.

Demissie G, Teshome A, Abakemal D, Tadesse A (2008). Cooking oils and "Triplex" in the control of *Sitophilus zeamais* Motschulsky (Coleoptera: Curculionidae) in farm-stored maize. J. Stored Prod. Res. 44(2):173-178.

Dubey RK, Rajesh K, Dubey NK (2007). Evaluation of *Eupatorium cannabinum* Linn. oil in enhancement of shelf life of mango fruits from fungal rotting. World J. Microbiol. Biotechnol. 23:467-473.

Emana G, Tsedeke A (1999). Management of maize stem borer using sowing date at arsi-negele. Pest Manag. J. Ethiopia. 3:47-51.

Fekadu G, Waktole S, Santiago DR (2012). Evaluation of plant powders and cooking oils against maize weevil, *Sitophilius zeamais* M. (Coleoptera: Curcurleonidae) under laboratory conditions. Mol. Entomol. 3:4-14.

Finney DJ (1971). Probit analysis, 3rd edn. Cambridge University Press, Cambridge, UK. pp. 1-333.

Gomez KA, Gomez AA (1984). Statistical Procedures for Agricultural Research. 2nd Edn., John Wiley and Sons Inc., New York, USA. P. 680.

Huang Y, Lam SL, Ho SH (2000). Bioactivities of Essential Oil from *Elletaria cardamomum* (L.) Maton. to *Sitophilus zeamais* M. and *Tribolium castaneum* (Herbst). J. Stored Prod. Res. 36:107-117.

Ileke KD, Oni MO (2011). Toxicity of some plant powders to maize weevil, *Sitophilus zeamais* (motschulsky) [Coleopteran: Curculiondae] on stored wheat grains (*Triticum aestivum*). Afr. J. Agric. Res. 6(13):3043-3048.

Kumar R, Mishra AK, Dubey NK, Tripathi YB (2007). Evaluation of *Chenopodium ambrosioides* oil as a potential source of antifungal, antiaflatoxigenic and antioxidant activity. Int. J. Food Microbial. 115(2):159-164.

Law-Ogbomo KE, Enobakhare DA (2007). The use of leaf powders of *Ocimum gratissimum* and *Vernonia amygladina* for the management of *Sitophilus oryzae* (L.) in stored rice, J. Entomol. 4:253-257.

Mahmoudvand M, Abbasipour H, Basij M, Hosseinpour MH, Rastegar

F, Nasiri MB (2011). Fumigant toxicity of some essential oils on adults of some stored-product pests. Chil. J. Agric. Res. 71:83-89.

Mbailao M, Nanadoum M, Automne B, Gabra B, Emmanuel A (2006). Effect of six common seed oils on survival, egg laying and development of the cowpea weevil, *Callosobruchus maculates*. J. Biol. Sci. 6(2):420-425.

Mekuria T (1995). Botanical insecticides to control stored grain insects with special reference to weevils (*Sitophilus* spp.) on maize. Proceedings of the Annual Conference of the Crop Protection Society of Ethiopia, May 18-19, 1995, Addis Abeba, Ethiopia. pp. 134-140.

Meseret B (2011). Effect of fermentation on quality protein Maize-soybean blends for the production of weaning food. M.Sc. Thesis, Addis Ababa University, Addis Ababa, Ethiopia.

Muhammad A (2009). Antixenotic and antibiotic impact of botanicals for organic management of stored wheat pest insects, Ph.D. Thesis, University of Agriculture, Faisalabad, Pakistan.

Mulungu LS, Lupenza G, Reuben SOWM, Misangu RN (2007). Evaluation of botanical products as stored grain protectant against Maize weevil, *Sitophilus zeamais*. J. Entomol. 4(3):258-262.

Negahban M, Moharramipour S, Sefidkon F (2007). Fumigant toxicity of essentail oil from *Artemisia siberi besser* against three stored product insects. J. Stored Prod. Res. 43(2):123-128.

Negeri A, Adisu M (2001). Hybrid maize seed production and commercialization: The experience of pioneer Hi-bred seeds in Ethiopia. Proceedings of the 2nd National Maize Workshop of Ethiopia, November 12-16, 2001, Addis Ababa, Ethiopia. pp. 166-169.

Odeyemi OO (1993). Insecticidal properties of certain indigenous plant oils against *Sitophilus zeamais* Mots. Appl. Eng. Phytopathol. 60:19-27.

Ogendo JO, Deng AL, Belmain SR, Walker DJ, Musandu AO, Obura RK (2004). Pest status of *Sitophilus zeamais* Motschulsky, control methods and constraints to safe maize grain storage in Western Kenya. Egenton. J. Sci. Tech. 5:175-193.

Oni MO, Ileke KD (2008). Fumigant toxicity of four botanical plant oils on survival, egg laying and progeny development of the dried yam beetle, *Dinoderus porcellus* (Coleoptera: Bostrichidae). Ibadan. J. Agric. Res. 4:31-36.

Paranagama P, Abeysekera T, Nugaliyadde L, Abeywickrawa K (2003). Effects of the essential oils of *Cymbopogon citratus*, *C. nardus* and *Cinnamonum zeylancium* on pest incidence and grain quality of rough rice (paddy) stored in an enclosed seed box. Food Agric. Environ. 134:134-136.

Parugrug ML, Roxas AC (2008). Insecticidal action of five plants against maize weevil, *Sitophilus zeamais* Motsch. (Coleoptera: Curculionidae). KMITL Sci. Tech. J. 8:24-38.

Rees D (2004). Insects of Stored Products, CSIRO Publishing, Australia. P. 49.

Sahaf BZ, Moharramipour S, Meshkatalsadat MH (2008). Fumigant Toxicity of Essential oil from Vitex pseudo-negundo against *Tribolium castaenum* (Herbst) and *Sitophilus orysae* (L.). J. Asia-Pacific Entomol. 11:175-179.

Selase AG, Getu E (2009). Evaluation of botanical plants powders against *Zabrotes subfasciatus* (Boheman) (Coleoptera: Bruchidae) in stored haricot beans under laboratory condition. Afr. J. Agric. Res. 4:1073-1079.

Sivakumar C, Chandrasekaran S, Vijayaraghavan C, Selvaraj S (2010). Fumigant toxicity of essential oils against pulse beetle, *Callosobruchus maculates* (F.) (Coleoptera: Bruchidae). J. Biopesticides 3:317-319.

Sori W, Ayana A (2012). Storage pests of maize and their status in jimma zone, Ethiopia. Afr. J. Agric. Res. 7(28):4056-4060.

Tapondjou LA, Adler C, Bouda H, Fontem DA (2002). Efficacy of powder and essential oil from *Chenopodium ambrosioides* leaves as post-harvest grain protectants against six-stored product beetles. J. Stored Prod. Res. 38(4):395-402.

Tripathi AK, Prajapati VAggarwal KK, Khanuja SPS, Kuman S (2000). Repellency and Toxicity of Oil from *Artemisia annua* to Certain Stored Product Beetles. J. Econ. Entom. 93:43-47.

14

Effect of delayed extraction and storage on quality of sugarcane juice

Krishnakumar T., Thamilselvi C. and Devadas C.T.

Department of Food and Agricultural Process Engineering, Tamil Nadu Agricultural University, Coimbatore-3, Tamil Nadu, India.

A study was conducted to determine the quality of sugarcane juice extracted from stored canes, as well as changes in quality of fresh juice stored at different temperatures. Cane stems were stored at 10 and 30°C, while the fresh juice was stored at 5 and 30°C. The parameters studied were juice yield, total soluble solids, total sugar content, titratable acidity, pH, viscosity, total microbial count and sensory evaluation for colour and flavor. Results showed that low temperature storage (10°C) of canes was able to maintain the quality of juice for 10 days, while low temperature storage (5°C) of juice could last for only 4 days. Spoilage of cane stored at 30°C occurred faster than that stored at 10°C. Fresh sugarcane juice became spoilt within a day when stored at 30°C. Microbial count (bacteria, yeast, fungi) especially lactic acid bacteria count increased, during storage of cane juice.

Key words: Cane stems, sugarcane juice, storage temperature, juice quality.

INTRODUCTION

Sugarcane (*Saccharum officinarum* L.) is one of the most important commercial crops in the world. As per Sugarcane Statistical Report (2008), India is the second largest producer of sugarcane in the world next to Brazil. In India, sugarcane is grown mainly for producing sweeteners such as sugar, jaggery and khandasari. The composition of sugarcane juice varies with variety, maturity, climatic and soil conditions and also the portion of the stalk from which it is extracted. Among the varieties grown in India, variety CoP 92226 is popular because of its high juice yield and sensory qualities (Chauhan et al., 2002). Sugarcane juice is a type of drink commonly found in Southeast Asia, South Asia and Latin America, and also in other countries where sugarcane is grown commercially. Sugarcane juice is very popular delicious drink and it is rarely available commercially in packaged form. It is extracted by crushing sugarcane between roller crusher and consumed with (or) without ice. Sugarcane juice contains water (75 to 85%), non reducing sugars (sucrose, 10 to 21%), reducing sugars (glucose and fructose, 0.3 to 3%), organic substances (0.5 to 1), inorganic substances (0.2 to 0.6) and nitrogenous bodies (0.5 to 1) (Swaminathan, 1995). Sugarcane juice has many medicinal properties, often it is used as a remedy for jaundice in traditional medicine (Subbannayya et al., 2007). Sugarcane juice, very useful in scanty urination, keeps the urinary flow clear and helps kidneys to perform their functions properly. Sugarcane juice of 100 ml provides 40 Kcal of energy, 10 mg of iron and 6 µg of carotene (Parvathy, 1983). Due to its commercial importance, it is envisaged that sugarcane juice production can become a profitable business provided efforts to be made to preserve its fresh quality during storage. In general, sugarcane juice is spoiled quickly due to presence of simple sugars (Krishnakumar and Devadas, 2006a). The quality of cane juice is also affected by chemical (acid) and enzymatic inversion (Singh et al., 2006). At present situation, harvested canes are often stored in the shed at ambient temperature before they are processed. Once the juice is extracted, it should be chilled and stored immediately at chilling temperature before distribution. The delay in extraction of harvested sugarcanes is reported to cause some changes in the juice quality (Densay et al., 1992). It has been observed that low temperature storage able to extend the shelf life of the juice for a few days. However, no study has been conducted to elaborate these parameters. Developing scientific knowledge of these aspects is very essential to the emerging industry. Therefore this study was sought to determine the effects of storing the canes on the quality of juice obtained, as well as to predict the physicochemical and microbiological changes in fresh sugarcane juice stored at different temperatures.

MATERIALS AND METHODS

Sugarcanes (10 ± 11 months old) were cut approximately 1 inch above the field surface and the stems were harvested at an average height of 20 inches. Harvested canes were immediately brought to the laboratory. The study was conducted in two parts. The first part of the study was to monitor changes in the quality of juice extracted from canes that was stored at 10 and 30°C. The second part of the study was to monitor the quality of juice stored at 5 and 30°C.

Quality of juice from stored sugarcanes

Upon arrival to the laboratory, sugarcane stems (CoP 92226) were cleaned and stored at 10 and 30°C for 15 days. Every day four canes were removed from storage for juice extraction. The physico-chemical tests carried out were yield of juice, total sugars, total soluble solids, titratable acidity, pH and sensory evaluation for colour and flavor (Ranganna, 1995). The total soluble solids were measured using an ERMA hand refractometer.

Extraction of sugarcane juice

Canes were cleaned, and cut into pieces of uniform length about 10 inches and washed them into fresh water so as to remove the dust and dirt particles. Extraction was done by three roller crusher. Juice was filtered through a four layer muslin cloth. The extracted juice was collected in a chilled container and chilled immediately before being analyzed. The chilling step is very essential to retain the original colour and flavour of the fresh sugarcane juice.

Quality of stored fresh sugarcane juice

Freshly extracted juice was used. The chilled juice were filled into pre sterilized glass bottles and stored at 5 and 30°C. The juices were subjected to similar physicochemical parameter analysis daily as above until they were considered no longer fit for consumption. Viscosity and microbial counts were also conducted. The viscosity of the juice was measured by digital viscometer (Brookfield Synchro-Lectric viscometer, USA) using spindle No.1 at 60 rpm. All the experiments were carried out in triplicates. Each reading was an average of three samples.

Microbiological analysis

The quality of sugarcane juice was based on the number and type of microorganism present which can be assessed by serial dilution and plating method for the differential enumeration of bacteria, yeast and fungi. Determination of total microbial counts (bacteria, yeast and fungi) for juice was carried out at 0 h and every 24 h. One milliliter of juice from each storage temperature was taken into a test tube containing 9 ml of sterile water. The mixture was homogenized. This homogenate represented 10^{-1} dilution. From here, serial dilutions of 10^{-2}, 10^{-3}, 10^{-4}, 10^{-5} and 10^{-6} were prepared. The plates were then incubated at room temperature for 48 h for bacteria and four days for fungi and yeast (Rao, 1986). Enumeration of bacteria was counted by nutrient agar media (Allen, 1953) with 10^{-6} dilutions. Yeast and fungi was counted by yeast extract agar media (Phaff, 1990) with 10^{-4} dilutions and martin's rose bengal medium (Martin, 1950) with 10^{-3} dilutions. The results (number of colony forming units) were obtained after the incubation time using the following formula:

$$Number\ of\ Colony\ Forming\ Units\ (CFU's)\ per\ gram\ of\ sample = \frac{Mean\ number\ of\ CFU's * Dilution\ Factor}{Quantity\ of\ sample\ on\ weight\ basis}$$

Sensory evaluation

Sensory evaluation of the juice extracted from stored canes was carried out by 20 panelists. The panelists rated the sample for colour, flavour, taste and acceptability using 9-point hedonic rating test method (1=dislike very much, 9=like very much) as recommended by Ranganna (1995). The evaluation of stored fresh sugarcane juice was carried out by 10 trained panelists using a triangle test. The panelists were asked to identify the odd sample in terms of colour, flavour and taste compared to a freshly extracted juice.

Statistical analysis

Data were subjected to statistical measurement of analysis of variance (ANOVA) using Agres package.

RESULTS AND DISCUSSION

Analysis of sugarcane juice from stored canes yield

Yield

The yield of juice obtained from stored canes seemed to decrease with increase in time of storage (Figure 1). The decrease in juice yield was more in canes stored at ambient temperature (30°C) than stored at 10°C. The yield started to decrease from the first day onwards until

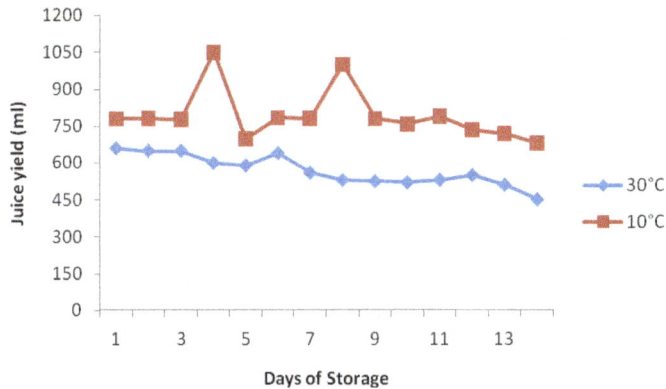

Figure 1. Juice yield from sugarcanes stored at 10 and 30°C.

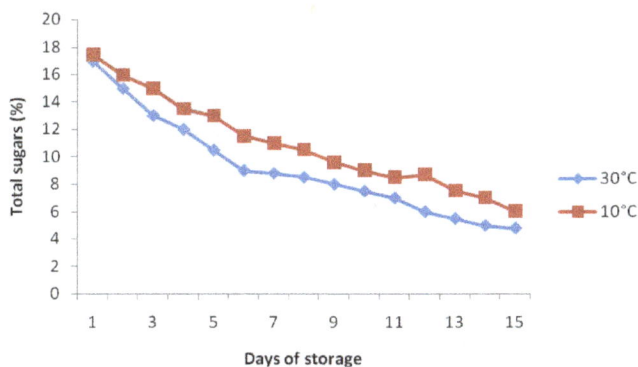

Figure 2. Total sugars contents of juice extracted from sugarcanes stored at 10 and 30°C.

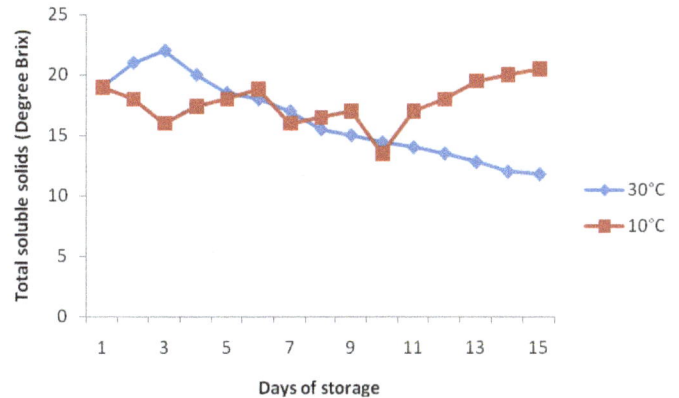

Figure 3. Total soluble solids contents of juice extracted from sugarcanes stored at 10 and 30°C.

for storage of sugarcane juice. Densay et al. (1992) reported that delayed extraction of the juice caused a loss in sugar content.

Total soluble solids (TSS)

There was an initial increase in the TSS content of juice extracted from canes stored at 30°C for the first 3 days of storage and after that it was decreased (Figure 3). This may indicate that, within the three days period, maturation of canes may have continued, resulting in an increase in sweetness. After three days, the TSS content decreased probably due to the onset of senescence. The canes stored at 10°C showed irregular values for TSS up to 9 days, but when the canes became stabilized the value began to increase as in the initial stage of storage at 30°C.

Titratable acidity

The titratable acidity of extracted juice increased during storage (Figure 4). The increase occurred on day 6 for both treatments. Sugarcane stems stored at 30°C recorded higher acidity increases compared to those stored at 10°C. This was perhaps caused by the utilization of sugar during respiration of the cane stem itself. The acidity increase was not identified by panelists until day 9. A similar result of high acidity in sugarcane juice was found out by Bhupinder et al. (1991).

pH

The pH value of juices extracted from canes stored at 30 and 10°C both decreased at about equal rates until day 9 Figure 5). After that the juice of cane stem stored at (30°C had a faster drop compared to the juice of cane stems stored at 10°C. Similar results were reported by

the end of the storage period of 15 days. After 15 days, the yield decreased by 44.5% for canes that were stored at 30°C compared to only 10.5% for those stored at 10°C. In India, the small or road side sugarcane juice sellers get their supply of canes on weekly basis. Before they are extracted, the bundles of canes stored under tree. This study shows that storage at the normal atmospheric temperature would incur great losses in juice yield and is certainly not a wise practice to continue. Meanwhile canes that were stored at 10°C only started to fluctuate in yield after three days of storage.

Total sugars

Storage of cane caused a decrease in total sugars content of the extracted juice (Figure 2). A significant decrease (p<0.05) in total sugar content of juice was observed at room and low temperature. The decrease in the sugar content of stored canes at 30°C was faster than at 10°C. The decrease in total sugars content may be due to breakdown of total sugars into reducing and other sugars. Similar results were reported by Sneh sankhla et al. (2012), Chauhan et al. (2002), Bhupinder et al. (1991)

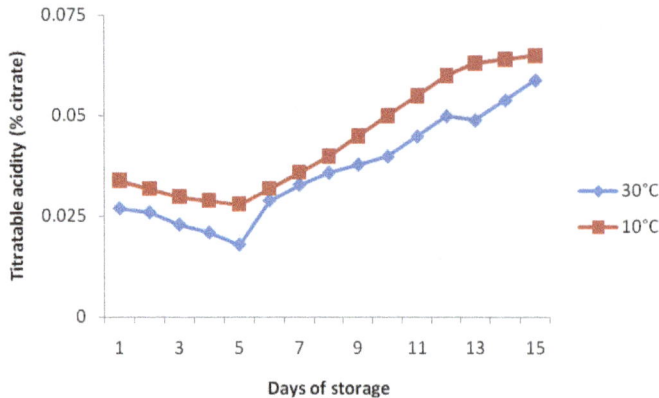

Figure 4. Titratable acidity values of juices extracted from sugarcanes stored at 10 and 30°C.

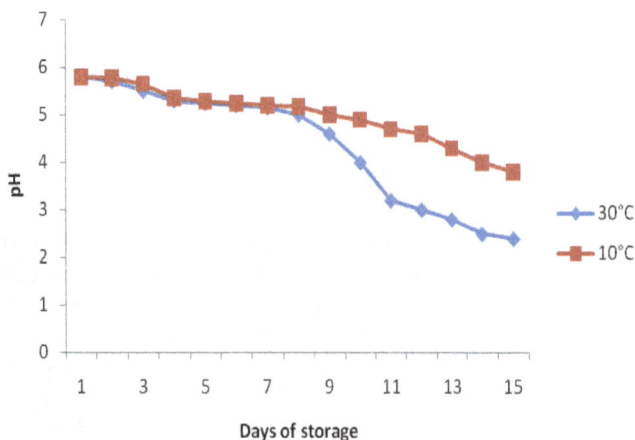

Figure 5. pH values of juice extracted from sugarcanes stored at 10 and 30°C.

Chauhan et al. (1997) for storage of sugarcane juice.

Sensory evaluation

The panelists did not indicate any significant change (p>0.05) in the colour of juices that were extracted from canes stored for 3 and 6 days at both storage temperature. However, on day 9 there was a significant difference (p<0.05) in the colour among the juices from both types of canes. The canes that were stored at 10°C produced juices that were rated higher in colour (6.8 out of 9) than cane stems stored at 30°C (3.9 out of 9.0). The mean ratings for flavor of the juice samples stored under the two temperature treatments were found to be significantly different also on day 9 (p<0.05). The flavor of the juice obtained from cane stems stored at 10°C for 9 days was high (7.53 out 9). This means that sugarcanes may be stored at low temperature (10°C for 9 days) without affecting the flavor of the juice while storage of canes at ambient temperature for 9 days results in

development of objectionable flavour. Similar results were reported by Sneh Sankhla et al. (2012) for storage of sugarcane juice using hurdle technology.

Analysis of stored fresh juice

Total sugars

Results indicated that, as storage time increased the total sugars content of sample decreased. The samples stored at 5°C decreased gradually within the storage period of 15 days while the ones stored at 30°C decreased sharply within 3 days (Table 1). The sharp drop in total sugars content that occurred in samples stored at 30°C was perhaps caused by microbes that utilized the sugars and in the end resulted in spoilage of juice. Chauhan et al. (1997) and Puspha Singh et al. (2002) also found similar results in sugarcane juice during storage. Microorganism present in juice leads to decrease total sugars content by formation of organic acid and ethanol (Sneh Sankhla et al., 2012). Major bacteria responsible for spoilage are *Leuconostoc,Enterobacter, Flavobacteruim, Micrococcus, Lactobacillus* and *Actinomyces* (Frazier and Westhoff, 1995).

TSS, titratable acidity and pH

There was an increase in the TSS content of juice stored at 5°C at the initial stage (up to 8 days), but the value decreased with time (Table 1). The juice sample stored at 30°C also showed an increase in TSS before it became spoilt. The increase was perhaps due to the breakdown of total sugars into simple sugars and acids during storage as a result of action of microorganism present in the juice. These observations are in agreement with the findings of Bhupinder et al. (1991). Acidity of juice increased with storage time (Table 1). Similar results were reported by Bhupinder et al. (1991) in sugarcane juice during storage. The increase in acidity caused a concomitant decrease in pH value (Table 1). A similar result reported by Abbo et al. (2006) revealed that there is a corresponding reduction in pH as the acidity increased in Soursoup juice. The reason for high acid and low pH could be due to acetic acid and lactic acid production.

Viscosity

The viscosity of juice stored at 5°C (Table 1) initially decreased within 10 days, after that it increased. Meanwhile the sample that was stored at 30°C recorded a decrease in viscosity. The increase in viscosity of juice stored at low temperature might be due to the development of dextran, that is, a gummy substance produced by bacteria such as *Leuconostoc mesenteroides.*

Table 1. Mean values for the physico-chemical characteristic of sugarcane juice stored at different temperatures.

Temperature (°C)	Storage duration (Days)	Total sugars (%)	Viscosity (cps)	TSS (°Brix)	Titratable acidity (%)	pH (%)
5	0	16.50^a	3.00^a	16.0^g	0.055^b	5.72^a
	1	16.10^{ab}	3.10^{ab}	17.4^{de}	0.045^b	5.79^a
	2	15.78^{ab}	3.08^{ab}	17.0^{ef}	0.052^b	5.76^{ae}
	3	14.82^{ab}	3.15^{ab}	17.7^{bcd}	0.053^b	5.76^a
	4	14.32^{ab}	2.72^{ab}	17.7^{bcd}	0.053^b	5.69^{ab}
	5	13.84^{abc}	2.49^b	18.0^{abc}	0.056^b	5.69^{ab}
	6	13.33^{abc}	2.29^{ab}	18.4^a	0.058^b	5.67^{ab}
	7	13.21^{abc}	2.40^{ab}	18.5^a	0.070^b	5.66^{ab}
	8	13.12^{abc}	2.50^{ab}	18.2^{ab}	0.063^b	5.63^{ab}
	9	13.03^{abc}	2.55^{ab}	17.5^{cd}	0.079^b	5.61^{ab}
	10	12.29^{abc}	2.55^{ab}	16.1^g	0.088^b	5.61^{ab}
	11	12.09^{abc}	2.61^{ab}	18.2^{ab}	0.094^b	5.49^{abc}
	12	12.06^{bc}	2.70^{ab}	18.0^{abc}	0.108^b	5.37^{abc}
	13	11.08^{cd}	2.77^{ab}	18.1^{ab}	0.058^b	5.15^{bcd}
	14	11.01^{cd}	3.11^a	17.4^{de}	0.054^b	5.05^{cd}
	15	08.78^d	3.20^a	16.8^g	0.149^a	4.67^d
30	0	16.50^a	3.00^a	16.0^b	0.055^a	5.72^a
	1	12.60^b	2.91^a	16.7^{ab}	0.170^{bc}	3.77^{bc}
	2	12.33^b	2.42^{ab}	16.10^b	0.200^b	3.80^b
	3	12.18^b	2.12^b	17.4^a	0.220^c	3.67^c

^{a}Means for each storage duration followed by the same letter are not significantly different p>0.05.

Table 2. Mean values for the microbial count of sugarcane juice stored at different temperature.

Storage duration	Temperature (°C)	Bacteria count (Cfu's)	Yeast count (Cfu's)	Fungi count (Cfu's)
1	5	3.27^a	0^b	0^b
	30	3.54^b	3.25^a	2.12^a
3	5	3.72^b	0^a	0^a
	30	4.42^a	3.46^b	2.16^b
6	5	3.74^a	0^a	2.40^b
	30	0.93^b	6.15^b	6.15^a
9	5	5.01^a	0^a	0^a
	30	5.45^b	4.41^b	4.53^b
12	5	2.35^a	0^a	-
	30	0.35^b	5.81^b	-
15	5	1.44^a	0^a	-
	30	0.27^b	5.98^b	-

^{a}Means for each storage duration followed by the same letter are not significantly different p>0.05.

This phenomenon was found by Lotha et al. (1994) who reported that the viscosity of mandarin juices increased during refrigerated storage and decreased when stored at ambient temperature.

Microbiological analysis

There were significant differences (p<0.05) in the bacteria, yeast and fungi of stored juice between storage temperature and duration (Table 2). Microbial count of the juice rises with time. There was an increase in bacteria and yeast count in juice stored at both temperature treatments. However the growth of fungi was detected in juice stored at 5°C on day 6 while no yeast was detected during the total storage period of 15 days. Presence of *Escherichia coli,* enterococci and other coliforms indicate faecal contamination of sugarcane juice, suggesting possible risk of infection involved with drinking such sugarcane juice (Pelczar et al., 1993). Richa Karmakar et al. (2011) reported that bacterial contamination may occur at different stages of juice processing such as by contamination of sugarcane, roller crusher, collecting vessel, ice, hands of the personnel and filter cloth. The low temperature storage may have retarded the growth of the organism. The growth of yeast and fungi in juice stored at 30°C increased significantly ($p < 0.05$) after 6 days, but bacteria were found to decrease at the later stage of storage.

Conclusion

The yield and quality of juice obtained are essential economic criteria in sugarcane juice business. Results of the study indicate that it is advisable to store sugarcanes at low temperatures to maintain the juice yield. The canes may be stored at 10°C for 9 days and still produce good quality juice, to a maximum of 11 days. Beyond that, the canes suffer chilling injury. The total sugar content of juice decreased after 3 days of storage. The colour and flavor of juice obtained were also superior to canes stored at high temperature. Storage of canes at 30°C for more than 4 days reduced the juice yield drastically. The colours of juice obtained were darker (more brown and less green) than the juices obtained from canes stored at 10°C and the flavour was also different. In the second part of the experiment, it was found that the freshly extracted unpasteurized juice could be kept at 5°C for only 4 days. Beyond that, the quality deteriorated, which could be observed by the colour and flavour change followed by increase in viscosity. The juice became spoilt within a day when kept at 30°C.

REFERENCES

Abbo ES, Odeyemi G, Glurius TO (2006). Studies on the storage stability of soursoup juice. Afr. J. Microbiol. 21(2):197 -214.

Allen DN (1953). Experiments in soil bacteriology (II edition). Burgees Publishing Co., Minneapolis, Minna. p. 127.

Bhupinder K, Sharma KP, Harinder K (1991). Studies on the Development and Storage Stability of ready-to- Serve Bottled Sugarcane Juice. Int. J. Trop. Agric. 9(2):128-134.

Chauhan K, Joshi VK, Lal BB (1997). Preparation and evaluation of a refreshing sugarcane juice beverages. J. Sci. Ind. Res. 56(4):220-223.

Chauhan OP, Dheer Singh, Tyagi SM, Balyan DK (2002). Studies on Preservation of sugarcane juice, Int. J. Food Propert. 5(1):217-229.

Densay JPS, Luthra R, Senthiya HL, Dhawan AK (1992). Deterioration of juice quality during post harvest storage in some sugarcane cultivars. Indian Sugar 42(2):92-95.

Frazier CW, Westhoff CD (1995). Food microbiology. Tata McGraw-Hill Publishing Company Limited, New Delhi, pp. 187-195.

Krishnakumar T, Devadas CT (2006a). Microbiological changes during storage of sugarcane juice in different packaging materials. Bever. Food World 33(10):82-83.

Lotha RE, Khurdiya DS, Maheswari ML (1994). Effect of storage on the quality of kinnow mandarin fruit for processing. Indian Food Packer 30(2):25 -32.

Martin JP (1950). Use of acid, rose Bengal and streptomycin in the plate method for estimating soil fungi. Soil Sci. 69:215.

Parvathy K (1983). Bottling of sugarcane juice. Proceedings of the scheme for studies on post harvest technology (ICAR), Coimbatore center, Annual Report, pp. 13-16.

Pelczar MJ, Chan EC, Krieg NR (1993). Microbiology of natural waters, drinking water and waste water. Chapter 29. In Microbiology: Concepts and applications, eds. M.J. Pelczar Jr, E.C. Chan, and N.R. Krieg Krieg International edition, McGraw-Hill Inc. New York. pp. 806-840.

Phaff HJ (1990). Isolation of yeasts from natural sources. In: Isolation of biotechnological organisms from nature (Eds.), McGraw-Hill publishing co., Newyork. p. 79

Puspha Singh, Shashi HN, Archnna Suman, Jain PC, Singh P, Suman A (2002). Sugarcane juices concentrate preparation, preservation and storage. J. Food Sci. Technol. 39(1):96-98.

Ranganna S (1995). Handbook of analysis and quality control for fruits and vegetable products. Tata –McGraw-Hill Publishing Company Limited, New delhi, India.

Rao NSS (1986). Soil microorganisms and plant growth (II edition). Oxford and IBH Publishing Co.Pvt.Ltd, New Delhi, pp. 290-309.

Richa k, Amit kG, Hiranmoy G (2011). Effect of pretreatments on physic-chemical characteristics of sugarcane juice. Sugar Tech. 13(1):47-50.

Singh I, Solomon S, Shrivastava AK, Singh RK, Singh J (2006). Post–harvest quality eterioration of cane juice: physiobiochemical indicators. Sugarcane Technol. 8(2&3):128-131.

Sneh sankhla, Anurag chaturvedi, aparna kuna, Dhanalakshmi K (2012). Preservation of sugarcane juice using hurdle technol. Sugar Tech. 14(1):26-39.

Subbannayya K, Bhat GK, Shetty S, Junu VG (2007). How safe is sugarcane juice. Indian J. Med. Microbiol. 25(1):73-74.

Swaminathan V (1995). Food science chemistry and experimental foods. Bangalore printing and publishing Co. Ltd, Bangalore.

15

Processing and preservation of African bread fruit (*Treculia africana*) by women in Enugu North agricultural zone, Enugu State, Nigeria

C. S. Ugwu and J. C. Iwuchukwu

Department of Agricultural Extension, University of Nigeria Nsukka, Enugu State, Nigeria.

The study examined methods used in processing and preservation of African bread fruit by women in Enugu North Agricultural Zone of Enugu State, Nigeria. A total of seventy two women were used for the study. Frequency, percentage and mean score were used for data analysis. Majority of the respondents engaged in processing and preservation of African bread fruit as their primary occupation and earned monthly income of ₦9,876 (about 61 US Dollars) on average. They engaged in these activities for mainly family consumption and their mean years of experience in the business was 31 years. Fermentation method of 7 to 14 days duration was the extraction method used by the respondents while the seeds were threshed with grinding or milling machine. Sun drying and keeping in bottle/air tight container without preservative was preservation method of choice. Water scarcity was a major problem encountered in processing while bad weather condition was a major problem encountered in preservation of African bread fruit. The study points out the need to encourage women to use good/modern methods of processing and preservation so as to get high quality seeds of African breadfruit that will attract more demand and income.

Key words: Processing, preservation, African breadfruit, rural women.

INTRODUCTION

Agriculture sector of a developing economy performs the primary role of provision of food to nourish the populace (Njoku, 2000). In Nigeria, agriculture employs about two-third of the total labour force, contributes 42.2% of the Gross Domestic Product (GDP) and provides 88% of non-oil earnings (World Bank, 2005). Crops contribute immensely to agricultural GDP. Apart from its contribution to the agricultural GDP, they are important to humans for it is a part in the food chain.

Some plants/crops resources have been widely exploited and used as food crops, while others, mainly the tree crops of which *Treculia Africana* (African breadfruit) is an important member have been under exploited and still harvested from the wild. The seed of african bread fruit (*T. Africana*) is variously named by different/ethnic tribes across the continent (Africa). In Nigeria, it is known as "ukwa" by the Igbo and the Yoruba refer to it as "bere – foo-foo" or "afou", "ediang" by Efik and Ibibio, "ize" by Bini and "bafafuta" by Hausa (Baiyeri and Mbah, 2006). It is referred to as "mwaya" in Swahili, "muzinda" by Lugadan and "brebretim" by Wolof (Enibe, 2006). It is widely cultivated in the southern states of Nigeria where it serves as low cost meat substitute for poor families (Badifu and Akubor, 2001).

According to Baiyeri and Mbah (2006), African breadfruit is an important natural resource for the poor, contributing significantly to their income and dietary intake. Hence African breadfruit does not only help to ensure food security by meeting the protein need of people but also provides income to rural poor households that produce, process and/or preserve this crop. The plant produces large, usually round compound fruit covered with rough pointed outgrowths. The seeds are buried in spongy pulp of the fruit (Keay, 1989; Osuji and Owei, 2010). The seeds are variously cooked as porridge alone or mixed with other food stuff such as sorghum (Onweluzo and Nnamuchi, 2009) or roasted and sold with palm kernel (*Elaeis guineensis*) as a roadside snack. The flour has high potential usage for pastries (Onyekwelu and Fayose, 2007). The seeds are highly nutritious and constitute a cheap source of vitamins, minerals, proteins, carbohydrates and fats (Osuji and Owei, 2010). Proximate analysis shows that the seed contain 17 to 23% crude protein, 11% crude fat and other essential vitamins and minerals (Akubor et al., 2000). The seeds are used in preparing pudding and as a thickener in weaning food for children (Onyekwelu and Fayose, 2007). As a tree crop, there is significant carbon sequestration benefit. It is expected that local sustainability will be bolstered by the increased food security that breadfruit stands to provide in the face of climate change (Pacific Agribusiness Research for Development Initiatives, 2011). Breadfruit is an important staple crop and makes substantive contributions to food security especially in rural communities. In countries like Nigeria, they serve as nutritious feed for livestock. In Malawi, blue monkey are very fond of the fruits and extract the seeds for food while in Tanzania the leaves are used as fodder (Enibe, 2006).

National survey in Nigeria showed that 40% of all household surveyed in all zones across the country and in all sectors were food insecure (Maziya-Bixon et al., 2004). There are classes of essential nutrients which must be combined in appropriate portion to ensure an adequate food intake. These include: Carbohydrate, proteins, fats and oil, vitamins and minerals (Mohammed, 2003). Therefore to achieve the goal of promoting good health, a cheap source of protein remains an ultimate step (FAO, 2002). African breadfruit serves as a cheap source of protein to the rural poor who cannot afford the luxury of buying meat or other sources of animal proteins. Also an attempt to achieve food security by increasing output requires a corresponding increase in processing and preservation of the food in order to avoid food losses.

Enugu North Agricultural Zone of Enugu State Nigeria has a favourable climate (low relative humidity, annual rainfall of 1680 to 1700 mm) for agriculture especially crop farming. Despite the socio-economic importance of African breadfruits in meeting the protein need as well as generation of income to a large population of the country, they are mainly found around homestead in the area as wild and protected crops. As a result of this and other factors,

demand for the crop has not been met. Processing and preservation of the crop become inevitable in order to increase its consumption among rural households especially during off-season. The ultimate goal of processing, however, is to preserve the nutrients in order to make them available to the consumers and to remove or reduce the levels of phytochemicals which interfere with nutrient digestion and absorption (Hassan et al., 2005). Processing and preservation help to supply wholesome, safe and nutritious food throughout the year for the maintenance of health as well as generation of income for the producers.

Poor processing and preservation leads to high post harvest losses. For example boiling and drying significantly reduced the selenium and iodine content of breadfruit seeds (Ijeh et al., 2010). They may also cause total loss of the seeds, imparts undesirable properties to the processed seeds such as offensive odour, variation in colour and in duration of cooking. These variation and undesirable qualities affect the nutritional and economic value of the crop and may not allow its preservation and storage for a long time. While minimally processed pulp has the appearance, texture, and taste of fresh breadfruit (Ragone, 2011).

In view of the aforementioned facts, the study sets out to characterize women involved in processing and preservation of African bread fruit in the area, identify methods used by them in processing and preservation of African breadfruit, ascertain reasons, sources of information and seeds they processed and preserved as well as problems they encountered in processing and preservation of African bread fruit.

METHODOLOGY

The study was carried out in Enugu North Agricultural Zone of Enugu State, Nigeria. The zone is located at the northern part of Enugu State and it is made up of eight blocks.

All women involved in processing and preservation of African bread fruit in the zone constituted the population for the study. Out of the 8 blocks in the zone, Nsukka 1 and 2 were purposively selected because of high level of involvement of the women in the area in processing and preservation of African bread fruit. Out of the eight cells in each of the blocks, three cells were randomly selected from each of the blocks giving a total of six cells. Twelve women who were involved in processing and preservation of African bread fruit were purposively selected from each of the cells making a total of seventy-two (72) respondents for the study.

Data were collected in July and August 2010 through the use of interview schedule. This was administered by the researcher and a research assistant to the respondents. Some of the variables contained in the instrument were: Marital status, level of education, primary occupation, reasons and frequency of processing and preservation of African breadfruit, sources of seed and information as well as processing and preservation methods used by these women in processing and preservation of African breadfruit. Respondents were requested to state their age in years, household size, years of experience in processing and preservation of African bread fruit, and their monthly income which were later classified/grouped.

Data on problems encountered in processing and preservation of

Table 1. Percentage distribution of the respondents base on their socio-economic characteristics.

Socio economic characteristic	Frequency	Percentage	Mean
Age (years)			
21-40	28	38.9	
41-60	36	50.0	47
61-80	7	9.7	
Above 80	1	1.4	
Marital status			
Married	45	62.5	
Widowed	22	30.6	
Divorced	4	5.5	
Single	1	1.4	
Household size			
1-5 persons	25	34.7	
6-10 persons	40	55.6	8
11-15 persons	7	9.7	
Level of education			
no formal education	24	33.4	
Primary education	20	27.8	
Secondary education	14	19.4	
Tertiary education	14	19.4	
Primary occupation			
Farming/processing of agricultural product	45	62.5	
Trading	21	29.2	
Artisan	4	5.5	
Civil service	2	2.8	
Monthly income (₦)			
1,001 - 10, 000 (6-62 US Dollars)	42	58.4	
10,001 - 20, 000 (63-125 US Dollars)	18	25.0	9,876 (61.7 Us Dollars)
>20,000 (125 US Dollars)	12	16.6	

Africa breadfruit were collected using a four point Likert type scale of 'not at, all'(0), 'occasionally' (1), 'often '(2) and 'very often' (3), with a mean of 1.5. Any variable with a mean equal or greater than 1.5 was regarded as a major problem, any variable with a mean less than 1.5 but greater than 1 was regarded as a minor problem while any variable with a mean equal or less than 1 was regarded as no problem to processing and preservation of African breadfruit. Some of the problems considered under processing were: Water scarcity, labourious nature of processing activities, bad odour and others. Bad weather condition, extra energy required in cooking preserved seeds, undesirable taste of preserved seeds, and others were problems considered under preservation. Data for the study were analysed using percentage and mean score. The statistical package for the social sciences (SPSS) version 16.0 was soft ware used for analysis.

FINDINGS AND DISCUSSION

Socio-economic characteristics of respondents

Data in Table 1 show that half (50%) of the respondents were within the age range of 41 to 60 years while 38.9% of them were between 21 and 40 years. The mean age was 47 years. This indicates that the respondents were in their middle age hence may be energetic to undertake task involved in processing and preservation of African breadfruit which according to Etoamaihe and Ndubueze, (2010) are labourious, time consuming and unhygienic in nature.

Majority (62.5%) of the respondents was married while 30.6% of them were widowed (Table 1). About 56% of the respondents had household size of 6 to 10 persons, 34.7% of them had household size of 1 to 5 persons while the mean house hold size was 8 persons (Table 1). This relatively large household size may be advantageous because they are likely to provide family labour for agricultural activities especially processing and preservation of African breadfruit.

Table 1 also shows that greater percentage (33.4%) of the respondents had no formal education while 27.8% of them had only primary education. This finding points at

Table 2. Percentage distribution of respondents based on processing and preservation characteristics of African bread fruit.

Factors in processing/preservation	Frequency	Percentage	Mean
Years of experience (years)			
1 - 10	14	19.4	
11 - 20	35	48.6	
21 - 30	10	13.9	
31 - 40	11	15.3	31
41 - 50	1	1.4	
51 - 60	1	1.4	
Reason for processing and preserving			
Income generation	35	48.6	
Family consumption	37	51.4	
Sources of seed			
Purchased	20	27.8	
Received as gift	19	26.4	
From own farm	33	45.8	
Frequency of processing and preservation			
Daily	1	1.4	
Twice a week	30	41.7	
Monthly	29	40.3	
As the need arises	12	16.6	
Sources of information			
Neighbor	30	41.7	
Friends	24	33.3	
Parents	11	15.3	
Extension agent	0	0	
Self	7	9.7	

the poor literacy level of respondents which is common in many rural communities especially among women. Thus, it hampers their socioeconomic status and accessibility to agricultural innovation irrespective of their enormous contribution to agriculture. It is also evident in Table 1 that majority (62.5%) of the respondents engaged in farming/processing of agricultural products as their primary occupation while 29.2% of them engaged in trading as their primary occupation. The table further reveals that greater proportion (58.4%) of the respondents had monthly income of ₦1,001 to ₦10, 000 (about 6 to 62 US Dollars) only, 25% of them had between ₦10,001 and ₦20,000 (about 63 to 125 US Dollars) as their monthly income while their mean monthly income was ₦9,876 (about 62 US Dollars).

This finding shows that these women are low income earners. Child bearing and caring as well as engagement of women in domestic works at home in developing countries might be a factor limiting the interest and deployment of women in activities that will earn them high income. In line with this, United Nations (1989) observed that female workers invariably predominate when production work involves mostly unskilled or semi skilled

jobs and relatively low wages.

Processing and preservation characteristics of African breadfruit

Years of experience

Entries in Table 2 show that 48.6% of the respondents had 11 to 20 years of experience while 15% of them had 31 to 40 years of experience in processing and preservation of Africa breadfruit. The mean years of experience were 31 years. The many years of experience implies that processing and preservation of the crop is not new in the study area and that the women are well experienced in the tasks. Experience is the first determinant of profitability (Yusuf, 2000) and perfection.

Reasons for processing and preservation of African bread fruit

The respondents indicated that their reasons for

engaging in processing and preservation of African breadfruit were: Family consumption (51.4%) and income generation (48.6%) (Table 2). This finding corroborates with Baiyeri and Mbah (2006) who noted that African bread fruit is an important natural resource for the rural poor households, contributing significantly to their income and dietary intake.

Sources of African breadfruit processed and preserved

Table 2 also indicates that 45.8% of the respondents sourced the African breadfruit from their own farm, 27.8% of them purchased while 26.4% of them received the one they processed and preserved as gift. Also greater proportion (41.7%) of the respondents processed and preserved African breadfruit twice a week, 40.3% processed and preserved on monthly basis while only about 1% of the respondents processed and preserved African bread fruit on daily basis. Since processing and preservation of African bread fruit were done for mainly family consumption and the fruits were gotten mainly from the farms of respondents, it is unlikely that the processing and preservation of this crop will be done often or on daily basis.

Sources of information on processing and preservation of African bread fruit

Table 2 further shows that respondents' sources of information on processing and preservation of African bread fruit were: Neighbors (41.7%), friends (33.3%), parents (15.3%) and self (9.7%). This means that there were no major sources of information on processing and preservation of African bread fruit. The few that existed were informal/interpersonal; hence, the knowledge generated is likely to be local or indigenous and may not be reliable. In view of this, the respondents may not keep pace with the modern or the scientific methods of processing and preservation of agricultural products especially African breadfruit when they rely solely on these information sources.

Methods used in processing African breadfruit

Extraction method

Table 3 reveals that majority (79.1%) of the respondents extracted African bread fruit using fermentation method while only 13.9% practiced fresh extraction method. Among these respondents that used fermentation method, 59.7% allowed a fermentation period of 7 to 14 days, 40.3% of them allowed 15 to 21 days, 6.5% allowed 6 to 7 days while only 1.6% allowed more than a month fermentation period before they extracted the

seeds (Table 3). In contrast with this finding, Ragone (2011) observed that fruits of breadfruit quickly ripen in just 1 to 3 days after harvest.

Reasons for using fermentation method

Table 3 also reveals that the reasons given by the respondents for using fermentation method were: It makes seeds easy to wash (54.8%), no reason (25.8%) and it reduces water requirement (9.7%). Reason given by the respondents for using fresh extraction method was that breadfruit seeds produced using the method gives better quality seeds (100%). This may be because of the cleanliness of the seeds produced through fresh extraction since the fruit did not undergo decaying or fermentation process. However, fermentation method has been known as a valuable preservation method because it does not only create more palatable food from less than desirable ingredients but produce vitamins through micro organism responsible for fermentation (FAO, 2002).

Table 3 further indicates that majority (80.5%) used clean water in washing their seeds, 11.1% used any available water, while 4.2% each accounted for respondents that washed their seeds with water from fermented cassava and water got from ground after rain. In as much as majority of the respondents used clean water in washing their seeds, it is important to note that for health reason, using dirty water as indicated by some of the respondents in washing the seeds after extraction or fermentation is not ideal as this predisposes consumers to diseases.

Parboiling of African breadfruit

In Table 4, greater percentage (50.0%) of the respondents indicated that they parboiled African bread fruit seeds for 10 min, while 40.3% parboiled them for 15 min after washing or before threshing. Also majority (97.2%) of the respondent did not add alum while only 2.8% added alum when parboiling the seeds. Reasons given by the respondents for not adding alum were: They dislike addition of alum (52.9%), it prolongs cooking time (22.8%), it changes the real taste (15.7%), and changes the colour (8.6%) (Table 4). These findings tend to contradict the fact that local producers claim that addition of alum into water used for parboiling breadfruit seeds increases the keeping quality of the product by extending the storage period and by leaving the cotyledons intact without breaking, thereby enhancing the appeal (Ihediohamma, 2009).

It is also obvious in Table 4 that the effects of parboiling on the seeds as indicated by the respondents were: Easy dehulling (65.3%) and prevention of the breakage of the cotyledons (30.7%). According to Nwabueze (2009) huge losses are encountered in most mechanical or traditional dehulling of parboiled African breadfruit seeds when the

Table 3. Percentage distribution of respondents based on extraction methods used in processing Africa bread fruit.

Methods	Frequency	Mean
Extraction processing methods		
Fresh extraction	10	13.9
Fermentation	57	79.1
Both methods	5	7.0
Duration of fermentation (n=62)		
6 - 7 days after cutting	4	6.5
7 - 14 days after cutting	37	59.7
15 - 21 days after cutting	25	40.3
More than a months	1	1.6
When to carryout fresh extraction (n=15)		
Immediately after cutting	13	86.7
2 days after cutting	2	13.3
Reasons for using fermentation method (n=62)		
Production of better quality of seeds	4	6.5
It requires less water	6	9.7
Makes washing of seed easy	34	54.8
Convenience	2	3.2
No reason	16	25.8
Reasons for using fresh extraction method (n=15)		
Quality of seed produced is better	15	100.0
Sources of water used for washing seeds		
Clean water	58	80.5
Water from fermented cassava	3	4.2
Any available water	8	11.1
Water got from ground after rain	3	4.2

Table 4. Percentage distribution of respondents based on parboiling and threshing methods.

Parboiling and threshing methods	Frequency	Percentage
Duration of parboiling (Min)		
5	6	8.3
10	36	50.0
15	29	40.3
No specific time	1	1.4
Addition of alum during parboiling		
Number that added alum	2	2.8
Number that did not add alum	70	97.2
Reasons for not adding alum (n = 70)		
It prolongs cooking time	16	22.8
It changes the colour	6	8.6
It affects the real taste	11	15.7
Dislike putting it	37	52.9

Table 4. Contd.

Threshing methods		
Manual (threshing on hard board/concrete floor)	11	15.3
Mechanical (using grinding/milling machine)	31	45.0
Both methods	30	41.7
Separating seeds from the hull		
Hand picking method (on hardboard/concrete floor/flat-broad plate)	2	2.8
Winnowing method (using flat-broad plate)	1	1.4
Both methods	69	95.8
Effect of parboiling		
Easy dehulling	47	65.3
Prevents breakage of the cotyledon	25	30.7

seeds are either over or under parboiled because of poor processing conditions. This points out the inefficiency of current methods of processing used by these women which may not afford them control over the temperature in which these seeds are parboiled.

Threshing of African breadfruit

After parboiling the seed, threshing is done to remove the hulls from the seeds before separating them. Greater proportions 43.0 and 41.7% of the respondents in their respective order threshed African bread fruit seeds with mechanical (using grinding/milling machine) and manual (threshing on a hard board, concrete floor using bottle, mortar) (Table 4). In Table 4 also, majority (95.8%) of the respondents separated seeds from the hulls by both hand picking (on hard board, concrete floor or flat-broad plate) and winnowing methods (using mainly flat-broad plates). The respondents might have combined these methods to ensure thorough and easy removal of the hull from the seeds.

Methods used in preservation of African breadfruit

Table 5 shows that greater percentage (45.8%) of the respondents preserved Africa breadfruit after processing by sun drying and keeping it in bottle/air tight container; 34.7% of them sun dried and kept in basin, 13.9% sun dried and kept in bag while only 4.2% sun dried and spread in open floor/mat. Also majority (88.9%) of the respondents asserted that the best duration of drying is when the seeds are completely dried while 11.1% asserted that it is when the seeds are moderately dried (Table 5). These findings are in line with what is found in rural communities especially in developing countries like Nigeria where preservation of agricultural products are done through traditional method of sun drying. Grains, legumes like African breadfruit are foodstuff that are preserved and stored using the technique FAO (2002). Also, rural women may have resorted to sun-drying because it is economical in the sense that they incur little or no cost in using this preservation method.

Table 5 also reveals that majority (99.6%) of the respondents did not use preservative in preserving African breadfruit. Some of the reasons given by them for not using these preservatives were: Preservatives are not necessary/needed (71.8%), preservative impart undesirable characteristic on the seeds (14.1%) while another 14.1% could not adduce any reason for not adding preservative (Table 5). This may mean that only few of the respondents could give tangible reason for not using preservatives. Thus, pointing at the ignorance of these rural women even in some of the technologies/practices they adopt. This may be attributed to their poor literacy level (Table 1) which will invariably affect their activities, output and income.

Effects of preservation (drying)

The effects of preservation (drying) on African breadfruit seeds as indicated by the respondents were: It makes it last longer (82.0%), makes it easy to cook (8.3%), and reduces its attack by insects/rodents and other micro-organisms (8.3%) (Table 5). In Table 5 also, it can be inferred that the life span of well preserved seeds was 3 to 7 months as indicated by majority (68%) of the respondents. Usually, processing and preservation of agricultural products like African bread fruits are done locally in rural households of developing countries like Nigeria without any preservative thereby making the preserved food to last for a limited time. In as much as these local/indigenous methods of processing and preservation of African breadfruit have disadvantages, there are positive attributes inherent from them due to the fact that food processed using these methods do not pre dispose consumers to diseases like cancer, heart problems, arthritis etc.

Table 5. Percentage distribution of the respondents based on methods used in preservation of Africa breadfruit.

Methods	Frequency	Percentage
Preservation methods		
Sun drying and keeping in bas	10	13.9
Sun drying and keeping in bottle/air tight container	33	45.8
Smoking and keeping above fire	1	1.4
Sun drying and keeping in basin	25	34.7
Sun drying and spreading on open floor/mat	3	4.2
Best duration of drying		
Till it is fully dried	64	88.9
Till it is moderately dried	8	11.1
Preservative used		
Alum	1	1.4
None	71	98.6
Reasons for not using preservatives (n = 71)		
Preservative are not necessary/needed	51	71.9
They impact undesirable characteristics on the seeds	10	14.1
No reason	10	14.1
Effects of preservation by drying		
Makes the seed last longer	56	82.0
Makes seeds easy to cook	6	8.3
Reduces attack by insect, rodent and other micro organisms	6	8.3
No effect	1	1.4
Life span of well preserved seeds		
12 months	11	15.3
1 month	12	16.7
3 to 7 months	49	68.0

Table 6. Mean score of major problems encountered in processing of Africa breadfruit (n=72).

Problems	Mean	S. D
Water scarcity	2.27	0.66
Laborious nature of processing activities	2.42	0.95
Takes long time	2.32	0.96
Difficulty in picking the seeds after threshing	2.28	1.1
Bad odour	1.93	1.17
Makes the environment dirty	1.86	1.04
Poor knowledge of improved processing methods	1.68	0.90
Lack of money to purchase seeds	1.64	1.13
Unavailability of processing equipment/machines	1.50	0.90
Poor storage facilities	1.40	0.73
Loss of seeds during processing	1.39	1.40
Decaying of seeds during fermentation	1.37	1.90
Long duration of fermentation	1.32	0.85
Poor price of finished products	0.96	1.02
Scarcity of energy for parboiling	0.86	0.86

Table 7. Mean score of problems encountered in preservation of African breadfruit (n=72).

Problems	Mean score	SD
Bad weather condition	2.62	0.78
Preserved seed require extra energy in cooking	2.01	1.01
Undesirable taste of preserved seed	1.97	0.96
Unacceptability of preserved seeds by consumers	1.81	1.21
Undesirable colour of preserved seed	1.79	0.95
Contamination preserved seeds by impunities	1.69	0.62
Poor knowledge on preservation methods	1.60	0.82
Poor storage facilities for preserved seeds	1.51	0.75
Attack of preserved seeds by insects and rodents	1.50	0.77
Poor return after preservation of seeds	1.29	0.19
Absence of good preservative	0.40	0.71
Lack of money for purchasing preservatives	0.39	0.80

Problems encountered in processing of African breadfruit

Data in Table 6 show that the major problems encountered by the respondents in processing African breadfruit were: Water scarcity (M = 2.71), laborious nature of the activities (M = 2.42), time consuming (M = 2.32), difficulty in picking the seeds after threshing (M = 2.28), bad odour produced by the seeds during processing (M = 1.93), makes the environment dirty (M = 1.86), poor knowledge of improved processing methods (M = 1.68), lack of money to purchase seeds for processing (M = 1.64) and unavailability of processing equipment and machine (M = 1.50). Water scarcity may be reason for using bad water in processing the fruit as indicated by some of the respondents in Table 3.The finding agrees with Nwabueze (2009) that traditional method of processing is very tedious, time consuming, wasteful, unhygienic and depends on climatic conditions. It also agrees with Onweluzo and Odume (2007) who asserted that traditional methods of extraction impacts characteristic offensive odour to the seeds.

Some minor problems facing processing of African breadfruit were: Poor storage facilities (M = 1.40), loss of seeds during processing (M = 1.39) and decaying of seeds during fermentation (M = 1.37). While poor price of finished products (M = 0.96) and scarcity of energy for parboiling of the seed (M = 0.86) were no problem to processing of African breadfruit in the area. The standard deviation for each of the constraints enumerated in the table was high (approximately one). Thus indicating disparity in the responses of the respondents which result to differences in the constraints they face in processing Africa breadfruit.

Problems encountered in preservation of African breadfruit

Table 7 shows that the major problems encountered by these respondents in preservation of African breadfruit were: Bad weather condition (M = 2.62), preserved seed require extra energy during cooking (M = 2.01), undesirable taste of preserved seed (M = 1.97), unacceptability of the products by consumer (M = 1.81), undesirable colour of preserved seeds (M = 1.79), contamination of preserved seeds by impurities (M = 1.69) , poor knowledge on preservation methods (M = 1.60), poor storage facilities for the preserved seeds (M = 1.51) and attack of preserved seeds by rodents and insects (M = 1.50). Truly bad weather condition can constitute problem to preservation of agricultural products like African bread fruit because farmers/ rural women can no longer predict weather of their environment due to climate change. In line with this, loss of crops under preservation and storage has also been perceived by extension workers in Anambra State Nigeria as one of the effects of climate change (Iwuchukwu and Onyeme, 2012). Secondly, although, processing and preservation of agricultural products aim at improving taste and consumer acceptance (Okafor, 2009), these benefits may be lost when these tasks are done in the wrong ways thereby constituting problems to marketing and consumption of the products.

A minor problem encountered by the respondents in preservation of Africa breadfruit was poor return after preservation of seeds (M = 1.29). Low standard deviation of 0.19 observed in 'Poor return after preservation of seeds' shows uniformity of responses of the respondents in relation to this variable as a constraint to preservation of African breadfruits which was not the case with other variables in the table that have high standard deviation of approximately one.

Conclusion

The study has revealed that majority of women involved in processing and preservation of African bread fruit in the area were married, middle aged with relatively large household size and poor educational background. They were also low income earners with long years of experience in processing and preservation of African breadfruit. These women sourced the bread fruit they processed and preserved from their own farm mainly for family consumption. There were no major sources of information on processing and preservation of African breadfruit in the area while the few that existed were informal.

Fermentation method was the processing method used by majority of the respondents. The seeds were washed with clean water after fermentation/extraction and parboiled for few minutes without alum so as to ensure easy dehulling of the seeds. Both manual and mechanical methods were used in threshing the seeds while separation of seeds from the hull was done by a combination of handpicking and winnowing.

Sun drying and keeping in bottle/air tight container was the preservation method used by greater percentage of the respondents. Also majority of the respondents did not preserve the seeds with preservative because they are not necessary or needed in preservation of African breadfruit. Water scarcity, laborious nature of the activities and the fact that processing of African breadfruit is time consuming were some major problems facing processing of African breadfruit. Bad weather condition, extra energy needed for cooking preserved African breadfruit seeds and undesirable taste of preserved seeds were some major problems facing preservation of African breadfruit in the area.

RECOMMENDATIONS

It is expected that the high level of experience of these women in processing and preservation of African breadfruit will be translated to high income but this is not true probably because processing and preservation of the crop was done at subsistent level (family consumption). There is need to boost quantity of this fruit head processed and preserved by encouraging these women not to rely solely on the fruit head gotten from their own farm but to purchase from market or people that have these fruit heads. Commercial processing and preservation of African breadfruit will invariably be translated to source of livelihood and higher income for these women.

Researches should be undertaken to invent better methods of processing and preservation of agricultural products generally and African breadfruit specifically. These methods should aim at improving upon the indigenous methods because of their inherent attributes. Extension through women in agriculture (WIA) programme should educate the women on these methods that will be less laborious and produce cleaner and higher quality seeds to attract more demand and income. Above all, social infrastructure/amenities

especially water; processing, preservation and storage facilities for African breadfruit should be provided by the government or through community development projects in the area. In a situation where this cannot be provided by the aforementioned bodies, women involved in processing and preservation of African breadfruit should form co-operative and pool their resources towards provision of these facilities. This will be of immense help in encouraging and enhancing processing, preservation, storage and even marketing of African breadfruit among these women, eliminate or reduce wastes/losses and invariably lead to availability of the food all year round at affordable price and reasonable profit.

REFERENCES

Akubor PI, Isolukwo PC, Ugbabe O, Onimawo IA (2000). Proximate composition and functional properties of African breadfruit Kernel and Wheat Flour Blends. Food Res. Int. 33:707-712.

Badifu GIO, Akubor PI (2001). Influence of PH and sodium chloride on selected functional and physical properties of African breadfruit (Treculia Africana Decen) kernel flour. Plant Foods Hum. Nutr. 56:105-115.

Baiyeri KP, Mbah BN (2006). Effect of soil-less and soil-based nursery media on stress of African breadfruit (Treculia Africana Decene). Afr. J. Biotechnol. 5:1405-1410.

Enibe SO (2006). Propagation, early growth, nutritional and engineering development project on African bread fruit (Treculia Africana Decne). J. Eng. 6:34-51.

Etoamaihe UJ, Ndubueze KC (2010). Development of motorized African breadfruit dehuller. J. Eng. Appl. Sci. 5:312-315.

Food and Agricultural Organization (FAO) (2002). Food and nutrition service. p. 13. No. 2 http://en.wikipedia/wiki/preservation.ng.

Hassan AB, Osman GA. Babiker EE (2005). Effects of domestic processing on antinutrients and availability of protein and minerals of lupin (Lupinus termis) seeds. J. Food Technol. 3:263-268.

Ihediohamma NC (2009). Effect of treatment with alum on the keeping quality of African breadfruit (Treculia Africana) seed. Nig. Food J. (27)2:129-135.

Ijeh IF, Ejike CE, Nkwonta OM, Njoku BC (2010). Effect of traditional processing techniques on the nutritional and phytochemical composition of African breadfruit (Treculiaafricana) Seeds. J. Appl. Sci. Environ. Manage. 14(4):169-173.

Iwuchukwu JC, Onyeme F N (2012). Awareness and Perceptions of climate change among extension workers of Agricultural Development Programme ADP in Anambra State, Nigeria. J. Agric. Exten. in press p. 16.

Keay RWJ (1989). Trees of Nigeria: A Revised Version of Nigerian Trees vol 1and2 Oxford, Standfield D.P, Clarendon Press.

Maziya-Dixon B, Akinyele IO, Oguntina EB, Nokoe S, Sanusi RA, Harris E (2004). Nigeria food consumption and nutrition survey 2001-2003. Summary International Institute of Tropical Agriculture (IITA), Ibadan. Nigeria.

Mohammed A (2003). Food insecurity vulnerability and information mapping system. A paper presented at the annual meeting of the inter-agency working group, Abuja, Nigeria, 1-3 October, 2003.

Njoku PCM (2000). Nigeria Agriculture and challenges of the 21st century agro-science. J. Trop. Agric. Food Environ. Exten. 5(1):22-23.

Nwabueze T (2009). Kernel extraction and Machine efficiency in dehulling parboiled African breadfruit (Trecullia Africana Decne) whole seeds. J. Food Qual. 32:669-683.

Okafor GI (2009). Economy of Nigeria: The role of food industries. In: Echezona, B.C (ed) General Agriculture: principles and practices. Nsukka, University of Nigeria press Ltd. pp. 93-99.

Onweluzo JC, Nnamuchi OM (2009). Production and Evaluation of porridge-Type Breakfast product from Treculia Africana and

Sorghum Bicolor flours. Pak. J. Nutr. 8(6):731-736.

Onweluzo LJC, Odume L (2007). Methods of extraction and demucilagination of *Treculia Africana*: effect on composition. Nig. Food J. 25(1):90-99.

Onyekwelu JC, Fayose OJ (2007). Effect of storage methods on germination and proximate composition of *Treculia africana* seeds. Conference on International Agricultural Research for Development. University of Kassel-I-Witzen-hausen and University of Gottigen.

Osuji JO, Owei SD (2010). Mitotic index studies on *Treculia africana* Decne in Nigeria. AJAE 1(1):25-28.

Pacific Agribusiness Research for Development Initiative(PARDI) (2011). Developing commercial breadfruit production in the South Pacific Islands Livai Tora livai@kokosiga.com.

Ragone D (2011) (revised). Farm and forestry production and marketing profile for breadfruit. In:Elevitch, C.R.(ed). Specialty Crops for Pacific Island AgroForestry. Permanent Agriculture Resource (PAR), Holualoa, Hawaii.http://agroforestry.net/scps.

United Nations (UN) (1989). 1989 World Survey on the Role of Women in Development. United Nations, New York.

World Bank, (2005). Getting agriculture going in Nigeria: Framework for a national strategies. Report P. 34618-NG.

Yusuf JF (2000). Economic analysis of millet production in Alkalari Local Government Area of Bauchi State. Unpublished M.Sc.Thesis, Abubakar Tafawa Belewa University Nigeria.

16

Effects of seed dormancy level and storage container on seed longevity and seedling vigour of jute mallow (*Corchorus olitorius*)

Ibrahim H., Oladiran J. A. and Mohammed H.

Department of Crop Production, School of Agriculture and Agricultural Technology, Federal University of Technology, Minna, Nigeria.

The study was conducted at the laboratory of the Crop Production Department, of the Federal University of Technology, Minna, Nigeria in 2009. The purpose was to determine the effect of hot-water steeping and storage container on seed longevity and seedling vigour of *Corchorus olitorius*. The study was also to determine if and when seed dormancy would be alleviated during the storage of the seeds of this crop. Samples of steeped (mildly dormant: md) and the unsteeped (strongly dormant: sd) seeds were packaged in glass bottles, and paper envelopes and stored at 30°C for 22 weeks. Seed germination, using the between paper method, seedling length and fresh seedling weight were determined at the onset of storage and subsequently at two weeks interval. Prior to storage, germination percentages of 72 and 28 were recorded for md and sd seeds respectively. There were some increases in germination percentages, seedling length and weight within the first six to eight weeks of storage and the level of increase depended on container and dormancy depth. A decline in the scores of all the parameters was subsequently recorded. Germination remained poor in the unsteeped seeds and viability was better preserved in glass bottles than in paper envelopes. Dormancy was not alleviated to any appreciable extent during the storage period. There were indications that seed death occurred even in the dormant state and that dormancy did not confer longevity superiority on the seed of this crop.

Key words: Jute mallow, *Corchorus olitorius*, dormancy, longevity, vigour.

INTRODUCTION

Jute mallow (*Corchorus olitorius*) is a leading leaf vegetable in many African countries and is cultivated in some Asian and Caribbean countries and Brazil. The two cultivars ("Amugbadu" and "Oniyaya") of this crop that are widely grown in Nigeria have been described by Akoroda (1988) and the seeds are known to possess dormancy (Fondio and Grubben, 2004). Dormancy normally results in non-uniformity in germination and seedling emergence, a problem that is a source of worry to vegetable growers (Gilberstone et al., 1981). However, Ndinya (2005) opined that seed dormancy may be a blessing as it may ensure the continuation of the seed over time and through periods of environmental stress. Seed dormancy in *C. olitorius* is caused by the hard seed coats (Schippers, 2000). According to Hartmann et al. (1997), germination in seeds that posses hard coats can be increased by any method that can soften or scarify the coverings. Seed coat-imposed dormancy has been described to be a survival strategy of many species by Kelly et al. (1992). Dark seeds of proso millet (*Panicum miliaceum* L.) which are known to have heavier seed coats have been shown to persist longer in soil than seeds with lighter coats (Khan et al., 1996). Debeaujon et al. (2000) have also shown that seeds of Arabidopsis with structural and/or pigmentation defects are less dormant. They also deteriorated faster in storage than the wild type which have complete testa layers and normal pigmentation and therefore less permeable. One of the techniques that have been developed to obtain uniform seed germination and seedling emergence is priming. This refers to controlled hydration followed by drying (Tarquis and Bradford, 1992; McDonald, 2000; Schwember and Bradford, 2010). However, whereas the technique has been reported to enhance longevity in some studies (Probert et al., 1991; Butler et al., 2009), the reverse was the case in others (Chojnowski et al., 1997; Schwember and Bradford, 2010).

To obtain high germination and uniform seedling emergence in *C. olitorius,* Schippers (2000) recommended the steeping of seeds in boiling water for five seconds followed by drying and planting. The effect this treatment might have on subsequent seed longevity has not been documented. Frequently used seed packaging containers include glass bottles, aluminum cans, laminated aluminum foil packets, etc (Rao et al., 2006). However, some farmers (especially in rural areas) may more readily have access to glass bottles and paper envelopes. Adebisi et al. (2008) listed glass bottles as one of the effective packaging containers in the storage of okra seeds. In seeds of some plant species, some period of after-ripening has been reported to alleviate dormancy (Probert, 2000; Steadman et al., 2003; Bair et al., 2006; Bazin et al., 2011). Information in this respect seems to be non-existing for *C. olitorius*. The aims of this study therefore, were (i) to determine the effects of hot water-steeping and packaging materials on the longevity of *C. olitorius* seeds. (ii) to study the changes in seed dormancy levels of untreated seeds as they aged during storage.

MATERIALS AND METHODS

The experiment was conducted in the laboratory of the Department of Crop Production, Federal University of Technology, Minna, Niger State (9° 40' and 6° 30' E), Nigeria. Some seeds of "Oniyaya" variety of *C. olitorius* were steeped in hot water at 97°C for five seconds to break dormancy as recommended by Oladiran (1986). They were subsequently spread on some layers of absorbent paper and dried in the shade to a moisture content of 8%. Samples of steeped (mildly dormant- md) and unsteeped (control referred to as strongly dormant- sd) seeds were then packaged in rubber-stoppered glass bottles and paper envelopes and stored at 30°C and 75% relative humidity for 22 weeks. Seed samples were drawn for germination test prior to storage and at an interval of two weeks thereafter. The between paper (BP) method was used for the germination test. At 22 weeks after storage (WAS), sd seeds packaged in stoppered-bottle were steeped as described above to

ascertain their real viability level. On each testing day, four replicates of 50 seeds each were counted and placed at equidistant spacing on two layers of moist kitchen towel. Two other layers of moist kitchen towel were then placed over the seeds. The set up was rolled carefully and arranged upright in plastic bowls and then incubated at 30°C. The kitchen towel and seeds were kept constantly moist with distilled water. Germination counts and measurement of seedling length and fresh seedling weight were taken after 14 days. To do this, the set up was normally carefully unrolled on a table and the covering layers of kitchen towel carefully lifted to prevent seedling damage. Germination counts were expressed as a percentage of the number of seeds sown. Seedling lengths and weights were thereafter recorded. The data collected on all parameters were subjected to analysis of variance (ANOVA) using Minitab (2003).

RESULTS

Figure 1 shows the survival curves of mildly (md) and strongly dormant (sd) seeds stored in glass bottle and paper envelope. Before storage, germination for the mildly dormant seeds was 72%; there was a general increase in germination up to 6/8 WAS followed by a decline. As from 8 WAS, the germination of the seeds packaged in paper envelope was generally significantly poorer than the values for seeds stored in glass bottles. The germination of strongly dormant seeds was poor all through the study period (a range of 10 to 35%). However, a slight improvement in germination up to about 4 and 8 WAS in paper envelope and glass bottled packages respectively was also recorded for them. When a sample of this class of seed (sd) packaged in glass bottle and stored for 22 weeks was steeped in hot water and the viability tested, a value of 64.5% was obtained as against 15% obtained for the unsteeped lot of the same age. The value is however, not significant from the 70% recorded for the mildly dormant seeds at age (22 WAS).

The seedling lengths from md and sd seeds were 6.08 and 7.53 cm, respectively prior to storage (Figure 2) with subsequent improvement within 2 to 8 WAS depending on treatments. This was followed by a decrease in value. Seedlings from the md seeds that were stored in glass bottle were generally significantly longer than those stored in paper envelope from the 2 to 20 WAS. Except at 14 and 16 WAS (when packaging in paper envelope gave better result), there were generally no significant differences in the length of the seedling from strongly-dormant seeds packaged in the different containers. An improvement in fresh seedling weight was also recorded within 2 to 6 WAS (Figure 3). Seedlings from mildly dormant (md) seeds packaged in glass bottle were generally superior to those from other treatments within the first 14 WAS.

DISCUSSION

The initial general increase observed for the parameters measured in this study could be linked to the loss of

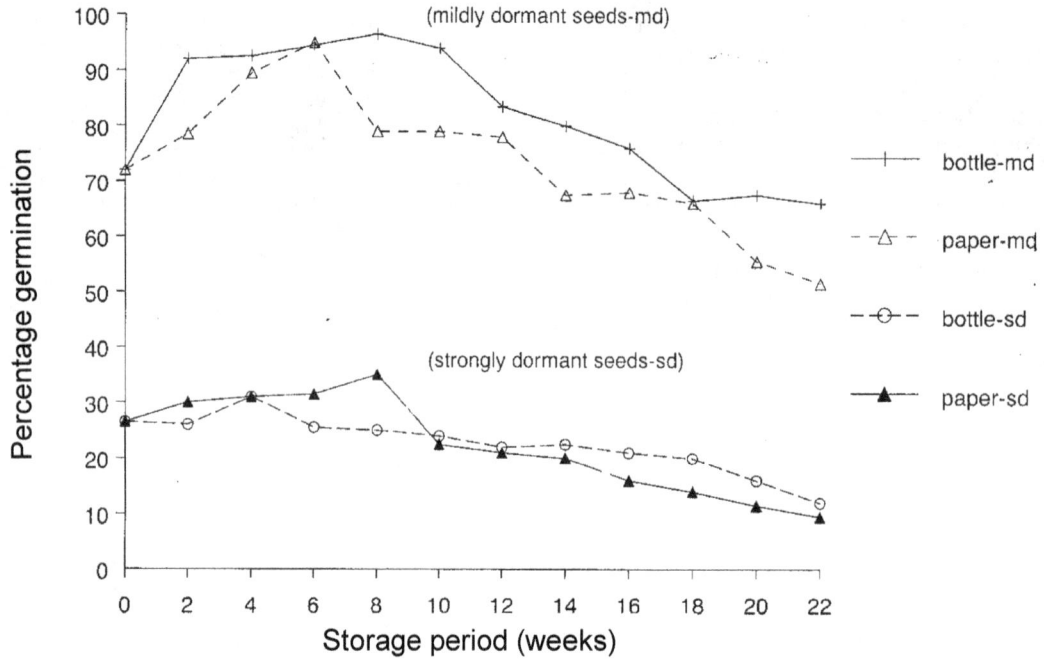

Figure 1. Effect of dormarncy status (mild and strong) packaging container (bottle and paper envelop) on seed longevity at 30°C.

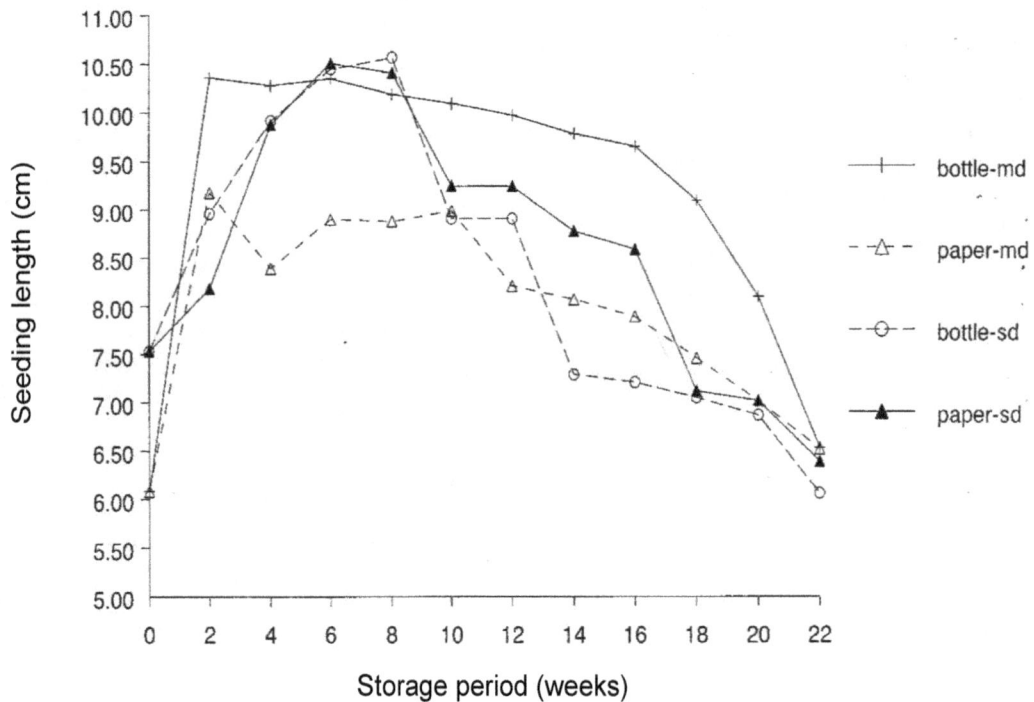

Figure 2. Effect of seed dormarncy status (mid-md or strong-sd) packaging materials on seeding length prior to and after storage at 30°C.

seed dormancy. Oladiran and Kortse (2002) reported that pepper seeds that were dormant when freshly harvested, gave higher germination values as storage progressed.

A situation of this type is an indication that some seeds of *C. olitorius* must have acquired the ability to germinate during dry storage, a process referred to as after-ripening

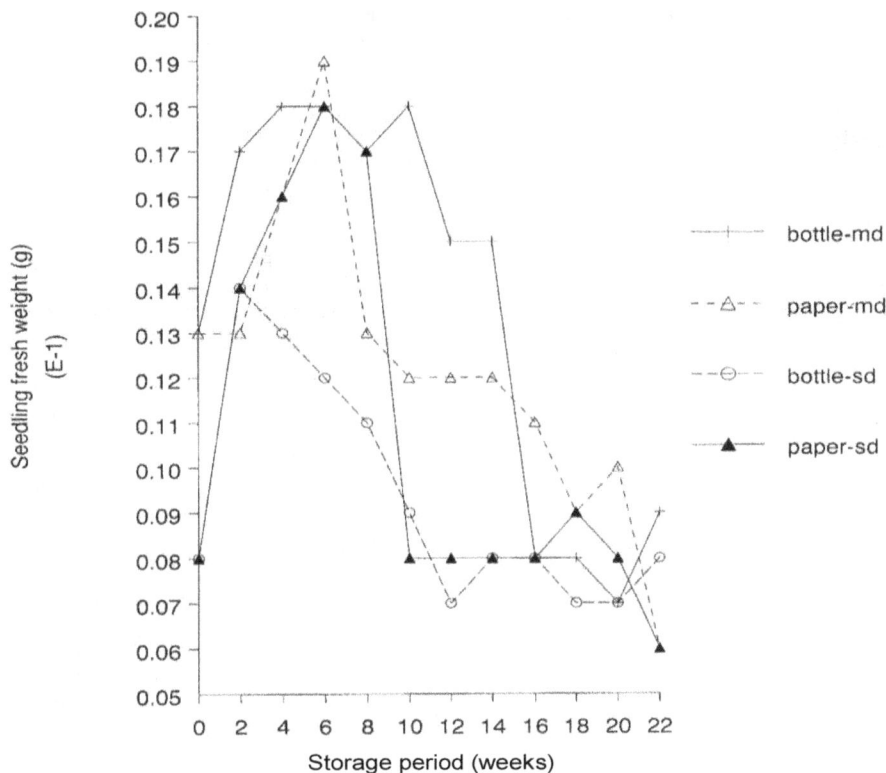

Figure 3. Effect of seed dormarncy status (mild-md or strong-sd) and packaging materials on seedling fresh weight prior to and after storage at 30°C.

(Bazin et al., 2011). Results of this study shows further that decline generally set in after the attainment of peak performance. This trend is normal in seed storage as has been shown to be the case by Ellis and Roberts (1980), Palma et al. (1995), Oladiran and Agunbiade (2000), who reported that as seeds aged they undergo gradual changes which lower their potential vigour and performance capability. According to Gilberstone et al. (1981), seedlings that emerge from old seeds show various cytological abnormalities such as chromosome breakage and the disturbance of mitotic spindle. Seedling growth reduction has also been shown to be recorded with age in storage by Mwai and Abukutsa-Onyango (2005). Results also revealed that strongly dormant seeds germinated poorly all through the study period. However, when a sample of the strongly dormant seeds stored in glass bottle for 22 weeks and which gave only 15% germination was steeped in boiling water and tested for germination, a value of about 65% was recorded in comparison to 70% obtained for mildly dormant seeds of the same age and in similar container. This is an indication that dormancy might not have conferred superior longevity on seeds in this crop. This is contrary to the findings reported by Khan et al. (1996) and Debeaujon et al. (2000) in which less dormant seeds deteriorated faster than more dormant ones. Furthermore, unlike the negative effect of priming on

seed longevity reported by Maude et al. (1994) and Chojnowski et al. (1997), hot water steeping as was practiced in this study, was not injurious to seeds of *C. olitorius*. The steeping of *C. olitorius* seed in hot water is normally done to alleviate dormancy which is due to hardseededness (Schippers, 2000; Fondio and Grubben, 2004). The improvement recorded in this study in respect of germination, seedling length and weight within the first 8/10 WAS (depending on treatment), may mean that dormancy may not be due to hard seed coat alone. This view agrees with Abukutsa-Oniyango (2007) who reported that some physiological aspects beside hardness of the seed could be contributing dormancy in this crop. Desal (1997) also reported that in some seeds, germination may be prevented by the presence to some inhibitory mechanisms within the seeds which must be removed before germination can occur. According to Bennett and Evans (2002), the need for after-ripening is due to a form of endogenous dormancy termed physiological dormancy. The use of heat and hot water have been reported to create cracks in seed coat and thereby alleviating dormancy (Dhillion and Singh, 1996). The failure of a large proportion of unsteeped sd seeds to germinate in this study even up to the point of viability decline after a long period of storage, is an indication that there might not have been any alteration in the seed coat structure that would have permitted germination. The ecological implication of this is that the annual bush fire, that is known to burn off seed coat, may be the only option to be relied on for the germination of dormant *C. olitorius* seeds in nature. The superiority of the glass bottle over paper envelope must, may no doubt, be due to the former's impermeability to moisture, making it an effective packaging material for seed conservation (Adebisi et al., 2003; Rao et al., 2006).

Conclusion

It is concluded that the steeping of *C. olitorius* seed did not reduce it longevity and that seeds of this crop could therefore be steeped in hot water, dried back to moisture content of about 8% and then packaged in glass bottles for future planting. However, if the temperature of the environment is as high as 30°C, seed viability may not be maintained at acceptable level (≥70%) for longer than about 16 weeks. Furthermore, the dormancy of the seeds of this crop may not be broken to any appreciable level during storage and viability may be lost without loss of dormancy.

REFERENCES

Abukutsa-Onyango M (2007). Seed production and support systems for African leafy vegetables in three communities in west Kenya. Afr. J. Agric. Nutr. Dev. 7(3):1-15.

Adebisi MA, Akintobi DCA, Oyekale KO (2003). Preservation of okro seed vigour by seed treatments and storage containers. Nig. J. Hort. Sci. 12:17-23.

Adebisi MA, Daniel IO, Ajala MO (2008). Storage life of soybean (*Glycine max* L. Merril) seeds after seed dressing. J. Trop. Agric. 42(1-2):3-7.

Bair NB, Meyer SE, Allen PS (2006). A hydrothermal after-ripening time model for seed dormancy loss in *Bromus tectorum* L. Seed Sci. Res. 16:17-28.

Bazin J, Batlla D, Dussert S, El-Maarouf-Bouteau H, Bailly C (2011). Role of relative humidity, temperature and water status in dormancy alleviation of sunflower seeds during dry after-ripening. J. Exper. Biol. 62(2):627-640.

Bennett MA, Evans AF (2002). Seed dormancy mechanisms in vegetable crop species. The Ohio State University Seed Production Seminar-Santiago, Chile. pp. 93-99.

Butler LH, Hay FR, Ellis RH, Smith RD, Murray TB (2009). Post-abscission, pre-dispersal seeds of *Digitalis purpurea* remain in a developmental state that is not terminated by desiccation *ex planta*. Ann. Bot. 103(5):785-794.

Chojnowski M, Corbineau F, Come D (1997). Physiological and biochemical changes induced in sunflower seeds by osmopriming and subsequent drying, storage and aging. Seed Sci. Res. 7(4):323-331.

Debeaujon I, Leeon-Kloosterziel KM, Koornneef M (2000). Influence of the testa on seed dormancy, germination and longevity in arabidopsis. Plant Physiol. 122(2):403-414.

Desal MH (1997). Association of seed colour with emergence and seed yield of snap beans. J. Am. Soc. Hort. Sci. 99(2):110-114.

Dhillion WS, Singh U (1996). Breaking of seed dormancy in different leguminous forage species. Int. Rice. Res. Notes 21(1):45-46.

Ellis RH, Robert EH (1980). Improved equation for the prediction of seed longevity. Ann. Bot. 45:13-30.

Fondio L, Grubben GJH (2004). *Corchorus olitorius* L. In: Grubben, GJH, Denton OA (Editors) Plant Resources of tropical Africa 2. Vegetables. PROTA Foundation, Wageningen, Netherlands/Baack-

huys Publishers, Leiden, Netherlands/CTA, Wageningen, Netherlands. pp. 217-221.

Gilberstone TL, Ferris ML, Brenner AC, Wilkins HF (1981). Effect of storage temperature on endogenous growth substances and shoot emergence in Freesia hybrid corns. J. Am. Soc. Hort. Sci. 112(4):641-644.

Hartmann HT, Kester DE, Davies Jr, FT, Geneve RL (1997). Plant propagation: Principles and Practices. Prentice-Hall, Inc., Englewood Cliffs, New Jersey, Sixth edition.

Kelly KM, Van Staden J, Bell WE (1992). Seed coat structure and dormancy. Plant Growth Regul. 11:201-209.

Khan MM, Cavers PB, Kane M, Thompson K. (1996). Role of the pigmented seed coat of proso millet (*Panicum millaceum* L.) in imbibitions, germination and seed persistence. Seed Sci. Res. 7:21-25.

Maude RB, Drew RLK, Gray D, Bujaiski W, Nienow AW (1994). The effect of storage on the germination and seedling abnormalities of leek seeds primed and dried by different methods. Seed Sci. Technol. 22:299-311.

McDonald MB (2000). Seed priming. In: Seed Technology and its Biological Basic (Eds. M. Black and J. Bewley). Sheffied Academic press Ltd. Sheffields. pp. 281-325.

Minitab Inc. (2003). Minitab Statistical Software, Release 14 for Windows, State College, Pennsylvania, USA.

Mwai GN, Abukutsa-Onyango M (2005). Potential salinity resistance in spider plant (*Cleome gynandra* L.). Afr. J. Food Agric. Nutr. Dev. 4:2.

Ndinya C (2005). Seed production and supply system of three African leafy vegetables in Kakamega District. In: Muriithin AN, Anjichi VE, Ngamau, K. Agong, SG, Frickle A; Hau B, Stutzel, H. (Eds). Proceedings of the third horticulture workshop on sustainable horticultural production in the tropics. Maseno University; Maseno. pp. 60-67.

Oladiran JA (1986). Effect of stage of harvesting and seed treatment on germination, seedling emergence and growth in *Corchorus olitorius* cv. "Oniyaya".

Oladiran JA, Agunbiade SA (2000). Germination and seedling development from pepper (*Capsicum annum*) seeds following storage in different packaging materials. Seed Sci. Technol. 28:413-419.

Oladiran JA, Kortse PA (2002). Variations in germination and longevity of pepper (*Capsicum annum* L.) seed harvested at different stages of maturation. Acta Agron. Hung. 50(2):157-162.

Palma B, Vog G, Neville P (1995). Endogenous factors that limit seed germination of *Acaccia senegalensis* wild Phyton. Int. J. Exp. Bot. 57:97-102.

Probert RJ, Bogh SV, Smith AJ, Wechsberg GE (1991). The effects of priming on seed longevity in *Ranunculus sceleratus* L. Seed Sci. Res. 1:243-249.

Probert RJ (2000). Role of temperature in the regulation of seed dormancy and germination. In: Fernner M. ed. Seeds. The ecology of regeneration in plant communities. Oxon: ABI Publishing. pp. 261-292.

Rao NK, Hanson J, Dulloo ME, Ghosh K, Nowell D, Larinde M (2006). Manual of seed handling in genebanks. Handbook for Genebanks No 8. Bioversity international, Rome, Italy.

Schippers RR (2000). African Indigenous vegetables. An overview of the cultivated species. Chattam, UK: Natural Resources Institute. P. 214.

Schwember AR, Bradford KJ (2010). Quantitative trait loci associated with longevity of lettuce seeds under conventional and controlled deterioration storage conditions. J. Exp. Bot. 61(15):4423-4436.

Steadman KJ, Craeford AD, Gallagher RS (2003). Dormancy release in *Lolium rigidum* seeds is a function of thermal after-ripenig time and seed water content. Funct. Plant. Biol. 30:345-352.

Tarquis AM, Bradford KJ (1992). Prehydration and priming treatments that advance germination also increase the rate of deterioration of lettuce seeds. J. Exp. Biol. 43:307-317.

17

Quality of low temperature heat-shocked green asparagus spears during short-term storage

Kai Ying Chiu[1] and Jih Min Sung[2]

[1]Department of Post-Modern Agriculture, Mingdao University, Peetow, Changhwa County, 523, Taiwan.
[2]Department of Food Science and Technology, Hung Kuang University, 34 Chung-Chie Rd, Sha Lu, Taichung City, 43302, Taiwan.

The present study evaluated the effects of heat-shock treatment (washing in 48°C distilled water for 5 min) on quality of green asparagus spears stored at 25 or 4°C for 7 days. The results showed that the quality characteristics including toughness, lignin and cellulose contents, and phenylalanine ammonia-lyase, cinnamyl alcohol dehydrogenase and peroxidase activities were increasing in both heat-shocked and control spears stored at 25°C; however, the heat-shocked spears exhibited lower degrees of increases than control spears. Similar increasing trends but with even smaller magnitudes in these characteristics were obtained from the heat-shocked spears stored at 4°C. Heat-shocked spears stored at 25 or 4°C also demonstrated lesser proliferations of total mesophilic aerobes, *Escherichia coli* and mold than their respective controls. Thus, a combination of 48°C heat-shock treatment and 4°C cold storage can be used to conserve the quality of stored green asparagus spears.

Key words: Green asparagus, heat-shock, lignification, microbial load, storage.

INTRODUCTION

Green asparagus (*Asparagus officinalis* L.) spear is a healthy but highly perishable vegetable, which has a very short shelf life (generally three to five days) with normal post-harvest handling at ambient temperature. Many physical and chemical changes such as the loss of water and the increase of toughness that reduce the quality of green asparagus spears occur during harvesting, handling and storage (Rodriguez-Arcos et al., 2002; Liu and Jiang, 2006). Among these changes, the increase of toughness is an important factor in de-grading the quality of green asparagus spears and consequently decreasing their market value. The toughness of green asparagus spears is reported to be related to the degree of lignification in their tissues, which is regulated by enzymes Phenylalanine Ammonia-Lyase (PAL), cinnamyl alcohol dehydrogenase (CAD) and peroxidase (POD) (Boudet, 2000; Chen et al., 2002; Cai et al., 2006; Li and

Zhang, 2006). Additionally, many pathogenic micro-organisms can contaminate the produced green asparagus spears during cultivation, harvest and post-harvest handling. The physical and chemical changes of green asparagus spears that occurred during storage would facilitate the proliferation of loaded microorganisms if the harvested green asparagus spears were not properly sterilized. As a result, the increased microbial load would further reduce the quality of green asparagus spears and shorten their shelf life (Sothornvit and Kiatchanapaibul, 2009). Many methods have been developed to ensure microbial safety and extending shelf life of green asparagus spears.

Washing asparagus with chlorine was reported to reduce the microbial load and maintained its quality (Sothornvit and Kiatchanapaibul, 2009). Aqueous ozone treatment plus a subsequent modified atmosphere packaging also proved effective in decreasing the toughness and extending the shelf life of treated asparagus (An et al., 2007). A combination of edible coating of chitosan and 2°C cold storage also exhibited a reduction of decay of green asparagus and extended its shelf life (Qiu et al., 2013). Nevertheless, heat treatment is still extensively used to eradicate pathogens and preserve the quality of various fresh and fresh-cut produces because of the increased concern about the safety of chemicals used in post-harvest treatments (Sivakumar and Fallik, 2013). The applied heating parameters (for example, time and temperature regime) can vary to accomplish almost any degree of microbial inactivation, ranging from limited reduction of microbial load to complete sterilization (Rajkovic et al., 2010). Heat treatment has also been used to preserve the quality of harvested asparagus spears (Lau et al., 2000). Heat-shock is a method which usually implies a washing step at a temperature ranging from 43 to 70°C for a few minutes (usually less than 5 min) (Fallik, 2004; Rico et al., 2007). This method is proved useful to preserve the quality of produces such as green onion (Cantwell et al., 2001), lettuce (Murata et al., 2004) and kiwi fruit (Beírãao-da-Costa et al., 2008). But it has not been tested for controlling the proliferation of microorganisms and extending the shelf life of green asparagus spear.

In the present study, attempts were made to characterize the physical and chemical changes in green asparagus spears that have been subjected to 48°C heat-shock, and then stored at 4 or 25°C conditions for seven days. The activities of PAL, CAD and POD in the stored asparagus spears were determined and compared. The effectiveness of heat-shock treatment on initial microbial de-contamination and proliferation during storage was also examined. Knowledge of these variations should help to identify the potential value of low temperature heat-shock treatment on safety control and quality preservation of green asparagus spears.

MATERIALS AND METHODS

Plant materials and chemicals

Spears of fresh green asparagus (*Asparagus officinalis* L. cultivar Tainan 12) were obtained from Tainan District Improvement Station (Tainan county, Taiwan). The spears were cut at ground level between 8:00 and 9:30 AM, then placed in crushed ice and transported to the laboratory within three h on the day of harvest. The spears that were 15 to 20 mm in diameter and 15 cm (from the tip) in length with closed bracts were used in the study. The selected and cut spears (15 cm in length) were washed with distilled water, then drained and cleaned with paper towels to remove the water. A preliminary heat shock trial by using several temperature regimes (ranged from 45 to 57°C) had shown that a heat shock at 48°C water for 5 min was capable of reducing the increases of toughness to some extent in the treated asparagus spears stored at ambient temperature for five days. Therefore, in this study, the 48°C heat-shock treatment was used for asparagus spear quality control. While a portion of selected asparagus spears was not heat-shocked (the control group), another portion of the

selected asparagus spears (5 kg) underwent the heat-shock treatment in 48°C distilled water for 5 min. The heated-shocked spears were then immersed in ice-cold water for 10 min. Following cooling, the spears were drained and dried with paper towels and then packed into an airtight plastic container (20 × 10 × 5 cm), and then stored in airtight chambers (with about 85% relative humidity) at 4 or 25°C for seven days. Spears from the control samples were also immersed in ice-cold water for 10 min and then drained and stored in the same way as the heat-shocked spears. The stored spears were sampled on day 0, 1, 3, 5 and 7, and the top portions of spears (first 12 cm from the tip) were used for physical and chemical analyses.

The polyvinylpyrrolidone, 2-mercaptoethanol, ethylenediaminetetraacetic acid (EDTA) sodium salt, and the chemicals including tris (hydroxymethyl) aminomethane-HCl, xylenol, L-phenylalanine, borate buffer, coniferyl alcohol, Nicotinamide adenine dinucleotide phosphate, guaiacol used for enzyme activity assays were purchased from Sigma-Aldrich (St. Louis, MO, USA). The growth media for microbial analyses were purchased from Chisso Corporation (Tokyo, Japan). All other chemicals used in this study were of analytical grade.

Fresh weight loss and texture toughness determinations

The fresh-weight loss of the spears (harvested asparagus spears 15 cm in length) was determined with a precision balance during the entire storage period. The tissue toughness measurements were performed in a Tahdi Texture Analyzer (Stable Micro-Systems Ltd, Godalming, UK) with a 500 N load cell and a 5 mm diameter probe. A single puncture measurement was made on each spear sample (10 mm depth of penetration) at a speed of 1.0 mm s^{-1}. The recorded breaking force (maximum peak force, N) was used as an indicator of texture toughness parameter.

Lignin, cellulose and H$_2$O$_2$ contents determinations

For lignin determination, five randomly selected spears (harvested asparagus spears 15 cm in length) were diced and frozen in liquid nitrogen and then stored at -20°C prior to analyses. Lignin was extracted and measured by using the procedures detailed by Bruce and West (1989). Different samples from the same frozen bulked and diced spears that were used for lignin analyses were also used for the cellulose assay following the procedures detailed by Oomena et al. (2004). The content of H$_2$O$_2$ content was determined by using the procedures detailed by Lin et al. (1988).

Enzymes activities determinations

For phenylalanine ammonia-lyase (PAL) activity assay, 5 g of diced and bulked frozen spears prepared from the harvested asparagus spears (15 cm in length from the tip) were homogenized for 2 min in 15 ml 0.1 M borate buffer (pH 8.8) containing 6 g polyvinylpyrrolidone, 5 mM 2-mercaptoethanol and 2 mM EDTA. The homogenate was centrifuged for 15 min at 14,000 g and the supernatant was collected for enzyme activity determination. The PAL activity was measured according to the method of Cheng and Breen (1991). For cinnamyl alcohol dehydrogenase (CAD) activity assay, 5 g of diced and bulked frozen samples were extracted by using 10 ml Tris-HCl buffer (0.2 M, pH 7.5). The homogenized mixture was centrifuged at 14,000 g for 15 min and the supernatant was used as enzyme extract. The CAD activity was assayed by using the procedures detailed by Goffner et al. (1992). For peroxidase (POD) activity determination, the enzyme was extracted from 5 g of diced and bulked frozen samples in 0.05 M phosphate buffer (pH 7.0). The homogenate was centrifuged at 14,000 g for 15

Table 1. The fresh weight loss (%) and toughness (N force cm^{-2}) measured in the control and heat-shocked green asparagus spears stored at 25 or 4°C for seven days.

Treatment	Storage temperature (°C)	Storage period(days)				
		0	1	3	5	7
			Fresh weight loss (%)			
Control	25	-	1.12±0.19^{bc}	1.62±0.37^{d}	3.94±0.18^{g}	10.01±1.61^{f}
Heat-shocked		-	0.71±0.09^{a}	1.05±0.13^{c}	1.84±0.28^{e}	7.41±0.63^{d}
Control	4	-	0.68±0.08^{a}	0.81±0.03^{b}	0.89±0.09^{b}	0.94±0.07^{b}
Heat-shocked		-	0.61±0.06^{a}	0.67±0.04^{a}	0.73±0.05^{a}	0.81±0.04^{a}
			Toughness (N force cm^{-2})			
Control	25	2.37±0.06	2.83±0.41^{a}	3.51±0.11^{c}	3.87±0.10^{c}	4.22±0.14^{c}
Heat-shocked			2.78±0.19^{a}	3.15±0.19^{b}	3.56±0.14^{b}	3.83±0.11^{b}
Control	4		2.61±0.34^{a}	2.73±0.10^{a}	3.07±0.32^{a}	3.74±0.12^{b}
Heat-shocked			2.53±0.13^{a}	2.67±0.23^{a}	2.94±0.26^{a}	3.28±0.30^{a}

Values for fresh-weight loss and toughness with different superscript letters within columns were significantly different at $P < 0.05$, respectively.

min at 4°C, and the supernatant was used to determine the POD activity by using the method described by Sheu and Chen (1991).

Microbiological analyses

The diced green asparagus spears samples (25 g) were added to 225 ml of 0.85% (w/v) NaCl solution, and then homogenized and diluted (6-fold dilution) with distilled water. Following dilution, 1 ml of diluted sample was surface-plated on media prepared following the manufacturer's instructions (Chisso Corporation, Tokyo, Japan) for the counting of total mesophilic aerobes, *Escherichia coli* and mold. All the plates were incubated at 35°C for 48 h, 35°C for 24 h or 25°C for 5 days for total mesophilic aerobes, *E. coli* or mold counting, respectively. After incubation, the colonies of microbes were enumerated and expressed as log colony forming units per g (log cfu g^{-1}).

Statistical analysis

The experiments were conducted in a completely randomized design with four replicates. Analyses of variance (ANOVA) were performed and means were compared using Duncan's multiple range tests at P < 0.05. Correlation analyses were also used to characterize the relationships between the contents of lignin, cellulose and H$_2$O$_2$, and toughness, as well as the related enzymes.

RESULTS AND DISCUSSION

Fresh weight loss and toughness at 25°C storage condition

The fresh-weight loss is a crucial parameter for stored green asparagus spears because the loss in fresh weight represents an economic loss. In the present study, the fresh-weight loss in control and heat-shocked green asparagus spears stored at 25°C increased progressively over the seven days of storage (Table 1). For non-treated control, the fresh-weight loss detected over seven days of storage was 10.01% of the initial weight. The fresh-weight loss of stored spears is mainly due to the loss of water (Albanese et al., 2007). The heat-shock treatment was effective in slowing down the fresh-weight loss of green asparagus spears (Table 1). The fresh-weight loss was 7.41% (26% less than control) in the heat-shocked spears stored at 25°C for seven days. The toughness of stored green asparagus spears is an important factor that determines the acceptance or rejection by consumers. As shown in Table 1, a steady increase of breaking force (an indicator of tissue toughness) was detected in the spears stored at 25°C for seven days. After seven days of storage, the measured toughness for non-treated control spears gradually increased to 4.22 N force cm^{-2}.

Increases in toughness were also obtained from the heat-shocked spears over seven days of storage. However, the extent of increases was much lower in the heat-shocked spears; the measured toughness only reached to 3.83 N force cm^{-2} at the end of seven days storage. These results suggest that the heat-shock treatment is effective in retarding the increased tissue toughness during 25°C storage.

Lignin, cellulose and H$_2$O$_2$ contents at 25°C storage condition

Lignin is a complex polymer of phenylpropanoid residues occurring in large quantity in the secondary cell walls of fibers, xylem vessels and tracheids. These vascular tissues of green asparagus spear are reported to continually produce lignin after harvest (Hennion et al., 1992; Waldron et al., 2003). Boudet (2000) reported that the changes in toughness of control and heat-shocked asparagus spears during 25°C storage were linked to the

Table 2. The contents of cellulose (% of fresh weight), lignin (% of fresh weight) and H_2O_2 (μmole g^{-1} fresh weight weight) measured in the control and heat-shocked green asparagus spears stored at 25 or 4°C for seven days.

Treatment	Storage temperature (°C)	Storage period(days)				
		0	1	3	5	7
		Lignin content (% fresh weight)				
Control	25	0.50±0.09	0.65±0.02[b]	0.76±0.05[b]	0.90±0.06[b]	1.21±0.11[d]
Heat-shocked			0.59±0.04[ab]	0.71±0.03[b]	0.84±0.07[b]	0.94±0.09[c]
Control	4		0.53±0.10[a]	0.57±0.05[a]	0.63±0.04[a]	0.76±0.04[b]
Heat-shocked			0.52±0.11[a]	0.55±0.04[a]	0.60±0.04[a]	0.67±0.03[a]
		Cellulose content (% fresh weight)				
Control	25	0.55±0.12	0.89±0.05[b]	1.48±0.05[d]	1.79±0.05[c]	2.24±0.12[d]
Heat-shocked			0.61±0.07[a]	1.32±0.10[c]	1.49±0.05[b]	1.89±0.08[c]
Control	4		0.60±0.09[a]	0.88±0.01[b]	1.36±0.08[b]	1.62±0.06[b]
Heat-shocked			0.58±0.16[a]	0.71±0.03[a]	1.14±0.06[a]	1.46±0.03[a]
		H_2O_2 content (μmol g^{-1} Fresh weight)				
Control	25	25.43±1.95	36.31±0.72[d]	42.55±2.15[c]	45.27±2.25[c]	56.13±3.09[d]
Heat-shocked			31.16±0.80[b]	36.42±3.08[b]	40.21±2.20[b]	49.24±2.59[c]
Control	4		28.82±0.61[a]	31.74±1.46[a]	31.73±1.53[a]	37.01±2.01[b]
Heat-shocked			27.21±0.66[a]	28.25±1.48[a]	30.73±2.24[a]	33.26±1.47[a]

Values for lignin, cellulose and H_2O_2 with different superscript letters within columns were significantly different at $P < 0.05$, respectively.

concomitant changes in their lignin contents. Therefore, in the present study, the contents of lignin, cellulose and H_2O_2 in the tested asparagus spears were determined and compared. As shown in Table 2, the lignin content in the control spears was increased from 0.53 to 1.21% over the seven days of 25°C-storage (Table 2). The lignin content in heat-shocked asparagus spears was also increasing, but the extent of lignin increases was lower than non-treated control spears (Table 2). These results indicate that the 48°C heat-shock treatment prior to storage could retard the lignin increases in green asparagus spears during seven days of storage at 25°C. The content of cellulose also plays a key role in the texture attributes of asparagus cell wall (An et al., 2007).

In the present study, the changes in cellulose content were also observed in control spears (increased from 0.55 to 2.04%) stored at 25°C for seven days (Table 2). The cellulose content in the heat-shocked spears was also increasing, but with the extent of cellulose increases lower than those in the controls (Table 2). H_2O_2 is an ubiquitous and dynamic constitute of plant cells. It serves as a substrate in several cell-wall formation reactions such as the apoplastic cross-linking of cinnamyl alcohols during lignin biosynthesis (Ros Barceló, 1998). Our results showed that the changing trend of H_2O_2 contents in the control and the heat-shocked spears stored at 25°C were basically the same, which increased gradually throughout the entire storage period (Table 2). However,

the heat-shock treatment did reduce H_2O_2 accumulations in the treated spears (increased to 49.24 μmol g^{-1} fresh weight) as compared to those in the control spears (increased to 56.13 μmol g^{-1} fresh weight) during seven days of 25°C-storage (Table 2). The reduced H_2O_2 contents indirectly supported the findings of decreased lignin formation in green asparagus spears heat-shocked at 48°C. Boudet (2000) reported that the increases in toughness of stored asparagus spears were coupled with increases in lignin and cellulose contents. The calculated positive correlations ($P < 0.01$) between toughness and lignin, cellulose and H_2O_2 contents (across control and heat-shocked spears) supported his findings (Table 3).

Phenylalanine ammonia-liase (PAL), cinnamyl alcohol dehydrogenase (CAD) and peroxidases (POD) activities at 25°C storage condition

The degree of lignification of plant tissues is reported to be enhanced by enzymes such as PAL, CAD and POD (Chen et al., 2002; Cai et al., 2006; Li and Zhang, 2006). PAL catalyzes the first reaction in the biosynthesis of plant phenylpropanoid products (Cai et al., 2006). CAD plays a down-regulating role in the last step of monolignol pathway, while POD catalyzes the polymerization of monolignol like p-coumaroyl, coniferyl, 5-hydroxyconiferyl and sinapyl alcohols to complete the process of

Table 3. The correlation coefficients between the tested physical and chemical characteristics of green asparagus spears (across control and heat-shocked spears) stored at 25 or 4°C.

Parameter	Lignin	Cellulose	H_2O_2
25°C			
Toughness	0.903**	0.913**	0.865**
PAL activity	0.571**	0.626**	0.593**
CAD activity	0.193	0.355*	0.201
POD activity	0.539**	0.529**	0.563**
4°C			
Toughness	0.923**	0.837**	0.892**
PAL activity	0.907**	0.815**	0.829**
CAD activity	0.445**	0.552**	0.510**
POD activity	0.735**	0.656**	0.639**

*, ** Values are significant at $P < 0.05$ and 0.01, respectively.

lignification (Holm et al., 2003). As shown in Table 4, an increase in PAL activity was observed in the control spears stored at 25°C for three days (increased from 16.81 to 54.47 units g^{-1} fresh weight), and then followed by a rapid decrease (reached to 26.01 units g^{-1} fresh weight) afterwards. Heat-shocked spears showed retardation and delay in the changing trend of PAL activity, with the PAL activity increasing from 16.81 (day 0) to 48.15 units g^{-1} fresh weight (day 5) and then declining (Table 3) afterwards. Both CAD and POD activities in the control and heat-shocked spears stored at 25°C followed the changing trends of PAL activity, but each peaked at different sampling time (Table 4). The reduced PAL, CAD and POD activities may explain in part why heat-shock treatment retards the lignifications of green asparagus spears and preserves their quality to some extent as compared to non-treated control spears. These notions were supported by the positive correlations (across control and heat-shocked spears) between the contents of lignin, cellulose and H_2O_2 and the activities of enzyme activities (except lignin or H_2O_2 vs. CAD) shown in Table 3.

Liu and Jiang (2006) pointed out that, in green asparagus, the plant hormone ethylene played a key role in the accumulation of lignification that resulted from the enhanced PAL, CAD and POD activities. Ethylene is produced in response to various kinds of environmental stress, including wounding, and the wound-induced ethylene is involved in plant lignification (Luo et al., 2007). The fresh-cut asparagus spear is considered as a wounded tissue; therefore, the observed lignification in fresh-cut spears can be a wound-induced ethylene effect. It appears that the wound-induced ethylene evolution of heat-shocked spears might be lower than that of non-treated control spears. Accordingly, the activities of PAL, CAD and POD are reduced to some extent, and result in a less lignin accumulation in heat-shocked asparagus

spears during 25°C storage (Falik, 2004).

Microbial loads at 25°C storage condition

Table 5 showed the populations of foodborne pathogens proliferated on the green asparagus spears during seven days of 25°C-storage. The initial levels of total mesophilic aerobes, *E. coli*, and mold measured in the green asparagus spears were 2.36, 2.07 and 1.06 log cfu g^{-1} fresh weight, respectively. After seven days of 25°C-storage, the populations of total mesophilic aerobes, *E. coli*, and mold were increased to 4.87, 3.36 and 2.37 log cfu g^{-1} fresh weight, respectively. The populations of total mesophilic aerobes, *E. coli*, and mold in heat-shocked asparagus spears were increased to 3.58, 3.19 and 2.17 log cfu g^{-1} fresh weight, respectively. The values were 24, 5 and 8% lower than the total mesophilic aerobes, *E. coli* and mold counts recorded on non-treated controls, respectively. These results are expected because the application of optimal pre-storage heat treatment to fresh-cut produces is shown to be beneficial in decreasing the proliferation of various microorganisms during storage (Fallik, 2004). Thus, it appears that the heat-shock treatment (washing in 48°C distilled water for 5 min) can be used to reduce the microbial load in fresh-cut green asparagus spears stored at 25°C.

Cold (4°C) storage effects on the tested physical and chemical characteristics

Temperature management is an important factor affecting the quality of fresh produce, and cold storage is generally recommended for highly perishable vegetables such as green asparagus (Albanese et al., 2007). In the present study, cold (4°C) storage period of seven days significantly reduced the fresh-weight losses in green

Table 4. The activities (unit g^{-1} fresh weight) of phenylalanine ammonia-lyase (PAL), cinnamyl alcohol dehydrogenase (CAD) and peroxidase (POD) measured in the control and heat-shocked green asparagus spears stored at 25 or 4°C for seven days.

Treatment	Storage temperature (°C)	Storage period(days)				
		0	1	3	5	7
		Phenylalanine ammonia-lyase activity (unit g^{-1} fresh weight)				
Control	25	16.81±0.93	28.61±1.26^c	54.47±3.19^d	38.13±2.61^c	26.01±1.82^a
Heat-shocked			25.60±1.43^b	40.41±1.89^b	48.15±2.16^d	38.90±2.11^c
Control	4		18.22±0.92^a	25.30±2.90^a	30.21±1.56^b	34.14±1.34^b
Heat-shocked			16.73±1.41^a	21.67±2.39^a	24.11±2.14^a	27.22±1.45^a
		Cinnamyl alcohol dehydrogenase activity (unit g^{-1} fresh weight)				
Control	25	4.20±0.24	5.14±0.11^b	9.37±0.50^e	5.71±0.24^b	3.45±0.44^a
Heat-shocked			4.70±0.16^a	7.57±0.72^c	6.77±0.42^c	4.92±0.99^b
Control	4		4.66±0.14^a	6.27±0.47^b	5.73±0.23^b	5.21±0.21^b
Heat-shocked			4.41±0.17^a	4.84±0.22^a	5.04±0.32^a	5.24±0.43^b
		Peroxidase activity (unit g^{-1} fresh weight)				
Control	25	17.42±0.83	24.42±0.72^b	38.62±2.91^{cd}	36.22±2.11^b	24.24±1.80^b
Heat-shocked			20.13±0.64^a	36.63±0.93^c	36.70±0.63^b	23.62±1.81^a
Control	4		18.57±0.73^a	25.91±2.23^b	23.74±1.05^a	21.48±1.24^a
Heat-shocked			18.10±1.63^a	21.82±1.46^a	20.58±0.67^a	22.19±1.26^a

Values PAL, CAD and POD with different superscript letters within columns were significantly different at $P < 0.05$, respectively.

asparagus spears (Table 1). Both control and heat-shocked green asparagus spears stored at 4°C showed considerably less fresh-weight losses than the spears stored at 25°C. The relatively less fresh-weight loss obtained from the spears stored at 4°C might be explained by the reduced water loss at low temperature. Cold (4°C) storage also significantly reduced the increasing trend of toughness in the stored asparagus spears (Table 1). Both control and heat-shocked spears stored at 4°C exhibited less toughness than the spears stored at 25°C at respective sampling time (Table 1).

The differences in toughness between control and heat-shocked spears were also significant, with the toughness values obtained from the heat-shocked spears lower than the values obtained from control spears. These results confirm that a combination of heat shocking and 4°C cold storage is effective in preserving the quality of green asparagus spears. As shown in Table 2, the contents of lignin, cellulose and H_2O_2 in control and heat-shocked green asparagus spears were also increasing during 4°C storage. But the increased lignin, cellulose and H_2O_2 contents within seven days of storage at 4°C were significantly lower than those of the spears stored at 25°C. Additionally, the heat-shocked spears contained less lignin, cellulose and H_2O_2 than their respective controls during seven days of 4°C-storage (Table 2).

As a result, positive correlation coefficients (across control and heat-shocked spears) between toughness and lignin, cellulose or H_2O_2 contents were also obtained

from the green asparagus spears stored at 4°C (Table 3).

Cold (4°C) storage effects on PAL, CAD and POD activities

The changes in PAL activities obtained from the spears stored at 4°C for seven days differed from the PAL activities obtained from the samples stored at 25°C (Table 4). Both the control and the heat-shocked spears stored at 4°C showed a steady increase in PAL activities throughout the cold storage period. The maximum PAL activities obtained from the spears stored at 4°C were lower than the maximum activities obtained from the spears stored at 25°C (Table 4). Additionally, the PAL activities obtained from the heat-shock spears were lower than the PAL activities obtained from the non-treated controls. The activities of CAD and POD obtained from the spears stored at 4°C were also lower than the activities of CAD and POD obtained from the spears stored at 25°C (Table 4). Minor differences in CAD and POD activities were detected between the control and the heat-shocked spears, with the heat-shocked spears generally having slightly lower CAD and POD activities developed lignification of asparagus spears during 4°C storage are probably due to their lower PAL, CAD and POD activities (Villanueva et al., 2005).

As shown in Tables 1, 2 and 4, it is noticeable that the green asparagus spears stored at 4°C maintain better physical and chemical qualities than that of the spars

Table 5. The counts of total mesophilic aerobes, *E. coli* and mold (log cfu g^{-1} fresh weight) measured on the bud segment of green asparagus subjected to various treatments and stored at 25 or 4°C for 7 days.

Treatment	Storage temperature (°C)	Storage period(days)				
		0	1	3	5	7
		Total mesophilic aerobes (log cfu g^{-1} fresh weight)				
Control	25	2.36±0.31	2.94±0.20^a	3.38±0.20^a	4.12±0.27^a	4.87±0.38^a
Heat-shocked			2.60±0.12^b	2.68±0.14^c	2.98±0.16^c	3.58±0.21^b
Control	4		2.87±0.10^a	3.02±0.11^b	3.25±0.11^b	3.56±0.18^b
Heat-shocked			2.58±0.09^b	2.78±0.16^c	2.99±0.10^c	3.04±0.17^c
		E. coli (log cfu g^{-1} fresh weight)				
Control	25	2.07±0.23	2.50±0.19^a	2.57±0.11^a	2.92±0.11^a	3.36±0.18^a
Heat-shocked			2.18±0.10^b	2.49±0.21^a	2.82±0.21^a	3.19±0.14^{ab}
Control	4		2.32±0.10^{ab}	2.58±0.18^a	2.79±0.24^a	2.91±0.19^b
Heat-shocked			1.86±0.21^c	2.01±0.17^b	2.22±0.26^b	2.47±0.21^c
		Mould (log cfu g^{-1} fresh weight)				
Control	25	1.06±0.02	1.16±0.12^a	1.56±0.11^a	1.93±0.07^a	2.37±0.13^a
Heat-shocked			1.08±0.23^a	1.38±0.09^b	1.71±0.13^b	2.17±0.11^a
Control	4		1.18±0.10^a	1.46±0.11^{ab}	1.72±0.08^b	2.11±0.18^{ab}
Heat-shocked			1.05±0.19^b	1.32±0.06^b	1.68±0.14^b	1.92±0.15^b

Values total bacterial, *E. coli* and mold counts with different superscripts were significantly different at the level of $P < 0.05$.

stored at 25°C. The improved physical quality of 4°C-stored green asparagus spears is possibly due to the relatively low ethylene production rate during 4°C storage (Zamorano et al., 1994).

Cold (4°C) storage effects on microbial loads

As shown in Table 5, both control and heat-shocked spears stored at 4°C showed steady increases in the populations of total mesophilic aerobes, *E. coli* and mold throughout the seven days of cold storage period. Nevertheless, the increased populations of total mesophilic aerobes, *E. coli* and mold in the spears stored at 4°C were significantly lower than those of the spears stored at 25°C. Additionally, the populations of total mesophilic aerobes, *E. coli*, and mold obtained from the heat-shocked spears were consistently lower than the populations of total mesophilic aerobes, *E. coli* and mold obtained from the non-treated controls. The asparagus spears subjected to heat shock treatment showed 1.83, 0.89 and 0.45 log cfu g^{-1} fresh weight reduction of total mesophilic aerobes, *E. coli* and mold comparing to non-treated control spears stored at 25°C, respectively, after seven days of 4°C cold storage (Table 5). Thus, both heat-shock and 4°C cold storage are useful approach for controlling the microbial proliferation on stored green asparagus spears. However, a combination of heat-shock treatment and 4°C cold storage is a desirable choice for further reducing microbial load on green asparagus spears.

Conclusion

In the present study, significant differences in physical and chemical parameters including the fresh-weight loss, toughness, the contents of lignin, cellulose and H$_2$O$_2$, and the activities of PAL, CAD and POD were found between the control and the heat (48°C)-shocked green asparagus spears that were stored at 25 or 4°C for seven days. The heat-shocked green asparagus spears tended to show less fresh-weight loss, show less toughness increase, accumulate less lignin, cellulose and H$_2$O$_2$, and exhibit less PAL, CAD and POD activities than the control spears. Heat-shock treatment also exhibited significant reductions in the proliferations of total mesophilic aerobes, *E. coli* and mold within seven days of storage. Cold (4°C) storage effectively preserved the quality of green asparagus spears by inducing significant declines in lignin, cellulose and H$_2$O$_2$ accumulations, decreasing the activities of PAL, CAD and POD, and decreasing the proliferations of total mesophilic aerobes, *E. coli* and mold. It appears that the 48°C heat-shock treatment can be considered for maintaining the quality of fresh-cut green asparagus spears to some extent. A combination of 48°C heat-shock and 4°C cold storage can further improve the quality of stored green asparagus spears.

REFERENCES

Albanese D, Russo L, Cinquanta L, Brasiello A, Di Matteo M (2007).

Physical and chemical changes in minimally processed green asparagus during cold-storage. Food Chem. 101:274-280.

An J, Zhang M, Lu Q (2007). Changes in some quality indexes in fresh-cut green asparagus pretreated with aqueous ozone and subsequent modified atmosphere packaging. J. Food Eng. 78:340-344.

Beírãao-da-Costa S, Steiner A, Correia L, Leitão E, Empis J, Moldão-Martins M (2008). Influence of moderate heat pre-treatments on physical and chemical characteristics of kiwifruit slices. Eur. Food Res. Technol. 226:641-651.

Boudet AM (2000). Lignins and lignificaton: selected issues. Plant Physiol. Biochem. 38:81-96.

Bruce RJ, West CA (1989). Elicitation of lignin biosynthesis and isoperoxidase activity by pectic fragments in suspension cultures of castor bean. Plant Physiol. 91:889-897.

Cai C, Xu C, Li X, Ferguson I, Chen K (2006). Accumulation of lignin in relation to change in activies of lignification enzmes in loquat fruit flesh after harvest. Postharvest Biol. Technol. 40:163-169.

Cantwell M, Hong G, Suslow TV (2001). Heat treatments control extension growth and enhance microbial disinfection of minimally processed green onions. HortScience 36:732-737.

Chen E-L, Chen Y-A, Chen L-M, Liu Z-H (2002). Effect of copper on peroxidase activity and lignin content in Raphanus sativus. Plant Physiol. Biochem. 40:439-444.

Cheng GW, Breen PJ (1991). Activity of phenylalanine ammonialyase (PAL) and concentrations of anthocyanins and phenolics in developing strawberry fruit. J. Am. Soc. Hortic. Sci. 116:865-869.

Fallik E (2004). Prestorage hot water treatments (immersion, rinsing and brushing). Postharvest Biol. Technol. 32:125-134.

Goffner D, Joffroy I, Grima PJ (1992). Purification and characterization of isoforms of cinnamyl alcohol dehydrogenase from Eucalyptus xylem. Planta 188:48-53.

Hennion S, Little ACH, Hartmann C (1992). Activities of enzymes involved in lignification during the post-harvest storage of etiolated asparagus spears. Physiol. Plant. 86:474-478.

Holm KB, Andreasen PH, Eckloff RG, Kristen BK, Rasmussen SK (2003). Three differentially expressed basic peroxidases from wound-lignifying Asparagus officinalis. J. Exp. Bot. 54:2275-2284.

Lau MH, Tang J, Swanson BG (2000). Kinetics of textural and color changes in green asparagus during thermal treatment. J. Food Eng. 45:231-236.

Li W, Zhang M (2006). Effect of three-stage hybobaric storage on cell wall components. Texture and cell structure of green asparagus. J. Food Eng. 77:112-118.

Lin ZF, Li SS, Lin GZ (1988). The relation of hydrogen peroxide accumulation and membrane lipid peroxidation in senescence leaf and chloroplast. Plant Physiol. 14:16-22.

Liu Z-Y, Jiang W-B (2006). Lignin deposit and effect of postharvest treatment on lignification of green asparagus (Asparagus officinalis L.). Plant Growth Regul. 48:187-193.

Luo Z, Xu X, Cai Z, Yan B (2007). Effects of ethylene and 1-methylcyclopropene (1-MCP) on lignification of postharvest bamboo shoot. Food Chem. 105:521-527.

Murata M, Tanaka E, Minoura E, Homma S (2004). Quality of cut lettuce treated by heat shock: prevention of enzymatic browning, repression of phenylalanine ammonia-lyase activity, and improvement on sensory evaluation during storage. Biosci. Biotechnol. Biochem. 68:501-507.

Oomena RJFJ, Tzitzikasa EN, Bakxb EJ, Straatman-Engelena I, Busch MS, McCann MC (2004). Modulation of the cellulose content of tuber cell walls by antisense expression of different potato (Solanum tuberosum L.) CesA clones. Phytochemistry 65:535-546.

Qiu M, Jiang H, Ren G, Huang J, Wang X (2013). Effect of chitosan coatings on postharvest green asparagus quality. Carbohyd. Polym. 92:2027-2032.

Rajkovic A, Smigic N, Devlieghere F (2010). Contemporary strategies in combating microbial contamination in food chain. Int. J. Food Microbiol. 141:S29-S42.

Rico D, Martín-Diana AB, Barat JM, Barry-Ryan C (2007). Extending and measuring the quality of fresh-cut fruit and vegetables: a review. Trends Food Sci. Technol. 18:373-386.

Rodriguez-Arcos RC, Smith AC, Waldron KW (2002). Mechanical properties of green asparagus. J. Sci. Food Agric. 82:293-300.

Ros Barceló A (1998). Hydrogen peroxide production is a general property of the lignifying xylem from vascular plants. Ann. Bot. 82:97-103.

Sheu SC, Chen AO (1991). Lipoxygenase as blanching index for frozen vegetable soybeans. J. Food Sci. 56:448-451.

Sivakumar D, Fallik AE (2013). Influence of heat treatments on quality retention of fresh and fresh-cut produce. Foor Rev. Int. 29:294-320.

Sothornvit R, Kiatchanapaibul P (2009). Quality and shelf-life of washed fresh-cut asparagus in modified atmospher packaging. LWT-Food Sci. Technol. 42:1484-1490.

Villanueva MJ, Tenorio MD, Sagardoy M, Redondo A, Saco MD (2005). Physical, chemical, histological and microbiological changes in fresh green asparagus (Asparagus officinalis L.) stored in modified atmosphere packaging. Food Chem. 91:609-619.

Waldron KW, Parker ML, Smith AC (2003). Plant cell walls and food quality. Compr. Rev. Food Sci. Food Safety 2:101-119.

Zamorano JP, Dopica B, Lowe AL, Wilson ID, Grierson D, Merodio CC (1994). Effect of low temperature storage and ethylene removal on ripening and gene expression changes in avocado fruit. Postharvest Biol. Technol. 4:331-342.

18

A review on sweet potato postharvest processing and preservation technology

M. O. OKE and T. S. WORKNEH

Bioresources Engineering, School of Engineering, University of Kwazulu Natal, Private Bag X01, Scottsville, Pietermaritzburg, Kwazulu Natal, 3209 South Africa.

Sweet potato (SP) is an important root crop grown all over the world and consumed as a vegetable, boiled, baked or often fermented into food and beverages. It could be a very good vehicle for addressing some health related problems and also serve as food security. The research into sweet potato processing has established the fact that there is a lot more in sweet potatoes than its starch. The review has established that the nutritional quality content in sweet potatoes can be enhanced by developing new varieties from available germplasm. Natural colourant and antioxidant present in purple- and red-flesh potatoes can be used for developing functional foods. Available evidence for Africa suggested that postharvest processing and subsequent storage of sweet potatoes need further research to explore the ways by which the new cultivars could be used for industrial and export purposes. Based on the report of the review, study of the combined effects of blanching and/or freezing pre-treatments with higher drying temperatures, determination of moisture diffusivity and activation energy during different drying conditions are recommended for future research.

Key words: Sweetpotato, processing, preservation, storage, postharvest.

INTRODUCTION

Sweet potato roots are bulky and perishable unless cured. This limits the distance over which sweet potato can be transported economically. It was established that in cases where countries are capable of generating surplus, it tends to be relatively localized but dispersed and this leads to a lack of market integration and limits market size (Katan and De Roos, 2004; FAO, 2011). Moreover, production is highly seasonal in most countries leading to marked variation in the quantity and quality of roots in markets and associated price swings. Most especially, in Africa, there is commercial processing into chips or flour, which could be stored for year round consumption for use such as in bread and cakes, or processing into fermented and dried products like fufu.

Sweet potato consumption has been adjudged to decline as incomes rise - a change often linked with urbanization, partly because it is perceived as a "poor man's food" but mostly because of the lack of post-harvest processing or storage (FAOSTAT, 2008; Centro Internacional de la Papa, 2009). The latter can lengthen the period for which sweet potato can be marketed but may also be relevant for subsistence oriented households to increase the period over which sweet potato can be consumed, particularly where there is a marked dry season. A sensible approach to achieve the goal of sweet potato product development would be to increase the nutritional content of this highly consumed crop. The aim of this review is to re-examine the information on the processing of sweet potatoes.

GLOBAL SITUATION OF SWEET POTATO

United Nation's Food and Agriculture Organization (FAO) (1990, 2011) reported that sweet potato (*Ipomea batatas* (L.) Lam.) is a very important crop in the developing world and a traditional, but less important crop in some parts of the developed world. According to FAO (2011), sweet potato is one of the seven crops in the world produce over 105 hundred million metric tonnes of edible food products in the world annually. Only potato and cassava, among the root and tuber crops, produce more. China alone produced 80 to 85% of the total sweet potato production in the world while the remaining countries in Asia have the next highest production and then, followed by Africa and Latin America (Centro Internacional de la Papa, 2009).

In another report, sweet potato is among the world's most important and under-exploited food crops (Scott and Maldonado, 1999; Grant, 2003). With more than 133 million metric tonnes in annual production, sweet potato currently ranks as the fifth most important food crop on a fresh-weight basis in developing countries after rice, wheat, maize, and cassava (Scott and Maldonado, 1999; Grant, 2003). Despite the fact that sweet potato commonly categorized as a subsistence, "food security" or "famine relief" crop, its uses have diversified considerably in developing countries over the last four decades (Grant, 2003). The major producing countries are China, Russian Federation, Indian, Ukraine and the USA with the annual world production of potatoes in 2009 of about 329,581 million tones (FAO, 2011).

Wheat which is suitable for bread-making cannot be grown satisfactorily in many countries, therefore, the FAO statistics has demonstrated that the importance of sweet potato in the area where wheat production is often disadvantaged due to climatic restraints, and utilization of indigenous crops could lead to reduction in importation of wheat or wheaten flour (Katan and De Roos, 2004). Sweet potato is a very efficient food crop and produces more dry matter, protein and minerals per unit area in comparison to cereals (Woolfe, 1992). Research have reported that sweet potatoes being the staple food in the developed countries account for 130 kcal of energy per person per day against 41 kcal in the developing countries where it is still considered as vegetable. Apart from being a rich source of starch, sweet potatoes contain good quantity of secondary metabolites and small molecules which play an important role in a number of processes (Friedman, 1997). Many of the compounds present in sweet potato are important because of their beneficial effects on health, therefore, are highly desirable in the human diet and functions as a functional food (Katan and De Roos, 2004).

The global situation of sweet potato as a commodity is

that it is widely grown throughout the world. However, only about one percent of production enters world trade with Canada, the United Kingdom, France and the Netherlands being the major importing countries (Katan and De Roos, 2004). The USA is the largest exporter of sweet potato accounting for 35% of world trade. The other exporters are China (12%), Israel (9%), France (7%), Indonesia (6%) and Netherlands/France (5%). The last two are also involved in re-exporting. Most of the product is used for table consumption with a small percentage going into industrial uses and animal feed. Sweet potatoes are grown throughout the world and are consumed in large quantities. One of the global health goals is to increase the availability of nutrients to a large population of the world. A reasonable and sensible approach to achieve this goal would be to increase the nutritional content of highly consumed crops (Katan and De Roos, 2004).

Sweet potato can, and does, play a multitude of varied roles in the human diets being either supplemental or a luxury food besides being a staple crop for some parts of the world (Papua New Guinea, some parts of Philippines, Tonga and Solomon Islands) (Sosinski et al., 2001). In Asia countries, sweet potato uses range from supplementary food of little status (Thailand) to a very important supplementary food (Ryukyu Islands, Japan) to rice and/or other root and tuber crops (Wanda, 1987). Sosinski et al. (2001) reported that in the United States and other developed countries, the role of sweet potato is strictly as a luxury food and in other parts of the world such as in Japan, where it plays its role as novel plant products and/or nutriceuticals.

NUTRITIONAL COMPOSITION OF SWEET POTATO

The nutritional composition of sweet potato which are important in meeting human nutritional needs including carbohydrates, fibres, carotenes, thiamine, riboflavin, niacin, potassium, zinc, calcium, iron, vitamins A and C and high quality protein (Tables 1 and 2). Sweet potato particularly provides energy in the human diet in the form of carbohydrates.

According to USDA (2009), besides carbohydrates, they are also rich in dietary fiber and have high water content and also provide 359 kJ energy with low total lipid content, which is only about 0.05 g per 100 g. In addition, sweet potatoes also are high in minerals such as potassium, calcium, magnesium, sodium, phosphorus, and iron (USDA, 2009). Because of the various roles that sweet potatoes play around the world, the concept of nutritional quality and its contribution must transform to meet specific roles in human diet. For instance, staple type diets could require high vitamin C, iron, potassium, protein and as well as high fibre. Similarly, supplemental types of sweet potato must have many of the same characters as staple types in terms of nutritional

Table 1. Sweet potato chemical composition (per Serving of one medium 5 inch long sweet potato; 130 g).

Nutrient	Unit	Composition
Calories	kJ/s	130
Calories from fat	g	0.39
Protein	g	2.15
Carbohydrate	g	31.56
Dietary Fiber	g	3.9
Sodium	mg	16.9
Potassium	mg	265.2
Calcium	mg	28.6
Folate	mcg	18.2
Vitamin C	mg	3.1 (excellent source)
Vitamin A	IU	18443 (excellent source)

Source: USDA, 2009.

Table 2. Nutritional value of raw sweet potato per 100 g.

Nutrient	Unit	Value per 100 g
Water	g	77.28
Energy	kJ	359.00
Protein	g	1.57
Total lipid (fat)	g	0.05
Ash	g	0.99
Carbohydrate	g	20.12
Fiber, total dietary	g	3.00
Calcium, Ca	g	30.00
Iron, Fe	mg	0.61
Magnesium, Mg	mg	25.00
Phosphorus, P	mg	47.00
Potassium, K	mg	337.00
Sodium, Na	mg	55.00
Vitamin C	mg	2.40
Pantothenic acid	mg	0.80
Vitamin B-6	mg	0.21
Vitamin A	IU	14187

Source: USDA (2009).

components. However, as they will not be major food component, the level of components may be more flexible and good.

PROCESSING TECHNIQUES OF SWEET POTATO

It was reported by Fellows (2000) that food processing entails combined procedures to achieve intended changes to the raw material and the processing technologies in the food industry. The processing is subdivided into two main groups, viz:

(i) Processing of foods with non-thermal methods (Lebovka et al., 2004, 2007) such as high pressure processing, pulsed electric field (PEF), electronic beams, and

(ii) Processing of foods with the application of heat (Yadav et al., 2006; Leeratanarak et al., 2006; Ahmed et al., 2010; Fernando et al., 2011; Singh and Pandey, 2012) such as blanching, pasteurization, sterilization, evaporation or concentration, drying or dehydration, microwave and infra-red heating.

PROCESSING OF FOODS WITH THERMAL METHODS AND HEAT

Heat treatment is one of the important methods used in food processing to extend the shelf life of foods either by destroying the enzymatic and microbial activity or by removing water to inhibit deterioration that results from higher water activity. Fellows (2000) enumerated the advantages of heat processing as:

(i) Simple control of processing conditions;
(ii) Production of shelf-stable products that need no refrigeration;
(iii) The destruction of anti-nutritional factors (e.g. trypsin inhibitor in some legumes); and
(iv) The enhancement of availability of nutrients for human consumption (e.g. improves digestibility of proteins and gelatinization of starches).

Processing by application of heat that can be used in product development from sweet potato can be carried out using four methods including:

(a) Heat processing with the use of hot air e.g. dehydration, baking, roasting (Ahmed et al., 2010; Doymaz, 2012).
(b) Heat processing with the use of water or steam e.g. blanching, pasteurization (Fernando et al., 2011).
(c) Heat processing with the use of hot oils e.g. frying (Troncoso et al., 2009).
(d) Heat processing using radiated and direct energy e.g. ohmic heating, di-electric heating, infrared heating (Zhong and Lima, 2003; Brinley et al., 2008; Wang et al., 2010).

PROCESSING OF SWEET POTATO INTO PRODUCTS

The traditional methods of processing sweet potato in most countries have been limited to washing, peeling and boiling. However, in some communities, the roots are washed, peeled, cut into small pieces and then lemon or tamarind juice sparingly added. The pieces are, then, dried in the sun and milled together with sorghum into flour that can be used in making porridge. Some farmers make chips, sun dry, store and later reconstitute by adding water then cook by boiling. Others dry the grated product, mill and then add to other flours to make composite flours. FAO (2011) developed improved processing methods to help overcome some of the problems associated with traditional method, in order to produce sweet potato flour (Figure 1) with improved odour, colour and nutritional qualities.

In cases where rare on-farm processing of sweet potato is done in sub-Saharan Africa, products made include flour which is mixed with sorghum to make porridge and mild alcoholic beverages from peeled, chopped, fermented and pounded sweet potato. This processing is only done when the crop has been harvested and there are no other immediate uses for the produce. In many other areas of the district, flour production was popular in the 60's but was abandoned in

favour of maize flour.

The development of processed products from sweet potato presents one of the most important keys to the expanded utilization of the crop. Just like white potatoes, sweet potatoes are multipurpose vegetables. The development in sweet potato research and development (R&D), has transformed the crop from a simple staple food to an important commercial crop with multiple uses such as a snack, ingredient in various foods and complementary vegetable. Lopez et al. (2000) reported that sweet potato flakes (called sweet potato buds) with an increased β-carotene content were produced in Guatemala to conquest vitamin A deficiency in children. Fresh-market sweet potatoes can be baked, microwaved, broiled, grilled, and baked. In some countries alcohol is distilled from sweet potatoes. They can also be used in plate garnishes, casseroles, sautéed vegetables, pasta sauces, dipping vegetables green salads, (fresh- cut sticks), soups, stir-fry, and stews (Dawkins and Lu, 1991). They can be processed as follows:

(i) Dried/dehydrated: flour, flakes, chips,
(ii) Frozen: dices, slices, patties, French fries, and
(iii) Canned: candied, baby foods, mashed, cut/sliced, pie fillings.

Sweet potatoes are also used as an ingredient in cakes, ice creams, icing, pie fillings, cookies, custards and various other bread products. As drying technology progressed, sweet potatoes began to be pureed and then dried to produce flakes, which can be easily reconstituted for direct use in various products like mashed sweet potato, pies and other products (Dawkins and Lu, 1991).

Dried sweet-sour sweet potato

Dried sweet-sour sweet potato was originally named Delicious-SP and it is a product that has the sweet and sour taste of dried fruits (Truong et al., 1998). The most acceptable product was made with boiled sweet potato slices 0.3 mm thick which were soaked in 60° Brix syrup containing 0.8 - 1.0% citric acid and dried at 65°C. Truong (1992) established that the Delicious-SP prepared from sweet potato variety VSP-1, which is a "moist" type sweet potato with low dry matter and starch content, obtained the highest sensory scores due to its attractive orange colour and soft texture. Dried sweet-sour sweet potato contains 13,033 I.U. of vitamin A per 100 g which is higher than both dried mango and dried apricot.

Sweet potato Catsup (Ketchup)

Sweet potato catsup consists of 32.3% (w/v) sweet potato, 42% water, 12.9% vinegar, 11.3% sugar, 1.0%

Fresh sweet potato roots

↓

Weighing

↓

Trimming

↓

Washing

↓

Peeling

↓

Slicing/shredding

↓

Soaking (bleaching/blanching)

↓

Pressing

↓

Spreading

↓

Drying

↓

Grinding

↓

Sifting

↓

Weighing

↓

Packaging

Figure 1. Sweet potato flour production.
Source: Okigbo (1989) and FAO (2011).

salt, 0.3% spices, and food colouring (references). The roots are washed, trimmed, chopped into chunks, and boiled. The boiled chunks are blended with water and other ingredients and boiled to the desired consistency before bottling. Various sweet potato cultivars having cooked flesh colours which range from yellow to orange and a "moist" texture can be used for catsup making. Sweet potato catsup had viscosity, pH, total soluble solids, and intermediate vitamin A content comparable to values found in banana catsup. In consumer acceptability tests, sweet potato catsup was ranked statistically equal to the leading brand of tomato and banana catsup in terms of colour, consistency, flavor, and general acceptability (Truong et al., 1990). Sweet potato catsup stored for four months at ambient temperature was given comparable sensory scores to that of freshly prepared samples.

Sweet potato jam

The sweet potato jam formula contains 20.7% (w/v) sweet potato, 45% sugar, 34% water, and 0% citric acid and this has proved most acceptable by the trained taste panel compared with other ratio. The initial steps in preparing sweet potato roots are similar to those for sweet potato catsup. The cooked chunks are blended with water, sugar, citric acid, and optionally with

flavourings. The slurring is then cooked until total soluble solids of 68° Brix was obtained (Truong et al., 1986). Due to the high starch content of sweet potato roots as compared to fruits, the proportions of sweet potato and sugar are different from the standard formula of 45% fruit and 55% sugar in fruit jams (Gross, 1979).

Sweet potato beverage

The processing steps for sweet potato beverage involve washing, peeling, trimming to remove damaged parts, steaming, extracting, and formulating with 12% (w/v) sugar, 20% (w/v) citric acid, and 232 mg/L ascorbic acid as vitamin C fortification (Truong and Fermentira, 1990). The formulated beverage is bottled in 150 ml glass containers and pasteurized at temperature of 90 to 95°C. Various sweet potato varieties were evaluated for their suitability in processing into the beverage. In general, the orange coloured beverage is preferred to other coloured products. Addition of the juice or pulp of different fruits, e.g., guava, pineapple, or Philippine lemon, at concentrations of 0.6 to 2.4% (w/v) significantly improved aroma scores. Similar to jam, incorporation of artificial orange flavouring also enhanced the aroma of sweet potato beverage. More than 85% of consumer respondents rated "like" for the sweet potato beverage, and 96% liked guava-flavoured sweet potato beverage (Truong and Fermentira, 1990).

Sweet potato leather

Steamed sweet potato chunks are blended with water, sugar, salt, citric acid, and optionally with artificial fruit flavours in processing sweet potato leather. The slurry is then thinly spread on plastic sheets and dried in a mechanical drier until the desired moisture content and texture of the product are obtained. A loading density of 4 kg slurry per m^2 produced the sweet potato leather which was rated with high sensory scores for thickness, texture, and general acceptability. The product also obtained scores of over 7.0 for colour, sweetness, and sourness on the 9-point hedonic scale. Addition of pectin at 0.05 to 0.15% w/w did not improve the texture of the product. Apparently the pectin content of sweet potato is sufficient to produce a leathery textured product (Truong et al., 1998).

OPPORTUNITIES FOR EXPANDING SWEET POTATO PROCESSING

Opportunities for expanding the use of sweet potato lie in three categories: (1) fresh and processed for human consumption, (2) fresh and dried for animal feed, and (3) starches and flours for food and non-food uses.

Table 3. Food crops and their vitamin-calorie contribution per capita/day.

Food crops	Calorie cost ($)	A (Si)	Vitamin		Minerals		
			B1 (mg)	C (mg)	Fe (mg)	Ca (mg)	Fe (mg)
Rice	2.75	0	0.69	0	20.82	486	2.78
Maize	1.38	1795	1.34	0	35.20	901	8.45
Cassava	0.94	3065	236	0.48	262.68	318	5.57
Sweet potato	1.34	78232	224	0.91	304.80	498	7.11

Source: Scott et al. (2000).

Human consumption

Sweet potato fulfils a number of basic roles in the global food system, all of which have fundamental implications for meeting food requirements, reducing poverty, and increasing food security. Sweet potato is a cheap calorie producer and is rich in vitamin A and C and minerals (Table 3). The production growth of sweet potato must be higher than the population growth for food security. World sweet potato production growth is projected at 1.45% and roots in fresh form generally have little competitive overlap on either the supply or demand side. The processed products made from sweet potato not only compete with cereals, but also with each other's processed products in terms of market and raw material. Declining availability of rice, population growth, modest absolute income levels for large segments of consumers, and declining farm size will contribute to a growing use of fresh roots, and in certain areas, of leaves for human consumption. It was established that consumers prefer processed products of roots such as noodles to fresh roots (Scott et al., 2000).

Consumption of fresh roots tends to decline as per capita income rises and consumers will switch to more preferred foods. Therefore, future research must investigate the feasibility of improving quality and lowering unit cost, or channelling output into emerging specialist markets such as the starch market for upstream industries. Future economic trends will also help determine whether shifts in relative prices and exchange rates, and pace of technological innovation, will change the market for this type of product either into a more regional market, or a highly localized one.

Sweet potato starch and flour can be processed into many food and non-food products (Figure 2). It is possible to develop flour and starch as strategic products for upstream industries. Expanding sweet potato for industrial uses must be backed up by innovative postharvest technologies. Physicochemical properties of sweet potato significantly differ among varieties. Therefore, suitable varieties for each processed product are needed (Lin, 2000).

Idris and Hasim (2000) reported that sweet potato starch can provide modified starch that is a raw material for processed products like sauces and fermented foods from sweet potato. This implies that there is an opportunity for expanding the uses of sweet potato in industry. The physical properties of starch differ widely across varieties and these differences markedly affect the quality of starch noodle produced (Collado and Corke, 2000; Panda and Ray, 2007; 2008; Panda et al., 2007, 2009). Some of the fermented foods include lacto-pickle, lacto-juice, sweet potato curd and yoghurt, wine and beer.

Lacto-pickles

Lactic acid (LA) bacteria influence the flavour of fermented foods in a variety of ways. In many cases, the most obvious change in LA fermentation is the production of acid and lowering of pH those result in an increase in sourness (Panda and Ray, 2007). Experimental work on pickling of β-carotene and anthocyanin-rich sweet potato by LA fermentation (sauerkraut process) using 5.20% (w.v) brine solution has been carried out at Regional Centre of CTCRI, Bhubaneswar, India (Panda et al., 2007, 2009). It not only produced LA which imparted taste and flavour to lacto-pickles, but also preserved ascorbic acid, phenols, and coloured pigments (β-carotene and anthocyanin); all these are considered as anti-oxidants (Shivashankara et al., 2004). Anthocyanin-rich sweet potato lacto-pickle had a pH (2.5 - 2.8), titratable acidity (TA) (1.5 - 1.7 g kg^{-1}), lactic acid (1.0 - 1.3 g kg^{-1}), starch (56 - 58 g kg^{-1}) and anthocyanin content (780 mg kg^{-1}) on fresh weight basis. Sensory evaluation rated the anthocyanin-rich sweet potato lacto-pickle acceptable based on texture, taste, aroma, flavour and aftertaste (Panda et al., 2009).

Lacto-juice

Lacto-juices processed by lactic acid fermentation bring about a change in the beverage assortment for their high nutritive value, vitamins and minerals which are beneficial to human health when consumed (Panda and Ray, 2008). Lacto-juice was prepared by fermentation of β-carotene and anthocyanin-rich sweet potato cultivars by inoculating LAB, *Lb. plantarum* MTCC 1407 (Panda and

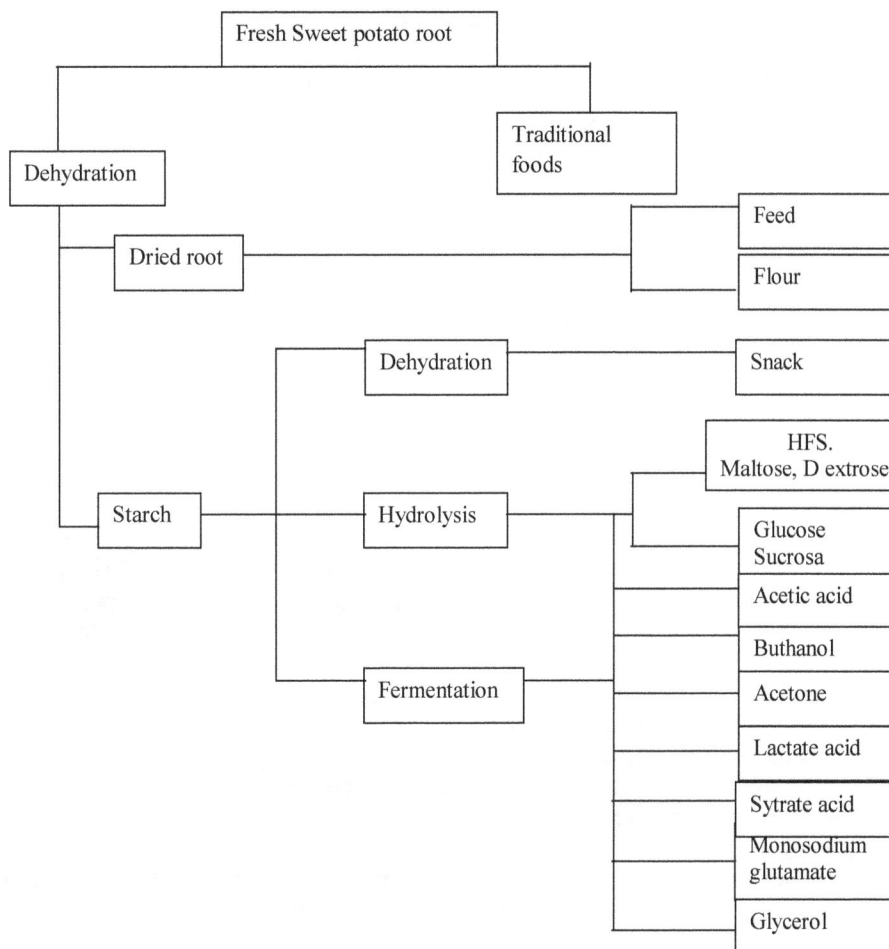

Figure 2. Flow chart on the use of sweet potato as starch and flour. Source: Lin (2000).

Ray, 2008; Panda et al., 2009). β-carotene-rich sweet potato roots (non-boiled/fully-boiled) were fermented with *Lb. plantarum* at 28 ± 2°C for 48 h to make lacto-juice.

During fermentation both analytical [pH, TA, LA, starch, total sugar, reducing sugar (g kg^{-1} roots), total phenol and β-carotene (mg kg^{-1} roots)] and sensory (texture, taste, aroma, flavour and aftertaste) analyses of sweet potato lacto-juice were evaluated. The fermented juice was subjected to panelist evaluation for acceptability. There were no significant variations in biochemical constituents (pH, 2.2 - 3.3; LA, 1.19 - 1.27 g kg^{-1} root; TA, 1.23 - 1.46 g kg^{-1} root, etc) of lacto-juices prepared from non-boiled and fully-boiled sweet potato roots except β-carotene concentration [130 ± 7.5 mg kg^{-1} (fully-boiled roots) and 165 ± 8.1 mg kg^{-1} (non-boiled roots) (Panda and Ray, 2008)].

Sweet potato curd and yoghurt

Yoghurt and curd are consumed by lactase-deficient individuals because much of the lactose in milk is converted to digestive LA by curd- or yoghurt-producing bacteria during fermentation. While the starter culture for curd is a mixture of undefined cocktail of LA producing micro-organisms, that is *Lb. bulgaricus, Streptococcus clemoris, St. thermophilus*, etc, the starter culture for yoghurt is the use of specific symbiotic or mixed culture of *Lb. bulgaricus* and *St. thermophilus*. In a recent study, a curd like product was prepared by co-fermenting boiled sweet potato pulp (8 to 16%) from β-carotene and anthocyanin-rich variety, sugar and curd inoculum (Panda et al., 2006; Mohapatra et al., 2007). Curd with 12 to 16% sweet potato pulp was most preferred by consumers' panelists (Ray et al., 2005). As this product is highly enriched with LAB, it has all the qualities to be addressed as 'probiotic' food.

Wine and beer from sweet potato

Yellow, red and black coloured beverages like beer (sparkling liquor) and wine are being sold in the Kyushu Province in Japan prepared from anthocyanin-rich sweet

Table 4. Total starch, α-amylase activity and trypsin inhibitor activity in fresh sweet potato roots at harvest.

Genotype	Dry matter (%, dry basis)	Total starch (%, dry basis)	α-amylase activity (Ceralpha unit/g, dry basis)	Trypsin inhibitor activity (U/mg, dry basis)
Hi-dry	33.5±0.9	73.6±0.5	0.41±0.01	16.5±1.84
Yan1	29.3±1.6	55.3±0.1	0.81±0.01	18.6±2.56
Chao1	22.6±0.6	46.8±2.0	1.73±0.06	3.90±0.18
Yubeibai	27.9±0.1	52.6±1.1	1.25±0.18	4.99±0.17
Guang7	26.9±1.2	57.6±3.4	1.14±0.04	8.74±0.89
Guang16	24.3±0.4	49.6±1.1	1.44±0.04	21.8±1.74
Mean	27.4	55.9	1.13	12.41
LSD (0.05)	2.59	4.8	0.20	3.50

Values are means of two replicates. Source: Zhang et al. (2002).

potato (Yamakawa, 2000).

Sweet potato as animal feed

BPS (2000) established that until 2000, the volume of fresh roots processed as feed in Indonesia was relatively low and the use of unmarketable fresh roots (very small size, damaged by pests/diseases) was most common in production areas. Moreover, sweet potato foliage as feed for livestock has been gaining importance. Cattle fed with it produce much manure which can be recycled as fertilizer in crop production. In a rice-sweet potato cropping system where rice is fertilized with cattle manure, root yield increases significantly (Wargiono et al., 2000). Good sweet potato growth means robust foliage for feed that, in turn, increases cattle manure to fertilize rice after growing sweet potato. Therefore, it is necessary to develop integrated crop management in production areas.

PREVIOUS WORK ON THE PROCESSING OF SWEET POTATO

Biochemical changes during storage of sweet potato roots

Changes during storage were investigated in carbohydrate level, digestibility, α -amylase, trypsin inhibitor activity and pasting properties of roots of six genotypes of sweet potato (*Ipomoea batatas* (L.) Lam) differing in dry matter content by Zhang et al. (2002). They reported that most genotypes showed a slight decrease in starch content during 0 to 180 days of storage, but in the genotype Hi-dry, it decreased substantially. Alpha-amylase activity increased during the first 2 months of storage, followed by a decrease with continued storage to a level similar to that at harvest. The decline in starch content was correlated with α -amylase activity in the first 60 days storage ($r=0.80$, $P=0.06$).

Trypsin inhibitor activity (TIA) in the fresh roots varied among genotypes from 3.90 to 21.83 U/mg (Table 4) and storage had little influence on TIA level. There was considerable genotypic variation in digestibility, with up to 27% reduction in digestibility after 120 days in storage. Glucose and sucrose concentration increased early in storage and then remained fairly constant. Storage reduced flour pasting viscosities, with up to nearly a 30% decline in peak viscosity.

Frying of sweet potato at different processing conditions

The research of Troncoso et al. (2009) studied the effect of different processing conditions on physical and sensory properties of sweet potato chips. Potato slices of Desire´e and Panda varieties (diameter: 30 mm; thickness: 3 mm) were pre-treated in the following ways: (i) control or unblanched slices without predrying; (ii) blanched slices in hot water at 85°C for 3.5 min and air-dried at 60°C until a final moisture content of 0.6 kg water/kg dry solid; (iii) control slices soaked in a 3.5 kg/m^3 sodium metabisulphite solution at 20°C for 3 min and pH adjusted to 3. Pre-treated slices were fried at 120 and 140°C under vacuum conditions (5.37 kPa, absolute pressure) and under atmospheric pressure until they reached final moisture content of w1.8 kg water/100 kg (wet basis). An experimental design was used to analyze the effect of pre-treatment, sweet potato variety, type of frying and frying temperature over the following responses: oil content, instrumental color and texture and sensory evaluation. Vacuum frying increased significantly ($p < 0.05$) oil content and decreased instrumental color and textural parameters. Sensory attributes, flavor quality and overall quality, were significantly improved using vacuum frying.

The higher frying temperature (140°C) increased DE, maximum breaking force, hardness and crispness and decreased L* and b* values. On the other hand, Panda sweet potato variety improved the color of the product. A great improvement on color parameters was obtained using sulphited potato slices instead of the other pre-treatments. Although, the better flavor was obtained for control sweet potato chips, no significant differences were found for overall quality between control and sulphited sweet potato chips. Significant correlations ($p < 0.01$) between sensory and instrumental responses were found.

Chemical and phytochemicals in sweet potato

Das et al. (2010) studied the chemical modification of sweet potato (*Ipomoea batatas*) by acetylation using vinyl acetate ranging from 4 to 10% and dual modification using propylene oxide at specific level of 7% followed by adipic acid anhydride at levels ranging from 0.05 to 0.12%. Degree of substitution ranges between 0.020 to 0.034% and 0.018 to 0.058% for dual-modified and acetylated starch samples, respectively. There was significant increase in water binding and oil-binding capacities, solubility, paste clarity, gel strength due to modification; however, rupture strength, gel elasticity and adhesiveness decrease in both modified starches. Analysis of SEM revealed that the modification altered starch morphology. Acetylation brought about slight aggregation or cluster formation of granules with deep groove in the central core region whereas in dual-modified starches there were present a number of aggregates of starch granules with development of few blister like appearances along with protuberances on their surfaces.

It was observed that morphology of sweet potato starch granule gets altered due to chemical modification and in food application; all these modified starches can be used for development of different products depending on the quality requirement of the product, which could be evaluated with specific applications on them.

Anti-oxidative activities by three methods in the sweet potato plant and in home processed roots were carried out by Jung et al. (2011). Total phenolic content was highest in the leaves. Eight root varieties were partitioned and analyzed for phenolics. The stem end of the root had significantly more phenolics. In all samples the predominant chlorogenic acids were 5-caffeoylquinic acid (5-CQA) and 3,5-diCQA. 3,4-diCQA was present in significant amounts in the leaves and the flower, and 4,5-diCQA in the leaves. Six home-processing/cooking techniques reduced phenolic content from 7% (baking) to -40% (deep frying/boiling). High correlations were observed between phenolic compounds determined by high-performance liquid chromatography (HPLC) and Folin-Ciocalteu, radical scavenging activity by 2,2-diphenyl-1-picrylhydrazyl (DPPH), and oxidative activity by ferric thiocyanate (FTC) and thiobarbituric acid (TBA) methods. The results show that there is a large variation in phenolics among sweet potato varieties and different

parts of the plant and that high-phenolic sweet potato leaves, widely consumed in Asian countries as a vegetable, should be considered for diets of other countries.

These observations on anti-oxidative effects of sweet potatoes complement and extend previous cited findings on the relationships between content of phenolic compounds and anti-oxidative activities in other sweet potato genotypes. Their findings emphasize the substantial variation found among sweet potatoes and suggest that consumers have a choice in selecting sweet potato varieties with a high content of phenolic compounds and high anti-oxidative effects. The studies also showed that sweet potato leaves contain considerably more phenolic compounds that the roots, indicating that perhaps leaves, widely consumed as vegetables in Asian countries, may merit inclusion in diets of Western countries.

In conclusion, the methods they used to obtain the cited data on the content and distribution of caffeoylquinic acids and of total phenolic content in sweet potato plant stems, roots, leaves, and flowers and in home processed roots should facilitate future studies designed to assess the role of sweet potato phenolic compounds in host-plant resistance, food microbiology, food chemistry, plant breeding, medicine and nutrition. The results may also help consumers select sweet potatoes with high levels of health promoting compounds to select cooking conditions that minimize losses and also for human use.

Ezekiel et al. (2013) reviewed the beneficial phytochemicals in sweet potato and reported that in addition to supplying energy, sweet potatoes contain a number of health promoting phytonutrients such as carotenoids, folates, flavonoids, anthocyanins, kukoamines, and phenolics. They established that phytochemicals content in sweet potatoes can be enhanced by developing new varieties from available germplasm high in these compounds. Antioxidant and natural colourant present in red-flesh- and purple sweet potatoes can be used for developing functional foods. Taking into consideration the large quantities in which sweet potatoes are consumed throughout the world, sweet potatoes could be a very good vehicle for addressing some health related problems. Available evidence suggests that postharvest storage of sweet potatoes do not significantly affect the content of phytochemicals, but antioxidant levels are generally higher in sweet potatoes grown in high-yielding environments, and increased during storage.

Since the cost of production of sweet potatoes is relatively low as compared to other horticultural crops, pigmented sweet potatoes may also serve as a potential source of natural anthocyanins for use in food industry. Furthermore, sweet potato is a high yielding crop in which cultural and storage practices are well established in its usage. However, for the economy of the whole process of pigment extraction, sweet potatoes with high

Table 5. Yields of sweet potatoes per unit area (t/ha).

Variety	d 100	d 130	d 160	Average
NS 88	21.75	26.25	27.38	25.13
XS 18	12.75	18.00	20.18	16.98
YZ 263	22.88	22.50	28.88	24.75
NS 009	20.63	15.75	22.88	19.76
NS 007	21.00	28.50	28.97	26.16
200730	27.38	31.50	34.01	30.96
SS 19	25.13	37.50	38.63	33.75
WS 34	19.50	29.25	31.13	26.63
2-l2-8	20.25	30.00	35.25	28.50
XS 22	18.00	26.25	34.13	26.13

Source: Jin et al. (2012).

Table 6. Feedstock consumptions (t) of the sweet potatoes to produce 1 t of anhydrous.

Variety	d 100	d 130	d 160	Average
NS 88	10.04	7.93	7.82	8.60
XS 18	7.25	6.45	6.61	6.77
YZ 263	8.12	7.03	7.34	7.50
NS 009	6.87	6.30	6.38	6.52
NS 007	6.57	6.02	5.99	6.19
200730	13.18	13.25	13.86	13.43
SS 19	8.07	7.10	7.61	7.59
WS 34	7.46	7.35	6.98	7.26
2-l2-8	8.52	7.34	8.57	8.15
XS 22	7.86	7.17	7.99	7.67

Source: Jin et al. (2012).

concentrations of anthocyanins are desirable. In general, cooking leads to losses of nutrients in sweet potatoes, however, phytonutrients are either not affected or sometimes increased due to increase in extractability and bioavailability. There is a need for further research to explore the ways by which losses in phytochemicals can be reduced such as co-pigmentation, and their stability can be enhanced. Sweet potatoes contain enough phytochemicals to justify the claim of being health promoters, therefore, should form a substantial part of our daily diet. Different pigmented sweet potato based foods needs to be developed and evaluated especially with respect to the antioxidant capacity and other health benefits.

Biofuel production from sweet potato

The performance in the ethanol production of 10 varieties of sweet potato was evaluated by Jin et al. (2012), and the consumption in raw materials, land occupation and fermentation waste residue in producing 1 ton of anhydrous ethanol were investigated. The comparative results (Tables 5 to 7) between 10 varieties of sweet potato at 3 growth stages indicated that NS 007 and SS 19 were better feedstock for ethanol production, exhibiting the highest level of ethanol output per unit area (4.17 and 4.17 ton/ha, respectively), less fermentation waste residue (0.56 and 0.55 tons/ton ethanol, respectively), the least land occupation (0.24 and 0.24 ha/ton ethanol, respectively), less feedstock consumption (6.19 and 7.59 tons/ton ethanol, respectively), and a lower viscosity of the fermentation culture (591 and 612 mPa S, respectively). In most varieties, the ethanol output speed at day 130 was the highest and therefore, NS 007 and SS 19 could be used for ethanol production and harvested after the period (130 days) of growth from an economic point of view. In addition, the high content of fermentable sugars and low content of fiber in sweet potatoes are criteria for achieving low viscosity in ethanol fermentation cultures.

In this work, among the 10 strains tested, the sweet potato strains NS 007 and SS 19 had the least land occupation, less feedstock consumption, and the highest

Table 7. Land occupation (ha) of the sweet potatoes to produce 1 t of anhydrous ethanol.

Variety	d 100	d 130	d 160	Average
NS 88	0.46	0.30	0.29	0.35
XS 18	0.57	0.36	0.33	0.42
YZ 263	0.35	0.31	0.25	0.31
NS 009	0.33	0.4	0.28	0.34
NS 007	0.31	0.21	0.21	0.24
200730	0.48	0.42	0.41	0.44
SS 19	0.32	0.19	0.20	0.24
WS 34	0.38	0.25	0.22	0.29
2-I2-8	0.42	0.24	0.24	0.30
XS 22	0.44	0.27	0.23	0.31

Source: Jin et al. (2012).

level and speed of ethanol output per unit area as well as a lower viscosity of fermentation culture and reduced fermentation waste residue. Although they did not have the lowest viscosity of fermentation cultures or the least fermentation waste residue, these strains could be used for ethanol production and harvested after growing for 130 days or act as parent crops to obtain hybrids that have ideal characteristics for ethanol production.

Drying process of Sweet potato for Human consumption

Yadav et al. (2006) reported the changes occurring in the characteristics of sweet potato flour as a result of processing. The pasting characteristics decreased due to gelatinization of starch during processing. The degradation of starch by amylases during hot air drying further lowered the total amylose and water binding capacity/viscosity and increased the digestibility compared to those of drum dried and native flour. Solubility and swelling power of the flours increased as a result of processing which subsequently increased with increase in temperature. Scanning electron micrographs of starch granules showed tendency of clustering, especially in drum dried samples. X-ray diffraction patterns showed alteration from Ca-type to V-type with a marked reduction in crystallinity index as a result of processing. The ^{13}C NMR spectra of processed starches showed reduced peak intensities and line widths due to depolymerizing effects, and also pointing to their change in crystallinity.

Hatamipour et al. (2007) worked on drying characteristics of six varieties of sweet potatoes in different dryers. In this work, which was designed for drying of agro-food products, six varieties of sweet potatoes (Santana, Marfona, Santea, Koncord, Diamant, and Renjer) were chosen as drying material. A fluidized bed dryer, with and without air circulation, and a pilot-scale tray dryer were used for performing drying

experiments. The experiments were performed with and without blanching. The changes in structure and colour of six varieties of potatoes were studied.

Temperature did not show significant effect on shrinkage, but blanching time and air circulation had significant effect on shrinkage as well as on the appearance of dried product. Less shrinkage occurred in Renjer and Diamant varieties at 80°C in comparison with other varieties. Santana (at 80°C), Santea (at 70°C) and Renjer (at 80°C) had better appearance and colour after free convection drying, whereas the appearance, colour and quality of Marfona variety was not acceptable at all. The quality and appearance of Marfona variety improved by using a tray dryer with air circulation. The quality and appearance of all varieties was very good in fluidized bed dryer. Blanching was effective in improving the colour of all dried varieties.

Ahmed et al. (2010) reported the effect of pretreatments with 1 w/v % sodium hydrogen sulphite ($NaHSO_3$) and 1 w/v % calcium chloride ($CaCl_2$) and drying temperatures (55, 60 and 65°C) on sweet potato flour. Flour treated with $CaCl_2$ had higher amounts of b-carotene and ascorbic acid (3.26–3.46 and 10.61–12.54 mg 100 g/L wet basis, respectively) than that treated with $NaHSO_3$ (3.05–3.43 and 9.47–11.47 mg 100 g/L wet basis, respectively). Water absorption index (wet basis) and total phenolic content were highest at 65°C when treated with $NaHSO_3$ (2.49 g g/L and 10.44 mg 100 g/L respectively) and $CaCl_2$ (2.85 g g/L and 9.52 mg 100 g/L respectively). Swelling capacity (wet basis) was highest at 55°C when treated with $NaHSO_3$ (2.85 g g/L) whereas when treated with $CaCl_2$ (2.96 g g/L) it was highest at 60°C. Freeze-dried samples treated with $CaCl_2$ had higher b-carotene and ascorbic acid while $NaHSO_3$-treated samples had higher lightness and total phenolic content. The results showed that good quality flour could be produced after soaking in $CaCl_2$ and dried at 65°C

The effect of pretreatments on drying characteristics of potato slices was investigated in a cabinet dryer by Doymaz (2012). The experiments were conducted on potato slices with thickness of 8 mm at 65°C with an air velocity of 2.0 m/s. Potato slices were pretreated with citric acid solution (1:25 w/w, 3 min, 20°C) or blanched hot water (3 min, 80°C) prior to drying while the untreated samples were dried as control. The shortest drying time was obtained with potatoes pretreated with citric acid solution. The drying data were fitted with ten mathematical models available in the literature. The results indicated that parabolic, logarithmic, modified Henderson and Pabis, two term, Midilli et al. and Verma et al. models were found better to describe the drying of potato slices. The values of effective moisture diffusivity were found to be range between 1.78×10^{-10} and 2.94×10^{-10} m²/s and influenced by pretreatments.

Fernando et al. (2011) studied the convective drying rates of thermally blanched slices of potato (*Solamum tuberosum*): Parameters for the estimation of drying rates. They concluded that thermal blanching is a pre-drying treatment applied to some food products in order to inactivate enzymes for preservation of colour and flavours. This process leads to modified surface layers in the products. Drying of such products involve moisture transfer across such layers leading to build-up of moisture concentration gradients within the material being dried. In this study microscopic investigation of surface layers of blanched potato confirms that the presence of such surface layers. Rates of drying of thermally blanched slices of potato (*Solanum tuberosum*) of different thicknesses are obtained experimentally in a laboratory convective tray dryer at 85 and 95°C. A model (Fernando et al., 2008) for estimation of dehydration rates of materials with surface layers which describe three parameters ($l/\sqrt{D}$), r and KpH related to the drying material and the process is employed in order to evaluate the respective parameters based on the experimental data with the aid of a Matlab optimization program. Consistent values of the parameters were obtained for different thicknesses indicating applicability of the parameters in the model for characterization of species for drying of the blanched materials. Drying rates regenerated using the averaged parameters showed compatibility with experimental data with a correlation coefficient of 0.992. It can therefore be seen that drying rates of blanched slices of potato could be evaluated using three parameters $l/\sqrt{D}$, r and KpH making it possible to estimate the drying rates of blanched slices using the model equation assumed.

The effects of drying conditions on the drying behavior of sweet potato (*Ipomoea batatas* L.) were investigated in a cabinet dryer by Singh and Pandey (2012). The convective air drying was carried out under five air velocities of 1.5, 2.5, 3.5, 4.5 and 5.5 m/s, five air temperatures; 50, 60, 70, 80 and 90°C, and three sweet potato cubes of 5, 8 and 12 mm thickness. The data generated were analyzed to obtain diffusivity values from the period of falling drying rate. Results indicated that drying took place in the falling rate period. Moisture

transfer from sweet potato cubes was described by applying the Fick's diffusion model, and effective moisture diffusion coefficients were calculated. The values of calculated effective diffusivity for drying at 50, 60, 70, 80 and 90°C of air temperature and 1.5–5.5 m/s of air flow rates ranged from 1.26×10^{-9} to 8.80×10^{-9} m²/s. Effective diffusivity increased with increasing temperature. An Arrhenius relation with an activation energy value of 11.38 kJ/mol expressed effect of temperature on the diffusivity. Two mathematical models available in the literature were fitted to the experimental data. The page model gave better prediction than the first order kinetics of Henderson and Pabis model and satisfactorily described drying characteristics of sweet potato cubes.

Sweet potato slices were dried using both low-pressure superheated steam drying (LPSSD) and hot air drying in the study of Leeratanarak et al. (2006). The effects of blanching as well as the drying temperature on the drying kinetics as well as various quality attributes of potato slices viz. color, texture, and brown pigment accumulation were also investigated. It was found that LPSSD took shorter time to dry the product to the final desired moisture content than that required by hot air drying when the drying temperatures were higher than 80°C. Drying methods had no obvious effect on the product quality except the browning index. Longer blanching time and lower drying temperature resulted in better color retention and led to chips of lower browning index. It was also established that blanching reduced the hardness and shrinkage of the product; however, the use of different blanching periods did not significantly affect the product hardness. A blanching time of 5 min followed by LPSSD at 90°C at an absolute pressure of 7 kPa was proposed as the best condition for drying potato chips in this study. These conditions gave puffed product, less hard with moderate browning index, which corresponded to less nutrients and other heat damages. These conditions also provided potato chips that had small changes of colors from their natural values and required shortest drying time. However, the best condition proposed still led to chips of inferior quality compared with the commercially available potato chips, especially in terms of hardness. The study of the combined effects of blanching and/or freezing pre-treatments with higher drying temperature is recommended for future work.

CONCLUSION

The research in sweet potato processing has established the fact that there is a lot more in sweet potatoes than starch. The nutritional quality content in sweet potatoes can be enhanced by developing new varieties from available germplasm. Natural colourant and antioxidant present in purple- and red-flesh potatoes can be used for developing functional foods. Considering the large quantities in which sweet potatoes are consumed throughout the world, sweet potatoes could be a very good vehicle for addressing some health related problems and also serve as food security. Available evidence suggests that postharvest processing and subsequent storage of sweet potatoes need further research to explore the ways by which the new cultivars could be used for industrial and export purposes in countries producing sweet potato. Based on this review, future studies of the combined effects of blanching and/or freezing pre-treatments with higher drying temperature, determination of moisture diffusivity and activation energy during different drying conditions are recommended.

REFERENCES

Ahmed M, Sorifa AM, Eun JB (2010). Effect of pretreatments and drying temperatures on sweet potato flour. Int. J. Food Sci. Technol. 45:726-732.

Badan Pusat Statistik (BPS) (2000). Statistik Indonesia 1999-2000: Produksi Ubijalar. Jakarta.

Brinley TA, Truong VD, Coronel P, Simunovic J, Sandeep KP (2008). Dielectric Properties of Sweet Potato Purees at 915 MHz as Affected by Temperature and Chemical Composition, Int. J. Food Ppties. 11(1):158-172.

Centro Internacional de la Papa (2009). Sweet Potato. (retrieved 06.10.12 from http://www.cipotato.org/sweet potato/)

Collado LS, Corke H (2000). Predicting the quality of sweet potato starch noodles. In: Cassava, Starch and Starch Derivatives. R.H. Howeler, C.G. Oates, and G.M. O'Brien (eds). Proceedings of the International Symposium held in Nanning, Guangxi, China Nov 11-15, 1996. Centro Internacional de Agricultura Tropical (CIAT).

Das AB, Singh G, Singh S, Riar CS (2010). Effect of acetylation and dual modification on physico-chemical, rheological and morphological characteristics of sweet potato (Ipomoea batatas) starch. Carbohy Polymers. 80:725-732.

Dawkins LN, Lu JY (1991). Physico-chemical properties and acceptability of flour prepared from microwave blanched sweet potatoes. J. Food Proc Pres. 15:115-124.

Doymaz I (2012). Drying of potato slices: effect of pretreatments and mathematical modeling. J. Food Proc Pres. 36:310-319.

Ezekiel R, Narpinder S, Shagun S, Amritpal K (2013). Beneficial phytochemicals in potato - A review. Food Res. Int. 50:487-496.

FAO (1990). Roots, Tubers, Plantains and Bananas in Human Nutrition. Rome: Food and Agriculture Organization.

FAO (2011). Statistical database. http://faostat.fao.org/site/567/DesktopDefault.aspx?PageID=567#ancor (accessed October 7, 2012).

FAOSTAT (2008). Production and area harvested statistics for sweetpotato for 2007. http://www.faostat.fao.org/site/567/default.aspx?PageID=567#ancor. Last accessed 16 September 2012.

Fellows PJ (2000). Food Processing Technology - Principles and Practice (2nd Edition) CRC Press, Boca Raton Boston New York Washington, DC Woodhead Publishing Limited, Cambridge England.

Fernando WJN, Ahmad AL, AbdShukor SR, Lok YH (2008). A study of intermediate drying rates of sliced agricultural products based on the build-up of moisture concentration gradients due to surface hardened layers: A case study for convective air drying of sliced papaya and garlic. J. Food Eng. 88:229-238.

Fernando WJN, Ahmad AL, Othman MR (2011). Convective drying rates of thermally blanched slices of potato (Solanum tuberosum): Parameters for the estimation of drying rates. Food Bioprod Process. 89:514-519.

Friedman M (1997). Chemistry, biochemistry, and dietary role of potato polyphenols. A review. J. Agric. Food Chem. 45:1523-1540.

Grant V (2003). Select markets for taro, sweet potato and yam. A report for the Rural Industries Research and Development Corporation (RIRDC). Publication No 0 3 /052 RIRDC project No UCQ-13A. Online: http://www.rirdc.gov.au.

Gross MO (1979). Chemical Sensory and Rheological Characterization of Low- Methoxyl Pectin gels, Ph.D. dissertation, Univ. Georgia, Athens.

Hatamipour MS, Hadji Kazemi H, Nooralivand A, Nozarpoor A (2007). Drying characteristics of six varieties of sweet potatoes in different dryers. Trans IChemE, Part C, Food Bioprod. Proc. 85(C3):171-177.

Idris K, Hasim P (2000). Modified sweet potato starches for sources. Starch Derivatives. Proceedings. China, the International Symposium Naning. pp. 224-234.

Jin Y, Fang Y, Zhang G, Zhou L, Zhao H (2012). Comparison of ethanol production performance in 10 varieties of sweet potato at different growth stages. Acta Oecologica, 30:1-5.

Jung J, Lee S, Kozukue N, Levin CE, Friedman M (2011). Distribution of phenolic compounds and antioxidative activities in parts of sweet potato (Ipomoea batata L.) plants and in home processed roots. J. Food Composition Anal. 24:29-37.

Katan MB, De Roos NM (2004). Promises and problems of functional foods. Critical Rev. Food Sci. Nutr. 44:369-377.

Lebovka NI, Praporscica I, Vorobiev E (2004) Effect of moderate thermal and pulsed electric field treatments on textural properties of carrots, potatoes and apples. Innov. Food Sci. Emerging Technol. 5:9-16.

Lebovka NI, Shynkaryk NV, Vorobiev E (2007). Pulsed electric field enhanced drying of potato tissue. J. Food Eng. 78:606-613.

Leeratanarak N, Devahastin S, Chiewchan N (2006). Drying kinetics and quality of potato chips undergoing different drying techniques. J. Food Eng. 77:635-643.

Lin L, Wheatley CC, Chen J, Song B (2000). Studies on the physicochemical properties of starch of various sweet potato varieties grown in China. In: Cassava, Starch and Starch Derivatives. R.H. Howeler, C.G. Oates, and G.M. O'Brien (eds). Proceedings of the 25 International Symposium held in Nanning, Guangxi, China Nov 11-15 1996. Centro Internacional de Agricultura Tropical (CIAT). pp. 216-223.

Lopez A, Iguaz A, Esnoz A, Virsed P (2000). Thin-layer drying behaviour of vegetable wastes from wholesale market. Drying Technol. 18:995-1006.

Mohapatra S, Panda SH, Sahoo SK, Sivakumar PS, Ray RC (2007). β-Carotene-rich sweet potato curd: production, nutritional and proximate composition. Int. J. Food Sci. Technol. 42:1305-1314.

Okigbo BN (1989). New Crops for Food and Industry. International Symposium on New Crops for Food and Industry, Southampton 1986 (G. Wickens, N. Haq, and P. Day, eds.), London, Chapman and Hall. P. 123.

Panda SH, Naskar SK, Ray RC (2006). Production, proximate and nutritional evaluation of sweet potato curd. J. Food Agric. Environ. 4:124-127.

Panda SH, Naskar SK, Sivakumar PS, Ray RC (2009). Lactic acid fermentation of anthocyanin-rich sweet potato (Ipomoea batatus L.) into lacto-juice. Int. J. Food Sci. Technol. 44:288-296.

Panda SH, Paramanick M, Ray RC (2007). Lactic acid fermentation of sweet potato (Ipomoea batatas L.) into pickles. J. Food Proc. Pres. 31:83-101.

Panda SH, Ray RC (2007). Lactic acid fermentation of bcarotene rich sweet potato (Ipomoea batatas L.) into lacto-juice. Plt Food Human Nutr. 62:65-70.

Panda SH, Ray RC (2008). Direct conversion of raw starch to lactic acid by Lactobacillus plantarum MTCC 1407 in semi-solid fermentation using sweet potato (Ipomoea batatas L.) flour. J. Scientific Industr. Res. 67:531-537.

Ray RC, Naskar SK, Sivakumar PS (2005). Sweet potato curd. Technical Bulletin Series: 39, Thiruvananthapuram, India: Central Tuber Crops Research Institute.

Scott GJ, Best R, Rosegrant M, Bokanga M (2000). Root and tuber crops in the global food system. A vision statement to the year 2020. Peru, Lima International Potato Center. pp 111-117.

Scott GJ, Maldonado L (1999). CIP sweet potato facts. A compendium of key figure and analyses for 33 important sweet potato producing countries. Peru, Lima International Potato Centre (CIP).

Shivashankara KS, Isobe S, Al-Haq MJ, Takenaka M, Shiina T (2004). Fruit- antioxidant activity, ascorbic acid, total phenol, quercetin, and carotene of Irwin mango fruits stored at low temperature after high electric field pretreatment. J. Agric. Food Chem. 52:1281–1286.

Singh NJ, Pandey RK (2012). Convective air drying characteristics of sweet potato (*Ipomoea batatas* L.) cube. Food Bioprod. Process. 90:317–322.

Sosinski B, He J, Cervantes-Flores R, Pokrzywa M, Bruckner A, Yencho GC (2001). Sweet potato genomics at North Carolina state University. Ames, T (ed). Proceedings of the first International Conference on sweet potato. Food and Health for the future, Acta Horticulture, 583:69-76.

Troncoso E, Pedreschi F, Zu´niga RN (2009). Comparative study of physical and sensory properties of pre-treated potato slices during vacuum and atmospheric frying. LWT- Food Sci. Technol. 42:187–195.

Truong VD (1992). Sweet potato beverages: Product development and technology transfer. In: Sweet potato Technology for the 21st Century, Hill, W.A., Bonsi, C.K and Loretan, P.A. (eds), Tuskegee University, Tuskegee, AL, USA. pp. 389-399.

Truong VD, Biermann CJ, Marlett JA (1986). Simple sugars, oligosaccharides, and starch concentrations in raw and cooked sweet potato. J. Agric. Food Chem. 34:421–425.

Truong VD, Fementira GB (1990). Formulation, consumer acceptability and nutrient content of non-alcoholic beverages from sweet potatoes. In: Proceedings of the Eighth Symposium of the International Society for Tropical Root Crops, Howler, R. H. (ed.). Bangkok, Thailand, pp. 589-599.

Truong VD, Guarte RC, De la Rosa LS, Cerna PF, Tabianan IC, Dignos AC (1990). Proceedings of the Eight Symposium of the International Society for Tropical Root Crops, Bangkok, ISTRC Department of Agriculture of Thailand, Bangkok, Thailand, P. 600.

Truong VD, Walter WMJr, Hamann DD (1998). Relationship between instrumental and sensory parameters of cooked sweet potato texture. J. Text Studies. 63(4):739-743.

USDA (U.S. Department of Agriculture), Agricultural Research Service. (2009). USDA National Nutrient Database for Standard Reference, Release 22. Nutrient Data Laboratory Home Page, http://www.ars.usda.gov/ba/bhnrc/ndl, Accessed 14 September 2012.

Wanda CW (1987). Genetic improvement for meeting human nutrition needs. Quebedeaux, B. and Bliss F (editors). Proceedings of the first International symposium on horticulture and human nutrition, Contributor of fruits and vegetables. Prentice Hall. pp. 191-199.

Wang R, Zhang M, Mujumdar AS (2010). Effects of vacuum and microwave freeze drying on microstructure and quality of potato slices. J. Food Eng. 101:131–139.

Wargiono J, Widowati S, Munarso J, Kartasasmita UG, Purba S (2000). Pengkajian dampak dan implementasi program satu hari setiap minggu tanpa konsumsi beras. AKT uslitbangtan, 16:53.

Yadav AR, Guha M, Tharanathan RN, Ramteke RS (2006). Changes in characteristics of sweet potato flour prepared by different drying techniques LWT-Food Sci. Technol. 39:20–26.

Yamakawa O (2000). New cultivation and utilization system for sweet potato toward the 21st century. In: Potential of Root Crops for Food and Industrial Resources (edited by M. Nakatani & K. Komaki). Twelfth Symposium of International Society of Tropical Root Crops (ISTRC), 10–16 Sept., 2000, Tsukuba, Japan. pp. 8–13.

Zhang Z, Wheatley CC, Corke H (2002). Biochemical changes during storage of sweet potato roots differing in dry matter content. Posthar. Biol. Technol. 24:317–325.

Zhong T, Lima M (2003). The effect of ohmic heating on vacuum drying rate of sweet potato tissue. Biores. Technol. 87:215–220.

19

Response of chips and flour from four yam varieties to *Tribolium castaneum* (Herbst) (Coleoptera: Tenebrionidae) infestation in storage

Zakka, U.[1], Lale, N. E. S.[1], Duru, N. M.[1] and Ehisianya, C. N.[2]

[1]Department of Crop and Soil Science, Faculty of Agriculture, University of Port-Harcourt, Choba, P. M. B. 5005, Rivers State, Nigeria.
[2]National Root Crops and Research Institute, Umudike, Abia State, Nigeria.

Chips and flour from four different yam varieties (Ame, Adaka, Nwopoko and Obiaturugo) obtained from the National Root Crops Research Institute (NRCRI), Umudike, Abia State, Nigeria were evaluated for responses to *Tribolium castaneum* (rust-red flour beetle) infestation in the laboratory under prevailing conditions (25-30°C and 70-90 RH) for a period of 90 days. 10 g of either chips or flour of each of the yam varieties were infested with 8 pairs of adult *T. castaneum* in 200 ml air-tight plastic containers. The experiment was set up in a completely randomized design (CRD) and experiments were replicated three times. The result showed that both forms of yam varieties were susceptible to *T. castaneum* but to varying degrees. The variation of susceptibility observed across the treatments showed that Nwopoko chips were the most susceptible and Nwopoko flour was the least susceptible. Development of *T. castaneum* was prolonged on flour than on the chips. Percentage weight loss was highest (3.33) in Adaka flour and weight gain was highest (3.67) in Nwopoko chips. The result also showed variations in the physicochemical and functional properties of the different yam varieties which played a role in pest performance and preference. *T. castaneum* has proved to be a cosmopolitan species that is able to colonize a wide range of substrates including different forms and types of yam cultivated in the Niger delta region of Nigeria. The outcome of this study necessitates the introduction of precautionary measures against this pest in order to prevent it from attaining pest status and causing economic damage to these yam products in storage.

Key words: *Tribolium castaneum*, yam, chips, flour, susceptibility index.

INTRODUCTION

In economic terms, yams (*Dioscorea* spp.) are the world's fourth most important tuber crop (Loko et al., 2013). Yams are perennial herbaceous vines cultivated for the consumption of their starchy tubers cultivated in most tropical countries where 95% of the world's output is produced (FAO, 2010; Babajide et al., 2010). It is a highly popular food crop especially in the yam zones of West Africa, comprising Cameroon, Nigeria, Benin, Togo, Ghana and Cote d'Ivoire (FAO, 2003; Izekor and Olumese,

2010). As a food crop, the place of yam in the diet of the people in Nigeria cannot be overemphasized. It contributes more than 200 dietary calories per capita daily for more than 150 million people in West Africa while serving as an important source of income to the people (Babaleye, 2003). Yam flour, dried cassava chips and yam chips constitute important sources of carbohydrates in the diet of the citizens of West Africa (Obadofin et al., 2013).

Yam is seasonal and the fresh tubers are highly perishable; postharvest losses can range from 30% to as high as 85% of the total production (Baco et al., 2004). Losses occur at every stage of the market chain ranging from storage to processing and marketing (Obadofin, 2013). Food commodities are usually liable to depredation by pest organisms such as micro-organisms, mites, insects, rodents and birds (Odeyemi, 2001). Insect infestation causes qualitative and quantitative losses of food commodities as they produce excrement and frass during their grain boring and oviposition activities (Rajendran, 2005).

In a developing country like Nigeria where a majority of the population is still dependent on forest and agriculture, storage of these products is a prime concern. Losses of yam in storage constitute a major challenge in storage. In Nigeria, yam is a major food and the tuber needs to be stored for consumption and as seed for the following year's crop. However, it is a highly perishable commodity and is easily contaminated by fungi and bacteria (Ikotun, 1983, 1989) or subject to sprouting due to increased metabolic activity (Ugochukwu et al., 1977). Insects often infest the chips during the process of drying and also in storage (Okigbo and Nwakammah, 2005). In Africa, about 1 million metric tons are lost in storage due to attack by insects and nematodes, which facilitate invasion of rotting organisms (Emehute et al., 1998). Infestation of stored products by storage pests poses nuisance to consumers as it limits the utilization and the economic/market value of products. In order to overcome this high perishability and irregularity of its availability throughout the year, processing yam into chips and flour (Akissoe et al., 2001 Babajide et al., 2008) or by peeling, pre-cooking and sun-drying (Vernier et al., 2005) is an alternative to the consumption of fresh tubers. Ekundayo (1986) reported that the conversion of yam tubers to flour is recommended as a suitable and convenient method of storing the crop to prevent postharvest losses encountered during storage. Vernier et al. (2005) are however of the view that yam chips and flour are often severely attacked by insects and reducing both the qualitative and quantitative values in a few months. This scenario makes availability of information in respect of infestation of stored yam chips and flour extremely necessary. This study assesses the postharvest losses of yam chips and flour attributed to *Tribolium castaneum* in storage so as to generate baseline information on this pest as it affects the quantity and nutritional quality of the products so that an effective means of management can be devised.

MATERIALS AND METHODS

Insect rearing

Adults of *T. castaneum* were obtained from infested wheat flour purchased from a local market in Choba, Rivers State, Nigeria. Wheat flour was sterilized using heat sterilization in a laboratory oven (GNP-9082) at 60°C for 90 min. The adults were subsequently maintained on the sterilized wheat flour in a 1-L Kilner jar and kept in a laboratory under prevailing laboratory conditions (25 to 30°C and 70 to 90% R.H.). On the seventh day, the adults were sieved out and eggs laid were allowed to develop to F_1 progeny in order to obtain adult *T. castaneum* of uniform age (Zakka et al., 2010).

Experimental materials

Dry chips of four varieties of yam namely: Adaka, Ame, Nwopoko and Obiaturugo were obtained from National Root Crop Research Institute (NRCRI), Umudike; half of each variety was ground into flour using an electric blender (Philips HR-2815 model); while the other half was left in chip form for the experiment.

Sexing of adult *Tribolium castaneum*

Adults of *T. castaneum* were sexed using their morphological characteristics; the males have a small patch of short bristles (sex patches) on the inside of the first pair of legs (Beeman et al., 2012) or hairy punctures on the ventral surface of the anterior femur which is absent in the female species (Dobie et al., 1984).

Infestation procedure

Ten grammes of each of the two forms of the different yam varieties (Adaka, Ame, Nwopoko and Obiaturugo) were weighed using a digital balance and kept in 200 ml plastic jars. Eight pairs of the newly emerged adult *T. castaneum* were introduced into each of the jars and left undisturbed on a work bench. The experiment was carried out in a completely randomized design (CRD) in which treatments were replicated three times. At the end of the experiment, the content of each jar was poured onto a transparent plastic tray and the numbers of teneral adults were counted taking note of living and dead insects. The immatures (larvae and pupae) were sieved out and counted. Weight of the different yam varieties and forms was taken in batches at termination of the experiment using an electronic sensitive balance J2003 model and the difference in weight was recorded and percentage weight loss was calculated by taking the difference between the initial and final weights and dividing it by the initial weight. The result was expressed in percentage as

$$\text{Percentage weight loss (PWL)} = \frac{C - T}{T} \times 100$$

Where, C, Initial weight (g); T = final weight (g) (*Jackai* and Asante, 2003).

The susceptibility indices were calculated according to the method of Dobie (1974) in order to determine the most susceptible and resistant varieties. The susceptibility index is given as

$$SI = \frac{Log Y}{T} \times 100$$

Where, SI, Susceptibility index; Log Y, log number of F_1 emerged adults; T, mean developmental period (days), estimated as the time from the day of infestation, to 50% of adult emergence.

RESULTS AND DISCUSSION

Table 1 shows that although more larvae developed in

Table 1. Mean number of *Tribolium castaneum* progeny that developed in chips and flour of four yam varieties.

Substrate	No. of larvae	No. of pupae	No. of immatures (larvae and pupae)	No. of adults	Total progeny
Adaka chips	0.00^b	1.33	1.33	2.33	3.67
Adaka flour	0.67^{ab}	1.00	1.67	1.67	3.33
Ame chips	0.67^{ab}	1.00	1.67	1.67	3.33
Ame flour	0.33^{ab}	1.00	1.33	3.00	4.33
Nwopoko chips	0.00^b	0.33	0.33	2.33	2.67
Nwopoko flour	1.00^{ab}	0.67	1.67	1.33	3.00
Obiaturugo chips	1.67^a	0.33	2.00	1.67	3.67
Obiaturugo flour	1.00^{ab}	0.33	1.33	3.67	5.00
		NS	NS	NS	NS

*Means with the same letters within columns are not significantly different at 5% level of probability.

Table 2. Mean weight loss in yam chips and flour, developmental period and mortality of *T. castaneum* bred on four yam varieties.

Substrates	Percentage weight		Developmental period (Days)	% Mortality	Susceptibility index
	Loss	Gain			
Adaka chips	-	1.67^{bc}	17.50	4.33	3.15
Adaka flour	3.33^a	-	16.17	4.67	2.05
Ame chips	-	2.00^{bc}	16.67	4.67	1.90
Ame flour	1.67^a	-	18.83	3.67	2.53
Nwopoko chips	-	3.67^c	16.17	5.33	3.35
Nwopoko flour	1.67^a	-	17.33	5.00	0.72
Obiaturugo chips	-	1.00^b	18.83	3.67	1.18
Obiaturugo flour	2.33^a	-	18.83	3.33	3.00
			NS	NS	

*Means with the same letters within columns are not significantly different at 5% level of probability.

Obiaturugo chips, there was no significant difference (P>0.05) in the number of *T. castaneum* larvae that emerged in the overall except for the Adaka and Nwopoko chips where there was no record of larvae at the end of the experiment. It also shows that number of pupae that developed was not significant in both forms of substrates; however, Nwopoko chips and Obiaturugo flour recorded the least number of pupae. Similar trends were obtained for the total immature stages of *T. castaneum*. Higher numbers of adults developed in Obiaturugo flour followed by Ame flour substrates and the least in Nwopoko flour though in the overall there was no significant difference among the treatments; a similar trend was observed in the mean number of total progeny (Table 1).

Table 2 shows that mean percentage weight loss was highest in Adaka flour, though, it was not significantly different (P≥0.05) from that in flour of the other yam varieties. While yam flour in each variety lost weight at the end of the experiment their corresponding chips gained weight with Nwopoko chips having the highest weight gain followed by Ame chips. There was no

significant difference in *T. castaneum* developmental period in both forms of substrates though it took longer days on Obiaturugo flour and chips and the shortest developmental period was recorded in Ame chips. There was no significant difference in percentage mortality across the various substrates. However, highest mortality was recorded in Nwopoko chips and the least in Obiaturugo flour. Table 2 shows that both chips and flour were susceptible to infestation but to varying degrees, nonetheless, Nwopoko chips were the most susceptible to *T. castaneum* infestation and its flour was the least susceptible.

The physicochemical and functional properties of the yam varieties indicate higher oil absorption capacity in Adaka and the least in Nwopoko. Percentage fat was relatively higher in Ame variety and least in Adaka, though it recorded higher percentage dry matter, swelling index and water absorption capacity. Percent Na and P were highest in Ame and Adaka varieties, respectively; Adaka also recorded the least % Na but highest in P and N (Table 3).

Although under standard storage conditions yam chips

Table 3. Physicochemical and functional properties of yam products from four varieties.

Yam variety	OAC	%fat	% CF	% Ash	WAC	SI	GT°C	BD	% MC	% DM	% Na	% K	% P	% N
Ame A	1.1	0.56	0.16	1.0	1.4	1.13	65	0.85	13.00	87.6	0.58	0.23	0.88	0.28
B	1.2	0.55	0.16	1.1	1.2	1.18	68	0.92	12.75	87.5				
Adaka A	1.9	0.36	0.24	0.43	1.6	1.20	64	0.85	12.40	87.8	0.23	0.33	1.55	0.77
B	1.6	0.37	0.20	0.4	1.7	1.19	65	0.86	12.50	88.0				
Nwopoko A	0.9	0.38	0.25	0.9	1.4	1.14	63	0.83	12.59	87.4	0.40	0.28	0.83	0.43
B	1.0	0.39	0.26	0.8	1.2	1.22	67	0.81	18.05	82.0				
Obiaturugu A	1.2	0.42	0.28	1.1	1.6	1.10	68	0.88	12.18	87.0	0.40	0.18	1.10	0.63
B	1.3	0.42	0.25	1.0	1.5	1.17	64	0.88	11.96	87.3				

OAC, Oil absorption capacity; WAC, water absorption capacity; DM, dry matter; BD, bulk density; SI, swelling index; MC, moisture content, GT, gelation temperature.

can be kept for up to two years without pest problem (Vernier, 2005), in this study both products (chips and flour) of the four yam varieties were susceptible to *T. castaneum* infestation. More *T. castaneum* progeny developed on average in flour than in chips. This agrees with the report of Loko et al. (2013) on farmers' knowledge of the insect pest damaging yam chip stocks and diversity assessment that beetles act by penetrating the chips and drastically reducing their parts into powdery waste. This confirms the biology of *T. castaneum,* a secondary pest that has preference for fine flour for development as compared to chips (Odeyemi, 2001; Lale 2002). Yam chips and flour under rural conditions or during trading are often severely attacked by insects (Vernier et al., 2005). It was also observed in Nwopoko that the number of immature stages (larvae and pupae) that developed in the substrates was more in the flour.

With the exception of Adaka yam variety, developmental period of *T. castaneum* in flour was longer in chips. This is a deviation from the normal biology of *T. castaneum*, which is known to be more fecund and to develop faster on flour than on solid substrates or broken crop products (Lale and Yusuf, 2001; Lale and Modu, 2003; Haines, 1991); *T. castaneum*, being a secondary pest ought to perform poorly or slowly on chips in storage. This could be attributed to the relatively high amounts of antibiotic compounds being made more available in the flour which might have negatively affected the biology of *T. castaneum*. The presence of phenolic compounds and saponins (secondary metabolites) in yams as an antibiotic against developing stages of *T. castaneum* was earlier reported by Degras (1993) and IITA (1995). Although all the chips were infested by *T. castaneum*, this form of yam substrate, increased in weight. The increase in weight might be attributed to the higher functional properties such as water absorption capacity (WAC) and swelling index (SI) which are known for having high propensity to increase the hygroscopic property of the substrates, causing it to draw more moisture from its surrounding environment and probably causing desiccation in the insects hence increasing the mortality rate of the insect.

The lower values of susceptibility index recorded across the various forms and types of all the yam varieties used might be attributed to the fact although the improved varieties may have been bred for other traits, they also showed relative resistance to *T. castaneum* infestation. Similar results were obtained by Ashamo (2002) who worked on improved maize varieties and attributed it to inherent traits such as resistance. In another study, Siwale et al. (2009) reported that the lower susceptibility indices they obtained were partly due to low moisture content of the substrates. Moisture content of the yam products was not determined in this study, the overall result suggests that this physical attribute may not have significantly influenced the outcome of the investigation.

The findings of this study have proved that *T. castaneum* is a cosmopolitan species and can colonize a wide range of substrates including different forms and types of yam cultivated in the Niger delta region of Nigeria. This insect is capable of posing a serious threat to yam in storage and consequently to food security, this is particularly significant because yam is one of the major staple foods in Nigeria. It is evident that steps must be taken to protect yams and storage houses from infestation by *T. castaneum* in yam growing areas of the tropics and subtropics.

REFERENCES

Akissoe N, Hounhouigan DJ, Bricas N, Vernier P. Nago CM, Olorunda AO (2001). Physical, chemical and sensory evaluation of dried yam (Dioscorea rotundata) tubers, flour and amala, a flour-derived product. Trop. Sci. 41(3):151-155.

Ashamo MO (2002). Relative performance of *T. castaneum* (Herbst) in different flour(s). Appl. Trop. Agric. 7:46-49.

Babaleye T (2003). West Africa; improving yam production technology ANA-BIA supplement Issue/Edition P. 463.

Babajide JM, Bello OQ, Babajide SO (2010). Quantitative dtermination of active substances (preservative) in *Piliostigma thonningii* and *Khaya ivorensis* leaves and subsequent transfer in dry yam. Afri. J. Food Sci. 4:(6) 382-388.

Babajide JM, Atanda OQ, Ibrahim TA, Majolagbe HO, Akinbayode SA (2008). Quantitative effect of 'abafe' (*Piliostigma thonningii*) and 'agehu' (*Khaya ivorensis*) leaves on the Microbial load of dry-yam 'gbodo'

Afri. J. Microbiol. Res. 2:292-298.

Baco MN, Tostain S, Mongbo RL, Dainou O, Agbangla C (2004). Gestation dynamique de la diversite varietale des ignames cultivees (*Dioscorea cayenesis – D. rotundata*) dans la commune de Sinende au nord Benin. (English Trans.) Plant. Genet. Resour. Newsletter 139:18-24.

Beeman RW, Haas S, Friesen K (2012). Beetle Wrangling Tips. An Introduction to the Care and Handling of *Tribolium castaneum*). Unites States Department of Agriculture (USDA),Agricultural Research Service (ARS). Retrieved from: http://www.ars.usda.gov/Research/docs.htm?docid=12892. Assessed, November, 2012.

Degras LM (1993). The Yam: A Tropical Root Crop. The Technical Centre for Agricultural and Rural Cooperation (CTA). *The Macmillan Press Ltd*, London, England. P. 408.

Dobie P (1974). The contribution of the tropical stored products centre to the study of insect resistance in stored maize. Trop. Stored prod. Inform. 34:7–22.

Dobie P, Haines CP, Hodges RJ, Prevett PF (1984). Insect and Arachnids of Tropical Stored Products: Their Biology and Identification. Trop. Develop. Res. Institute. P. 237.

Ekundayo CA (1986). Biochemical changes caused by mycoflora of yam sclices during sun drying. Microbios Letters 32:13-18.

Emehute JKU, Itokun T, Nwauzor EC, Nwokocha HN (1998). Crop protection In: Orkwor GC, Asiedu R and Ekanayake IJ (Editors). Food yams: Advances in research at IITA and NRCRI Nigeria. pp. 141-186.

FAO (2003) Website: www.faosat database Food and Agricultural Organization Rome, Italy, 2003 http://www.fao.org/

FAO (2010) Website: www.faosat database, Food and Agricultural Organization Rome, Italy, 2010 http://www.fao.org/

Haines CP (1991). Insects and Arachnids of Tropical Stored Products: Their biology and identification (a training manual). *National Resource Institute* (NRI), Chatham, Maritime, U.K.

Ikotun T (1983). Post-harvest microbial rot of yams in Nigeria. Fitopatologia Brasiliera 8(1):1-7.

Ikotun T (1989). Diseases of yam tubers. Int. J. Trop. Dis. 7(1):1-21.

International Institute of Tropical Agriculture (IITA) (1995). Annual report, Ibadan, Nigeria.

Izekor OB, Olumese MI (2010). Determinants of yam production and profitability in Edo State, Nigeria. Afri. J. General Agric.. 6(4):205-210

Jackai LEN, Asante SK (2003). A case for standardization of protocols used in screening cowpea for resistance to *Callosobruchus maculatus* F. (Coleoptera: Bruchidae) J. Stored Prod. Res. 39:251-263.

Lale NES (2002). Stored Product Entomology and Acarology in Tropical Africa. First Edition Mole Publications (Nig.) Maiduguri.

Lale NES, Yusuf BA (2001). Potential of varietal resistance and *Piper guineense* seed oil to control infestation of stored millet seeds and processed products by *Tribolium castaneum* (Herbst). J. Stored Prod. Res. 37:63–75.

Lale NES, Modu B (2003). Susceptibility of seeds and flour from local wheat cultivars to *Tribolium castaneum* infestation in storage. Trop. Sci. 43(4):174-177.

Loko YL, Dansi A, Tamo M, Bokonon-Ganta AH, Assogba P, Dansi M, Vodouhe R, Akoegninou A, Sanni A (2013). Storage insects on yam chips and their traditional management in Northern Benin. The Scientific World J. 484536:11.

Obadofin A., Joda AO, Oluitan JA (2013). Market survey of insect infestation of dried root and tuber products across Nigeria-Benin land border. Int. J. Appl. Sci. Technol. 3(5):72-77.

Odeyemi OO (2001). Biology, Ecology and Control Of Insect Pests Of Stored Processed Cereals And Pulses. In: Ofuya TI and Lale NES (Editors) Pests of Stored Cereals and Pulses in Nigeria. Biology, Ecology and Control. Dave Collins Publishers, Nigeria, P. 171.

Okigbo RN, Nwakammah PT (2005). Biodegradation of white yam (*Dioscorea rotundata* Poir) and Water yam (*D. alata* L.) slices dried under different conditions *KMITL* Sci. Technol. J. 5(3):577-586.

Rajendran S (2005). Detection of insect infestation in stored foods. Advan. Food. Nutr. Res. 49:165-232.

Siwale J Mbata K, Microbert J and Mareck JH (2009). Comparative resistance of improved maize genotypes and landraces to maize weevil. Afri. Crop Sci. J. 17(1):1-16.

Ugochukwu EN, Anosike EO, Agogbua IO (1977). Changes in enzymes activity of white yam tuber after prolonged storage. Phytochemistry 16:1159-1162.

Vernier P, Georgen G, Dossou R, Letourmy P, Chaume J (2005). Utilization of biological insecticides for the protection of stored yam chips. Outlook on Agriculture 34(3):173-179.

Zakka U, Lale NES, Ehisianya CN. (2010). Infestation of two products from processed dry tubers of sweet potato by the rust-red flour beetle *Tribolium castaneum* (Herbst) (Coleoptera: Tenebrionidae) in a humid environment. Nig. J. Plant Protect. 24:48-53.

20

Genetic divergence and cluster analysis studies of different apple genotypes using D^2 statistics

Girish Sharma[1], Nirmal Sharma[2], Rubiqua Bashir[1], Amit Khokhar[2] and Mahital Jamwal[2]

[1]Department of Fruit Breeding and Genetic Resourses, Dr. YSP UHF, Solan, HP, India.
[2]Division of Fruit Science, SKUAST-J, Jammu and Kashmir, India.

Medium quantum of genetic divergence was observed among sixteen apple genotypes under the present study. All the genotypes, on the basis of total variability were grouped into four distinct clusters. Maximum number of cultivars were accommodated in Cluster IV (Fuji, Gala, Jonadel, Jonagold, Red Fuji, Royal Gala and Spijon) followed by Cluster I (Arlet, Ruspippin, Sinta and Summerred), Cluster III (Crimson Gold, Elstar and Neomi) and Cluster II ('Spartan' and 'Quinte'). Cluster IV had highest intra cluster value so was most divergent and Cluster I having least intra cluster value was least divergent. Highest value for inter cluster distance was recorded between Cluster I and II while it was lowest between Cluster III and IV. Cluster means were maximum in Cluster II followed by Clusters I, III and IV. Neomi is best cultivars for fruit yield/plant, fruit length, fruit diameter, fruit weight, total sugars and non-reducing sugars. However, Jonagold is best for TSS. Cultivars Spartan, Elstar, Royal Gala, Jonagold and Summerred would prove best for different vegetative characters.

Key words: Apple, cluster analysis, D^2 statistics, genetic divergence.

INTRODUCTION

The cultivated apple (*Malus × domestica* Borkh.) is a member of family rosaceae and sub family pomoideae, have originated in south western Asia, Asia Minor, the Caucasus mountains of Russia, central Asia and the Himalayan region of India and Pakistan (Juniper et al., 1999). It is an important temperate fruit crop of India with respect to acreage, production, economic value and above all popularity among the consumers. In India it is a prime commercial fruit crop of Himachal Pradesh, Jammu and Kashmir and Uttranchal and some parts of north eastern states including Arunachal Pradesh, Sikkim, Nagaland, Meghalaya and Nilgiri hills of Tamil Nadu (Awasthi and Chauhan, 2001). Apple productivity has

gradually declined since 1975 till date and on the basis of low production and productivity, India is now ranked 10th in the world apple cultivation scenario (Sardana, 2012). The important factors which are responsible for low productivity are age old varieties, inappropriate sites, irregular bearing, poor soil conditions, lack of suitable adaptable cultivars and poor selection of pollinizers and their inadequate proportion. Therefore urgent need is felt for development/introduction of new improved varieties which could help in elevating the apple productivity in India. For the success of any breeding programme the basic requirement is the variability found within the members of the population. It is this variation which if heritable could be used for cultivar improvement, as improved cultivars are the backbone of any orchard system. Therefore, prior to initiation of any breeding programme they should be tested and extent of variability present must be adequately assessed so that the breeding programme could yield the desired results. To use or exploit the available variability present in the genetic material in the form of some specific groups or classes, the divergence studies based upon some desirable/suitable parameters is of very essential and of utmost significance. Keeping in view the above the genetic divergence and cluster analysis using D^2 statistics was undertaken with the objectives to assess the variability present among the sixteen apple genotypes and potential use of this variability for hybridization programmes. Use of Mahalanobis D^2 statistics to estimate or evaluate the net/total divergence in breeding for crop improvement has been indicated by number of workers in different fruit crops (Saran et al., 2007). The use of genetically divergent parents in hybridization under transgressive breeding programme is dependent upon categorization of breeding material on the basis of appropriate criteria (Santos et al., 2011). Apart from providing requisite assistance or help in selection of divergent parents in hybridization, D^2 statistics also adequately assists in the measurement of diversification and the contribution of the relative proportion of each component trait towards the total genetic divergence or variation.

MATERIALS AND METHODS

Studies were carried out in the Department of Fruit Breeding and Genetic Resources, Dr Y.S. Parmar University of Horticulture and Forestry, Nauni, Solan, Himachal Pradesh, India on sixteen apple genotypes viz. Arlet, Crimson Gold, Elstar, Fuji, Gala, Jonadel, Jonagold, Neomi, Quinte, Red Fuji, Royal Gala, Ruspippin, Sinta, Spartan, Spijon and Summered. Plant height and spread (North-South and East-West) were measured with the help of measuring pole and were expressed in meters. The trunk girth was measured at a height of 9 cm above graft union with the help of measuring tape and was expressed in centimeters. Shoot length was measured by selecting twenty uniform shoots on periphery of each tree and recording the length of shoots with the help of measuring tape and expressed in centimeters. Internodal length was worked out by dividing the shoot length with number of nodes. Spur frequency was recorded by selecting ten uniform branches in each tree of 1 m length more than one year old and number of spurs were counted and mean was worked out dividing number of spurs with length of branch taken multiplied by 100. Number of days from the date of opening of first flower to the date of opening of last flower was taken as the duration of flowering. Fruits retained in all the genotypes were recorded one week before harvesting of fruits and expressed in percentage by dividing number of fruit retained with total number of flowers multiplied by 100. The dates on which fruits were harvested was recorded as date of harvesting. The crop load of apples harvested from each plant was recorded and the results were expressed in yield per plant in kilograms. The length and diameters of ten fruits was measured with the help of digital vernier calliper and mean was worked out and expressed in centimeters. The fruit weight was worked by weighing ten fruits

selected randomly from each tree and weighed on a single pan kitchen balance and mean was expressed in grams (g). Flesh firmness of fruit was measured after removing the skin (0.8 cm) and using effigy penetrometer (model FT 327) with plunger of 11 mm dia. The results were expressed in kg/cm². Specific gravity was measured by dividing the weight of the fruits by their volume. In counting number of seeds per fruit chaffy and shrivelled seeds were discarded. Total soluble solid contents of five uniformly ripened fruits of each tree were determined with an Erma hand refractometer (0 to 32°Brix) by placing few drops of juice on the prism and reading was taken. Titratable acidity, total sugars, reducing sugars and non -reducing sugars were determined as per the method suggested by AOAC (1990). Mahalanobis D2 statistic was used for assessing the genotypic divergence between populations (Mahalanobis, 1936). The generalized distance between any two populations is given by formula:

$$D^2 = \Sigma\Sigma\lambda_{ij}\ \sigma_{ai}\sigma_{aj}$$

Where, D^2 = Square o generalized distance; λ_{ij} = Reciprocal of the common dispersal matrix; σ_{ai} = (μ_{i1} - μ_{i2}); σ_{aj} = (μ_{j1} - μ_{j2}); μ = General mean.

Since, the formula for computation requires inversion of higher order determinant, transformation of the original correlated unstandardized character mean (Xs) to standardized uncorrelated variable (Ys) was done to simplify the computational procedure. The D^2 values were obtained as the sum of squares of the differences between pairs of corresponding uncorrelated (gs) values of any two uncorrelated genotype of D^2 value. All n (n-1) / 2 D2 value were clustered using Toucher's method described by Rao (1952). The intra cluster distances were calculated by the formula given by Singh and Choudhary (1997):

Square of the intra cluster distance = ΣD^2_i/n

Where, ΣD^2_i is the sum of distance between all possible combinations of the entries included in a cluster and n is number of all possible combinations.
    The inter cluster distances were calculated by the formula described by Singh and Choudhary (1997):

Square of the intra cluster distance = ΣD^2_i/n_in_j

Where, ΣD^2_i is the sum of distances between all possible combinations (n_in_j) of the entries included in the clusters under study. n_i is number of entries in Cluster I and n_j is number of entries in cluster j. The criterion used in clustering by this method was that any two genotypes belonging to the same cluster, at least on an average, show a small D^2 value than those belonging to two different clusters.

RESULTS AND DISCUSSION

The clustering pattern of sixteen cultivars of apple on tree, shoot, flowering and fruit characters are presented in Table 1. The genetic divergence in the present study observed among the sixteen cultivars is of medium quantum. The sixteen cultivars on the basis of net variability were grouped into four distinct clusters. Maximum number of cultivars (7) were accommodated into Cluster IV (Fuji, Gala, Jonadel, Jonagold, Red Fuji, Royal Gala and Spijon) while the minimum number (2) were in Cluster II which included Spartan and Quinte.

Table 1. Clustering pattern of sixteen cultivars of apple on the basis of genetic divergence.

Cluster number	Number of cultivars	Cultivars included
I	4	Arlet, Ruspippin, Sinta and Summered
II	2	Spartan and Quinte
III	3	Crimson Gold, Elstar and Neomi
IV	7	Fuji, Gala, Jonadel, Jonagold, Red Fuji, Royal Gala and Spijon

Table 2. Intra and inter cluster distance (D^2).

Clusters	I	II	III	IV
I	8.201	30.331	16.428	14.598
II		8.410	24.595	21.102
III			9.156	9.994
IV				9.321

Inter and intra cluster divergence values (D^2) between and within four clusters are presented in the Table 2. The intra cluster distance was maximum (9.32) for Cluster IV and minimum (8.20) for Cluster I. Highest value (30.331) for inter cluster distance was recorded between Cluster I and II while it was lowest (9.994) between Cluster III and IV. On the basis of results it is inferred that subsequent hybridization between the genotypes having broad genetic base should result in maximum heterotic performance and eventually the desirable transgressive recombinants, as broad genetic base is a fundamental requirement for any crop improvement programme. The Cluster IV accommodating cultivars Fuji, Gala, Jonadel, Jonagold, Red Fuji, Royal Gala and Spijon were more divergent, followed by Cluster III having three cultivars namely Crimson Gold, Elstar and Neomi. Wide diversity in the progeny is expected when hybridization is attempted within cultivars which are more divergent. Since inter cluster distance is maximum (30.331) between Clusters I and II so maximum variability will be achieved when hybridization between the cultivars accommodating these clusters is attempted. The cluster means of the various tree, shoot, flowering and fruiting characters are presented in Table 3. The average cluster means revealed highest values for the characters like fruit weight (8.81 gm), fruit set after 50 days (64.66) fruit retention (49.88%), TSS (11.79°B) in Cluster I. The Cluster II had better mean performance for the traits like fruit yield per plant (39.08 kg), duration of flowering (19.25 days), flesh firmness (12.34 kg/cm^2), fruit diameter (6.59 cm), fruit length (5.74 cm) and plant spread (4.92 m). In similar way, Cluster III revealed superior mean performance for the characters like trunk girth (46.76 cm), followed by spur frequency (12.66%), total sugars (7.62%), reducing sugars (6.78%). The characters like shoot length (9.67 cm), plant height (4.95 m), plant

spread (3.89 m) had higher values in Cluster IV. The character fruit set after 50 days (64.66) in Cluster I, yield per plant (39.78Kg) in Cluster II, trunk girth (46.76 cm) in Cluster III and shoot length (19.47 cm) in Cluster IV showed the highest values. Pereira et al. (2003) clustered apple genotypes into different groups on the basis of traits like internodal length, spur frequency, spur coefficient, number of long shoot. In the present study the above said characters revealed variation but in a narrow sense, probably the cultivars in the present study were not much divergent in respect of these traits. Saran et al. (2007) grouped 35 ber varieties into 7 clusters using D^2 statistics. Linoaiah et al. (1998) after canonical variate analysis in apple found that plant height, stem girth, flowering shoot per square meter and percentage flowering per square meter contributed much towards the genetic diversity. These characters though not very significant in our study still they could be effectively exploited in future crop improvement. While studying the clonal variability in mango Manchekar et al. (2011) reported substantial variation after applying D^2 statistics. Hence it is concluded that genotypes of cultivars with wide genetic variation accompanied with useful characteristics could be effectively employed in intra specific crosses with the hope that this would lead to the transmission of higher genetic gain for different putative traits major being yield from practical utility point of view. On the basis of the performance of different cultivars and the cluster analysis, the sixteen apple cultivars have been identified for different characters (Table 4), which are potential parents for hybridization programmes. Neomi is best cultivars for fruit yield/plant, fruit length, fruit diameter, fruit weight, total sugars and non-reducing sugars. However, Jonagold is best for TSS. Cultivars Spartan, Elstar, Royal Gala, Jonagold and Summerred would prove best for different vegetative characters.

Table 3. Cluster means of sixteen apple cultivars.

Character	Clusters			
	I	II	II	IV
Plant height (m)	4.06	4.76	4.61	4.95
Plant spread (NS)	2.96	3.76	3.76	3.89
Plant spread (EW)	4.06	4.92	4.26	4.67
Trunk girth (m)	31.46	41.96	46.76	42.33
Shoot length (cm)	17.71	16.22	17.08	19.47
Internodal length (cm)	2.88	3.06	2.82	2.94
Spur frequency (%)	12.04	12.25	12.66	11.66
Duration of flowering (days)	17.54	19.25	18.56	17.83
Fruit set after 50 days (%)	64.46	44.50	60.00	59.33
Fruit retention (%)	49.88	32.25	46.94	46.52
Fruit yield per plant (Kg)	29.47	39.08	28.13	36.59
Fruit length (cm)	5.73	5.74	5.38	5.68
Fruit diameter (cm)	6.36	6.59	6.17	6.28
Fruit weight (g)	8.81	8.46	8.48	8.68
Flesh firmness (kg/cm^2)	11.56	12.34	11.16	11.61
TSS (^{o}B)	11.79	11.34	11.40	11.57
Acidity (%)	0.69	0.72	0.60	0.64
Total sugar (%)	7.20	7.59	7.62	7.48
Reducing sugar (%)	6.40	6.50	6.78	6.43
Number of seeds per fruit	6.96	7.00	5.55	5.83

Table 4. Promising cultivars of apple for different characters.

Characters	Highest cultivars	Lowest cultivars	Promising cultivar at par with highest
Plant height (m)	Spartan	Fuji	Royal Gala, Jonadel, Neomi
Trunk girth (cm)	Elstar	Summered	Fuji, Royal Gala
Plant spread NS (m)	Royal Gala	Neomi	Spartan
Plant spread EW (m)	Jonadel	Summered	Royal Gala, Elstar
Shoot length (cm)	Jonagold	Quinte	Arlet, Gala
Internodal length (cm)	Summered	Elstar	Crimson Gold
Spur frequency (%)	Fuji	Jonagold	-
Duration of flowering	Neomi, Ruspippin	Crimson Gold	-
Fruit retention (O.P.)	Ruspippin	Spartan	Gala and Summered
Date of harvest			
Fruit yield/plant (Kg)			
Fruit length (cm)	Neomi	Crimson Gold	-
Fruit diameter (cm)	Neomi	Crimson Gold	-
Fruit weight (g)	Neomi	Crimson Gold	-
Flesh firmness (kg/cm^2)	Red Fuji	Fuji	Arlet, Elstar
Number of seeds/fruit	Ruspippin	Sinte	Fuji, Spartan, Spijon and Summered
Total soluble solid (^{o}B)	Jonadel	Quinte	
Fruit acidity (%)	Spijon	Neomi	Crimson Gold
Total sugar (%)	Neomi	Crimson Gold	
Reducing sugar (%)	Neomi	Spijon	

REFERENCES

AOAC (1990). Official Methods of Analysis. 15th edition. Association of official Agricultural chemists, Washington, D.C.

Awasthi RP, Chauhan PS (2001). Apple. In: *Handbook of Horticulture.* K. L. Chadha, (ed.). Indian Council of Agricultural Resaerch, New Delhi, pp. 119-131.

Juniper BE, Watkin R, Harries SA (1999). The origin of apple. Acta Hort.

484:27-30.

Linoaiah HB, Reddy NS, Kulkarni RS, Thomas KK (1998). Genetic divergence in cashew (*Anacardium occidentale* L.) genotypes. In: Development in plantation crop research.

Mahalanobis PC (1936). The generalized distance in statistics. Proc. Nat. Acad. Sci. (India). 2:79-85.

Manchekar MD, Mokashi AN, Hegde RV, Venugopal CK, Byadgi AS (2011). Clonal variability studies in alphonso mango (*Mangifera indica* L.) by genetic divergence (D^2) analysis. Karnataka J. Agric. Sci. 24(4):490-492.

Pereira LS, Ramos CAM. Ascasibar EJ and Pineiro AJ (2003). Analysis of apple germplasm in North Western Spain. J. Amer. Soc. for Hort. Sci. 120(1):67-84.

Rao CR (1952). Advanced statistical methods in biometrical research. John Wiley and Sons, Inc. New York. pp. 357-363.

Santos CAF, Corrêa LC, Costa SR (2011). Genetic divergence among *Psidium* accessions based on biochemical and agronomic variables. Crop Breed. App. Biotech. 11:149-156.

Saran PL, Godara AK, Dalal RP (2007). Biodiversity among indian jujube (*Ziziphus mauritiana* Lamk.) Genotypes for powdery mildew and other traits. Not. Bot. Hort. Agrobot. Cluj. 35(2):15-22.

Sardana MMK (2012). Towards increasing productivity and improving post harvest management in apple cultivation in Himachal. http://isidev.nic.in/pdf/DN1003.

Singh PK, Choudhary RD (1997). Biometrical Methods. In. Quantitative Genetic Analysis, Kalayani Publishers, New Delhi, pp. 178-185.

21

Importance of designer eggs for the Nigerian population

Dike I. P.

Department of Biological Sciences, Covenant University, Ota, Ogun State, Nigeria.

Epidemiological studies have led to recommendations that people should consume at least two servings of fruit and three servings of vegetable daily for healthy living but majority of Nigerians falls well short of meeting these guidelines. Hence, there is a need to make up for this shortage. Through studies, it has been established that in addition to being a natural functional food, the egg's nutrient content can be altered by designing the feed given to the chickens thus resulting in the production of designer eggs. A crucial feature of these designer eggs is the synergistic combination of healthy Omega-3 fatty acids with major antioxidants, Vitamin E and lutein, as an important approach to the improvement of the human diet. These eggs will not be able to replace vegetable and fruits as a major source of natural antioxidants and fish products but can substantially improve the diet, especially in a country like Nigeria, significantly contributing to the recommended daily intake of essential nutrients. Thus, this study reviews the importance of designer eggs in the Nigerian context.

Key words: Vitamin E, designer egg, omega-3 fatty acids, antioxidants, lutein, Nigeria.

INTRODUCTION

Eggs have been described as "Nature's original functional food" (Hasler, 2000) packed with various important vitamins and minerals. Eggs are said to contain the highest quality protein, when compared to other animal protein sources and they are inexpensive when compare to other protein sources.

Chicken's eggs have been used as a food by human beings since antiquity. Compared with the hen's egg, no other single food of animal origin is eaten by so many people all over the world and none is served in such a variety of ways. Its popularity is justified not only because it is so easily produced and has so many uses in cookery, but also because of its nutritive excellence.

The nutrient value of one egg have been tabulated in Table 1 and per the World Health Organization (WHO) recommendation, 2 eggs per day are required for optimum growth.

Eggs contain a number of beneficial nutrients, some of which have functions that are currently being studied. Egg yolks provide an excellent, highly bio-available source of the carotenoids, lutein and zeaxanthin (Handelman et al., 1999). Recent research demonstrated the link between these dietary compounds and the macular pigment of the retina of the eye (Landrum and Bone, 2001). Lutein and zeaxanthin are the primary carotenoids found in the macular region. Sufficient quantities of these nutrients in the diet are thought to reduce the risk of age-related macular degeneration, a

Table 1. Nutritional value of one large egg.

S/N	Nutrient	Amount
1	Calories	70 kcal
2	Total fat	4.5 g
3	Saturated fat	1.5 g
4	Polyunsaturated fat	0.5 g
5	Monounsaturated fat	2.0 g
6	Cholesterol	213 mg
7	Sodium	65 mg
8	Potassium	60 mg
9	Total carbohydrate	1 g
10	Protein	6 g

Source: Hargis (1988).

leading cause of blindness in the elderly.

In addition to possibly reducing the risk of macular degeneration, lutein has been associated with a protective effect for early atherosclerosis. Dwyer et al. (2001) reported that increased amounts of dietary lutein from green leafy vegetables and egg yolks could be protective against atherosclerosis by slowing the progression of atherosclerotic lesions in humans and animals. Early arteriosclerosis was inversely related to levels of plasma lutein which were affected by dietary intake indicating an inverse relationship between dietary lutein and arteriosclerosis development.

Choline is a nutrient naturally found in eggs that has been identified as contributing to fetal memory and brain development. Choline is found in the form of phosphatidylcholine and sphingomyelin, which are types of phospholipids. Choline's chief function in the body is as an important part of cellular compounds such as the neurotransmitter acetylcholine and lecithin, a naturally occurring emulsifier present in cell membranes and bile. One large egg contains approximately 300 mg choline. Eggs are good sources of choline since the recommended daily intakes range from 425 to 550 mg for adults, including pregnant and lactating women, according to the National Academy of Sciences (Yu and Sim, 1987).

Eggs naturally contain essential and functional nutrients to promote health. In addition, the nutrient content of eggs can be modified to provide nutrients above and beyond what is normally found in generic shell eggs.

Surprisingly, even with the clear picture of the advantages of eggs and its low cost when compared to other protein sources like meat, fish and other animal proteins, the consumption of eggs per day has been found to be on the decline in recent years.

DESIGNER EGGS

Designer eggs are eggs produced when hen is fed with special feed prepared to suite the nutrients one desires to be present in the egg produced. The benefits of the designer eggs are manifold; the one of highest order is the benefit it offers in terms of the fatty acid content.

Fatty acid content

Genetic selection of hens for lowered cholesterol has not been successful in lowering the egg cholesterol content. Thus, research into lowering egg cholesterol has centered mostly on diet and pharmacological intervention (drugs). Drugs have been successful in lowering egg cholesterol by as much as 50% (Sim et al., 1973). Drugs lower cholesterol in the egg by either inhibiting the synthesis of cholesterol in the hen or by inhibiting the transfer of cholesterol from the blood to the developing yolk on the ovary. But, the drugs which have shown promise in lowering cholesterol are not yet approved by the Food and Drug Administration (FDA) for commercial use.

Research has also shown that the most effective way to lower egg cholesterol content is alter the diet of the hen. Thus, introducing the concept of designer eggs, a designer egg is and egg laid by chicken feed with a special diet of feed.

Clinical and epidemiology research has proved that the consumption of small quantities of Omega-3 fatty acids (0.5 g/day) over a long period of time decreases the coronary heart disease mortality rate (Sim, 1990).

Altering the total fat content in the diet of the hen has little effect on the total fat content of the egg yolk. However, the fatty acid profile (or the ratios of the different types of fatty acids) of egg yolk lipid can easily be changed, simply by changing the type of fat used in the diet.

Consumption of polyunsaturated fatty acids has been reported to reduce the risk of atherosclerosis and stroke. Consumption of these fatty acids has also been shown to promote infant growth. Different feeds, such as flaxseed (linseed) (Caston and Leeson, 1990; Jiang et al., 1992; Nowokolo and Sim, 1989; Sim, 1990), safflower oil, perilla oils (Shrimpton, 1987), marine algae (Hargis, 1988) fish, fish oil (Hargis et al., 1991; Yu and Sim, 1987) and vegetable oil have been added to chicken feeds to increase the Omega-3 fatty acid content in the egg yolk.

The use of Omega-3 fatty acid rich eggs, showed a reduction in plasma and liver total cholesterol produced by 20 and 40%, respectively (Sim et al., 1973).

Omega-3 fatty acid-rich eggs may provide an alternative food source for enhancing intake of these 'healthy' fatty acids. Studies confirmed that designer eggs rich in Omega-3 fatty acids though rich in cholesterol do not provoke plasma cholesterol production instead suppress the low-density lipoprotein (LDL)-cholesterol production and increase the high-density lipoprotein (HDL)-cholesterol production (Sim et al., 1973).

Studies of the eggs during storage indicated that the

shelf life of the enriched eggs was comparable to that of typical eggs (Sim et al., 1973).

Many Omega-3 fatty acid-enhanced eggs are available in the U.S. market under various brand names such as Gold Circle Farms, Egg Plus, and the Country Hen Better Eggs (Sim, 1990).

Omega-3 fatty acid-enriched eggs taste and cook like any other chicken eggs available in the grocery store. However, they typically have a darker yellow yolk.

Additional benefits

Mineral content

There has been very little success in changing the calcium and phosphorus content of the albumen and yolk. It is possible, however, to increase the content of selenium, iodine and chromium (Yu and Sim, 1987). This has been done through dietary supplementation of the hen. These three minerals are important in human health.

Vitamin content

Designer eggs that have been produced contain higher concentrations of several vitamins. Two vitamins, A and E, are receiving the most interest as components of designer eggs. The vitamin content of the egg is variable and is dependent on the dietary concentration of any specific vitamin.

Pigment content

The colour of the yolk is a reflection of its pigment content. In addition, the type of pigment in the egg and its concentration are directly influenced by the dietary concentration of any particular pigment. Yolk colors can be achieved by using only natural pigments obtained from natural raw materials. Natural sources can be from plants such as marigold, chili, or corn. The high protein blue-green algae known as Spirulina has also been shown to be a very efficient pigment source for poultry skin and egg yolk (Landrum and Bone, 2001).

Recent research has shown that eggs may be beneficial in preventing macular degeneration; a major cause of blindness in the elderly. A recent study indicated that higher intake of carotenoids reduced the risk of age-related macular degeneration. The most effective carotenoids were lutein and zeaxanthin, which are commonly found in dark-green leafy vegetables, such as spinach and collard greens. Most of the carotenoids in egg yolk are hydroxyl compounds called xanthophylls. Lutein and zeaxanthin are two of the most common xanthophylls found in egg yolk. Lutein and zeaxanthin are high in pigmented feed ingredients such as yellow corn,

alfalfa meal, corn gluten meal, dried algae meal, and marigold-petal meal (Landrum and Bone, 2001).

Fortunately, both lutein and zeaxanthin are efficiently transferred to the yolk when these various feed ingredients are fed to laying hens. With a growing problem of macular degeneration in the elderly, the egg industry may want to seize this opportunity to produce lutein and zeaxanthin rich eggs.

WHY INTRODUCE DESIGNER EGGS IN NIGERIA

Traditionally, food products have been developed for taste, appearance, value, and convenience for the consumer. Epidemiological findings, supported by animal studies, have led to recommendations that people should consume at least two servings of fruit (like apples, grapes, bananas etc.) and three servings of vegetable (carrots, green peas, cabbage, tomatoes etc.) daily. Majority of Nigerians falls well short of meeting these guidelines. The population face severe issues with malnutrition most of which goes unnoticed until complications arise and thus can benefit greatly from the consumption of eggs.

Recent reports of increase in cardiac problems among the Nigeria population have be documented, indicating an increase in cholesterol in the population this may be due to the high animal fat consumption in the country. The Nigeria populations are inclined to the consumption of animal products which consists of more than 60% of total lipids, 70% saturated fats and 100% cholesterol. A large egg contains approximately 200 to 220 mg of cholesterol (Simopolous, 1991). Thus, it is of importance to look into means of combating the resultant heart diseases and cardiac problems faced by the population. Thus it may be of interest for Nigeria to tap into the benefits offered by designer eggs.

Consumption of polyunsaturated fatty acids has been reported to reduce the risk of atherosclerosis and stroke. Consumption of these fatty acids has also been shown to promote infant growth (Simopolous, 1991). For a long time the only dietary source of Omega-3 fatty acids was from fish and studies have shown that the consumption of these fatty acids protect against cardiovascular and inflammatory diseases as well as certain kinds of cancers (Beare-Rogers, 1991; Simopolous, 1991).

Other benefits include reduction in plasma triglycerides, blood pressure, platelet aggregation, thrombosis and atherosclerosis (Simopolous, 1991).

Different feeds, such as flaxseed (linseed) (Caston and Leeson, 1990; Jiang et al., 1992; Nowokolo and Sim, 1989; Sim, 1990), safflower oil, perilla oils (Shrimpton, 1987), marine algae (Hargis, 1988) fish, fish oil (Hargis et al., 1991; Yu and Sim, 1987) and vegetable oil have been added to chicken feeds to increase the Omega-3 fatty acid content in the egg yolk.

Recent research has also shown that eggs may be

beneficial in preventing macular degeneration, a major cause of blindness in the elderly. Thus, the development of eggs rich in lutein and zeaxanthin will also be of interest to the elderly population.

Thus, designer egg enriched in Vitamin E, lutein and Omega-3 fatty acids cannot be only a good nutritional product but also a good vector for the delivery of four essential nutrients vital for human health. A crucial feature of these designer eggs is the synergistic combination of healty fatty acids with major antioxidants, Vitamin E and lutein, as an important approach to the improvement of the human diet. These eggs will not be able to replace vegetable and fruits as a major source of natural antioxidants and fish products but can substantially improve the diet, especially in a country like Nigeria, significantly contributing to the recommended daily intake of essential nutrients.

Since most of the substances or raw materials required for the production of the designer feed such as yellow corn, dried algae meal, dried fish meal and alfalfa meal which can be used for the production of designer eggs are commonly found in Nigeria, its production in house can be easily carried out. But in order to implement the production of designer eggs it is important that studies be conducted using different combinations of feed in order to attain the desired nutrient values which will suite the Nigerian population requirements.

Conflict of Interests

The author(s) have not declared any conflict of interests.

REFERENCES

Beare-Rogers J (1991). Nutrition recommendation in Canada. Bureau of Nutritional Sciences. Food Directorate, Health Protection Branch, Health and welfare. AOCS, Annual Meeting at Chicago. Information 2(4):326.

Caston L, Leeson S (1990). Research Note: Dietary flaxseed and egg composition. Poult. Sci. 69:1617-1620.

Dwyer JH, Navab M, Dwyer KM, Hassan K, Sun P, Shircore A, Hama-Levy S, Hough G, Wang X, Drake T, Merz NB, Fogelman AM (2001). Oxygenated carotenoid lutein and progression of early atherosclerosis. The Los Angeles Atherosclerosis Study. Circulation 103(24):2922-2927.

Farrell DJ (1998). Enrichment of hen eggs with n-3 long-chain fatty acids and evaluation of enriched eggs in humans. Am. J. Clin. Nutr. 68:538-44. PMid:9734728

Handelman GJ, Nightingale ZD, Lichtenstein AH, Schaefer EJ, Blumberg JB (1999). Lutein and zeaxanthin concentrations in plasma after dietary supplementation with egg yolk. Am. J. Clin. Nutr. 70: 247-251. PMid:10426702

Hargis PS, van Elswyk ME, Harris MM (1991). Dietary modification of yolk lipid with menhaden oil. Poult. Sci. 70:874-885 http://dx.doi.org/10.3382/ps.0700874; PMid:1908579

Hargis PS (1988). Modifying egg cholesterol in the domestic fowl: A review, World's Poult. Sci. 44:17-29.

Hasler CM (2000). The changing face of functional foods. J. Am. Coll. Nutr.19:499S-506S. http://dx.doi.org/10.1080/07315724.2000.10718972; PMid:11022999

Jiang Z, Ahn DU, Ladner L, Sim JS (1992). Influence of feeding full fat flax and sunflower seeds on internal and sensory qualities of eggs. Poult. Sci. 71:378-382. http://dx.doi.org/10.3382/ps.0710378

Landrum JT, Bone RA (2001). Lutein, Zeaxanthin, and the macular pigment. Arch. Biochem. Biophys. 385:28-40. http://dx.doi.org/10.1006/abbi.2000.2171; PMid:11361022

Nowokolo E, Sim JS (1989). Barley and full-fat Canola seed in layer diets. Poult. Sci. 68:1485-1489. http://dx.doi.org/10.3382/ps.0681485

Shrimpton DH (1987). The nutritive value of eggs and their dietary significance. Egg quality-current problems and recent advances. Butterworth and Co. Ltd. Lodon. pp. 11-25.

Sim JS (1990). Flaxseed as a high energy/protein/omega-3fatty acid ingredient for poultry. Proceedings of the 53rd Flax Institute of the United States, NDSU, Fargo. N.D. pp 65-71.

Sim JS, Hudgon GS, Bragg DB (1973). Effect of dietary animal tallow and vegetable oil on fatty acid composition of egg yolk, adipose tissue and liver of laying hens. Poult. Sci. 52:51-57 http://dx.doi.org/10.3382/ps.0520051; PMid:4709787

Simopolous AP (1991). Omega-3 fatty acids in health and disease and in growth and development. Am J. Clin. Nutri. 54:433-435

Yu MN, Sim JS (1987). Biological incorporation of N-polyunsaturated fattyacids into chicken eggs. Poult. Sci. 66:1955.

22

Dairy production and marketing in Uganda: Current status, constraints and way forward

J. Ekou

Department of Animal Production and Management, Faculty of Agriculture and Animal Sciences, Busitema University, P. O. Box 236, Tororo, Uganda.

Dairy production in Uganda is mainly based on low-input traditional pasture production systems. This makes Uganda one of the few countries in the world that are low cost producers of milk. Dairy farming could play a greater role in the economy considering its strong potential to provide rural employment and regular income to the many resource-poor households. However, milk production is still largely subsistence. There is therefore a huge potential to increase dairy production and productivity. The aim of this article was to review the current status of the dairy sector in Uganda, to identify the major constraints to dairy production and productivity and to suggest possible areas of intervention so as to enable Uganda exploit its competitive advantage in dairy production for socio-economic development. The article identified and discussed the following as major constraints to dairy production in the country: breed factors, feed resources, climatic factors, particularly high ambient temperature, socio-cultural factors, and dominant informal sector in milk marketing. Selective crossbreed utilization, feed resource development, specific disease prevention and control strategies, support for pastoral production systems, and establishment and support for dairy co-operative societies were recommended for improvement of dairy production and marketing in the country.

Key words: Dairy production, Uganda, current status, constraints

INTRODUCTION

Uganda's economy is still dominated by agriculture. More than 80% of Uganda's workforce is engaged in agriculture based primarily on smallholder farms that are on average only 2 ha in area (Bahiigwa, 1999; RoU, 2004; FAO, 2010). The share of agriculture in the national gross domestic product (GDP) is about 14.6%. Although the contribution of the livestock sub-sector to the national GDP decreased from 1.5% in 2005 to 1.3% in 2010, the share of livestock in the agricultural GDP increased from 8.4 to 8.9% over the same period

(MAAIF, 2010). The dairy industry is estimated to contribute more than 50% of the total output from the livestock sub-sector, making it the second major agricultural activity contributing to the national GDP after cereal products (RoU, 2004; Grimaud et al., 2007a; DDA, 2010; Balikowa, 2011). The livestock sub-sector in Uganda is evolving in response to rapidly increasing demand for livestock products that is largely driven by human population growth, income growth and urbanization (Delgado et al., 1999; Faye and Alary, 2001;

Thornton, 2010).

Dairy production in Uganda is mainly based on the low-input traditional pasture production system. This makes Uganda to be one of the lowest cost producers of milk globally, an advantage that should be exploited (Hemme, 2007). Dairy farming could play a greater role in the economy, considering its strong potential to provide employment and regular income to the many resource-poor rural households. Uganda is one of the few African countries that has attained self-sufficiency in the production of milk (Hemme, 2007; FAO, 2010; Balikowa, 2011). However, milk production is still largely subsistence with the attendant inefficiencies and quality problems commonly associated with such production systems (Balikowa, 2011). Although total annual national milk production grew from 365 million litres in 1991 to over 1.5 billion in 2008 (Wozemba and Nsanja, 2008; DDA, 2010), the observed growth in milk production has been attributed mainly to growth in the cattle population rather than increased milk productivity per cow. Higher productivity per cow is still hindered by low adoption of improved technologies and management practices (Elepu, 2006). Therefore, there is considerable potential to increase dairy production and productivity in Uganda. The aim of this article is to identify the major constraints to increased dairy production in Uganda and to suggest possible areas of intervention so as to enable the country to exploit its competitive advantage in dairy production for socio-economic development.

CURRENT STATUS OF THE DAIRY SECTOR IN UGANDA

Annual total milk production was about 1.5 billion litres in 2008 (DDA, 2010). However, actual milk output in the country would be more than the recorded 1.5 billion litres if supplies were added from the Karamoja sub-region; the milk production from which is currently excluded from data collection due to hygiene consideration. About 85% of milk in Uganda is produced from indigenous cattle, mainly Ankole (Aliguma and Nyoro, 2004; Wozemba and Nsanja, 2008). The Ankole breed, a genetic intermediate between *Bos indicus* and *Bos taurus* and related to the Sanga cattle (Grimaud et al., 2007a), is not a dairy breed per se. Each Ankole cow produces on average 2 L of milk per day and graze over a wide expanse of land often in search of fresh forage and water. Approximately 70% of total milk production is marketed, with the balance consumed by producing households and their neighbours. The annual consumption of milk per person in Uganda was 54 L in 2009 (DDA, 2010) which is 73% less than the 200 L per capita consumption recommended by the FAO. However, the potential for expansion is high given the natural resources available for dairy production in Uganda. Seventy-five percent of the land (18 million square kilometers) could be used for

crops or grazing. Currently only 5 million hectares is used as pasture and grazing land.

Dairy production in Uganda takes place under any of four systems (Wozemba and Nsanja, 2008):

(1) Communal grazing which involves pastoral grazing on communal land owned by clan. Although discouraged, it is still practiced in Northern and Eastern parts of Uganda. It is deeply rooted in the culture of these communities who are either pastoralists or agro-pastoralists historically.
(2) Free range grazing; where cattle are grazed by moving them all over the farm. It is a traditional practice in the extensive grasslands in the Southern part of Uganda. The farmland is often not paddocked, but the boundaries are fenced with local hedgerow plants.
(3) Fenced/paddock grazing which involves grazing cattle in paddocks, with supplemental feeding with feed concentrates, is a common farming practice in areas where the land holdings are fairly small. This type of grazing requires land clearing and improved pasture. It Is largely practiced by farmers of hybrid and cross-breed cattle.
(4) Zero grazing where animals are confined in a small enclosure and fodder, feed concentrates, and water are brought to the animals. According to a study by Mbabazi (2005), at least 20% of low income households in Ankole sub-region in Western Uganda have received a zero grazing cow from either government or from such non-governmental organizations as Send a Cow (UK) and Heifer International.

Uganda is divided into six milk-sheds or dairy regions that are defined by agro-ecological and milk production factors as well as their dairy market situation (Figure 1). Milk producers in the different agroecological areas use different means of dairy herd management, from pastoral and extensive systems to agro-pastoral and agricultural intensive zones (Grimaud et al., 2007a). In the intensive systems, herds often include exotic cows, with a predominance of the Friesian-Holstein breed (Grimaud et al., 2004).

The six dairy regions exhibit significant differences in terms of milk production, cattle numbers, market dynamics, and dairy infrastructure, among other factors (DDA, 2004; Balikowa, 2011). These have largely been influenced by the dominant dairy production systems, with the North and Eastern regions (including Karamoja) lagging behind in terms of milk production. A significant part of the national milk production is provided by the Mbarara area, located in the South-West milk-shed (DDA, 2004).

The infrastructure for rural milk collection is still limited. Where it exists, chilled milk is delivered to processing plants and raw milk markets in insulated milk transport tankers. Uganda has a very limited capacity to process milk into value-added products. There are over twenty

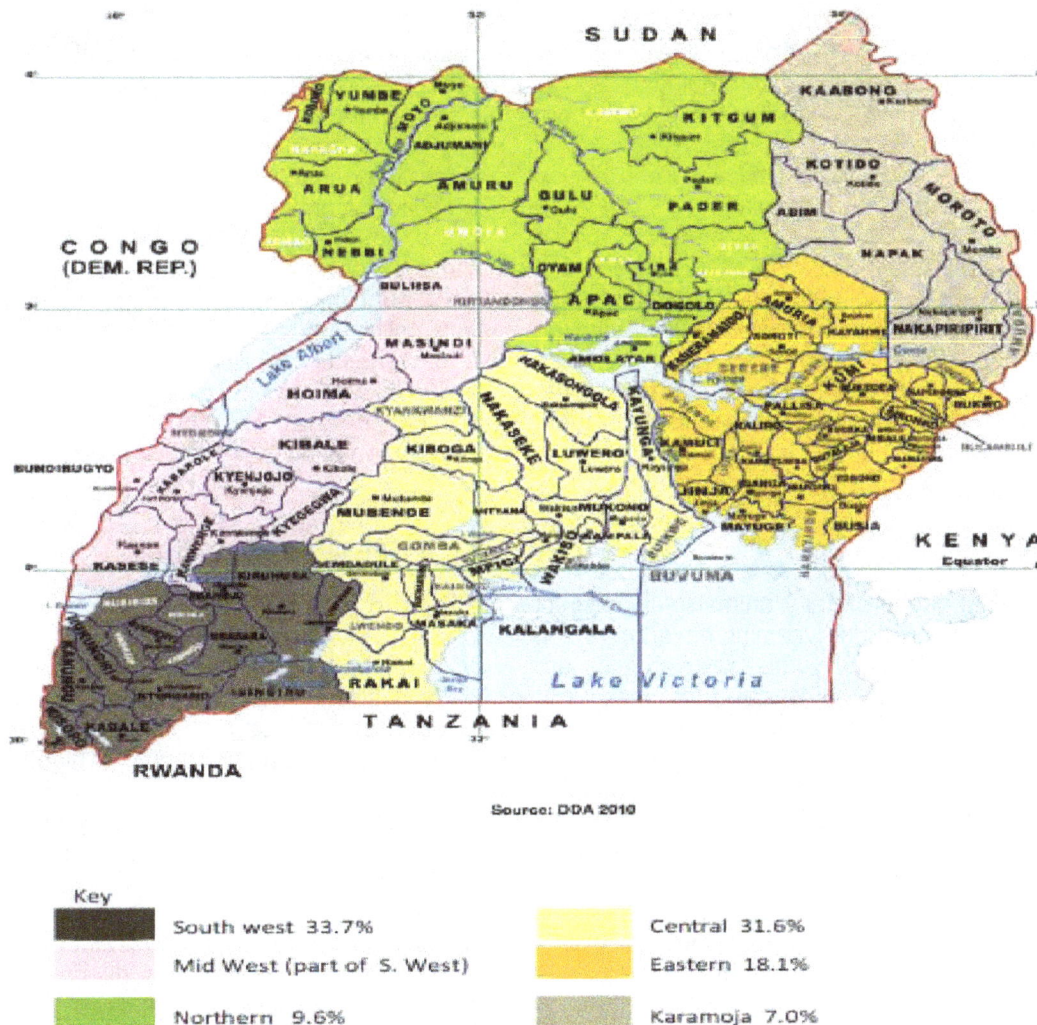

Figure 1. The dairy regions (milk sheds) of Uganda.

dairy processing companies, thirteen of which are milk processing plants and mini dairies. The combined installed capacity for processing plants was 463,200 L per day in 2008 (DDA, 2010). Only 10 to 20% of milk is currently marketed through the regulated formal market. The informal milk market takes up 80 to 90% of the marketable milk produced (Mpairwe, 2005; Grillet et al., 2005; DDA, 2010). The total quantity of milk and milk products imported into Uganda has been declining progressively from about 6,200 metric tons in 2003 to 800 metric tons in 2007. As a result of the decline in the imports, the amount of money spent on importing milk and milk products has experienced a steady decline from about $20 Million in 2001 to $1.6 Million in 2007. Between 2000 and 2008, Uganda exported an average of 380 metric tons of milk per year (DDA, 2010). UHT milk is the main dairy product exported to regional markets, including Rwanda, Kenya, Tanzania, DR Congo, Southern Sudan and Mauritius. Since May 2008 when

one Ugandan milk processor began producing milk power, some milk powder has also been exported. Informal dairy trade however goes on across all of Uganda's borders, but the volume traded is generally not significant.

MAJOR CONSTRAINTS TO DAIRY PRODUCTION AND MARKETING IN UGANDA

Breed factors

Cattle are the major source of milk in Uganda. The population of dairy goats and other milk animals (buffalo and camel) is insignificant (Balikowa, 2011). However, most regions in Uganda have cattle breeds that are slow growing and have low feed conversion ratios with very low milk yield and market weights. Currently the average adult live weight in the predominantly Zebu herds of Teso sub-region, for example, is only 180 to 350 kg per head, and this requires about five years to attain even if adequate pasture is available (Jain and Muladno, 2009). Of the 11.4 million cattle in Uganda, the Ankole longhorn (Sanga) breed is the most common, comprising 50% of the population. The small East African Zebu breed follows with 30% of the total population (MAAIF and UBOS, 2009). The Nganda intermediate breed represents 16% of the total population. Exotic breeds and their crosses make up only 4% of the total cattle population. The indigenous breeds generally have limited genetic potential for milk production and remain mediocre producers (500 to 1500 kg per lactation) even when the best possible husbandry conditions are available to them (Pagot, 1992).

Feed resources

While the nutritional needs of dairy animals with respect to energy, protein, minerals and vitamins have long been known and refined over many years, dairy animals are highly sensitive to changes in feeding regimes, and production can fall dramatically with small variations (Thornton, 2010). As understanding of the science of animal nutrition continues to expand and develop, most of the world's livestock, particularly ruminants in pastoral and extensive mixed systems in many developing countries, suffer from permanent or seasonal nutritional stress (Bruinsma, 2003). Natural and planted pastures are the major components in the diet of both indigenous and improved dairy cattle in Uganda. Because Uganda has many agro-ecological zones, the common naturally occurring pasture species vary from one region to another (Grimaud et al., 2007b). In the traditional cattle corridor, common sources of forages include grasses such as *Hyparrhenia rufa, Chloris gayana, Brachiaria decumbens, Cynodon dactylon, Themeda triandra, Digitaria* spp., *Hyparrhenia filipendula, Panicum maximum, Paspalum dilatatum* (Balikowa, 2011).

Most of the milk in Uganda is produced by smallholder producers that rely almost entirely on rain-fed natural pastures. However, a severe decline in the quantity and quality of pastures occurs during the dry season. This is often accompanied by widespread invasion of unpalatable grasses (mainly *Cymbopogon afronardus* and *Sporobolus pyramidalis)* as well as bush encroachment, with subsequent overgrazing of the palatable species, mainly *Brachiaria brizantha and Themeda triandra* (Grimaud et al., 2007b; Balikowa, 2011). Only a small number of households keeping improved dairy cattle make the effort to plant improved pastures. Consequently, very few of farms with improved dairy cattle produce enough fodder to meet the needs of their herds throughout the year (Balikowa, 2011). In addition, shortage of grazing land is becoming a serious constraint to dairy farming in most regions of Uganda.

The human population has been increasing rapidly, resulting in increased demand and competition for arable land. Households give priority to production of food crops. Land available for grazing is steadily dwindling in most regions. Extensive grazing of cattle, which has always been the most common management system, and is steadily becoming less popular except in the traditional cattle corridor (Balikowa, 2011). Hence, most animals thrive on sub-optimal energy levels for most of the year. Poor nutrition is one of the major production constraints in smallholder cattle systems of Uganda. Research on improvement of quality and availability of feed resources, including work on sown forages, forage conservation, use of multi-purpose trees, fibrous crop residues and strategic supplementation is available (Thornton, 2010).

Livestock diseases

Efforts to increase milk production in Uganda started in the 1950s with the importation of temperate dairy cattle (*Bos taurus*). However, the susceptibility of improved dairy cattle to local diseases and parasites, particularly tick-borne diseases and trypanosomiasis, and the high management costs remain the biggest impediment to development of commercial dairy farming in Uganda. Improved dairy cattle are still very unpopular among the poor farmers. The total population of improved dairy cattle was estimated at 5.57% of the national herd (MAAIF and UBOS, 2009; Balikowa, 2011). Tick-borne diseases remain a major constraint to the improvement of dairy production in Uganda (Norval et al., 1992; Bell-Sakyil et al., 2004). The cost of controlling ticks and tick-borne diseases is estimated to constitute about 85.6% (pastoral) and 73.8% (ranches) of total disease control costs (Ocaido et al., 2009).

The major tick-borne diseases in Uganda are anaplasmosis, babesiosis, cowdriosis and East Coast fever (ECF). Together, these diseases constitute the most important constraint to livestock production in Uganda (Ekou, 2013). In 1984, the government of Uganda stopped importing and distributing subsidized veterinary drugs and chemicals to farmers. Since then, government only imports certain veterinary products for use in control programs of particular endemic diseases such as Foot and Mouth Disease, Contagious Bovine Pleural Pneumonia, Lumpy Skin Disease, rabies and vectors such as tsetse flies. Instead, from 1994, it has been encouraging veterinarians to leave public service and set up private veterinary practice. To date, Government is still the major provider of animal health services (Wozemba and Nsanja, 2008; Balikowa, 2011). It has been reported, however, that routine strategic vaccinations often are not carried out, targeted vaccinations during outbreaks are delayed, and that in instances where vaccines were availed for targeted vaccinations, the vaccines did not cover all the livestock population in affected areas resulting in livestock becoming susceptible to preventable livestock diseases (RoU, 2009).

Climatic factors

Popularity of the high yielding temperate stock among the local farmers has been curtailed by their inability to withstand the tropical conditions (Balikowa, 2011). Numerous experiments have shown that a prolonged period in which temperatures are more than 25°C, particularly in humid air conditions, leads to a reduction of dry matter intake by milking cows and, as a consequence, a drop in their production. Experiments have shown that a fall in appetite due to heat is the principal factor in the depression of production. High ambient temperatures have another depressive action on milk production by reducing the fertility of the cows, thus lengthening the interval between lactations (Bligh, 1976; Pagot, 1992; Igono and Aliu, 1982). Heat stress significantly impacts animal production and profitability in dairy cattle by lowering feed intake, milk production and reproduction (Chase, 2006). Climate change will likely worsen the situation. It is predicted that climate change will have severely deleterious impacts in many parts of the tropics and subtropics, even for small increases in the average temperature (IPCC, 2007). It will undoubtedly increase livestock production risks as well as reduce the ability of farmers to manage these risks (Thornton, 2010).

Socio-cultural factors

Exploitation of cattle for milk is one of the features of pastoral communities that have lived in symbiosis with their cattle for millennia. Herding people are often nomadic or transhumant and do not practice agriculture. Pastoralism, an economic and social system well adapted to dryland conditions and characterized by a complex set of practices and knowledge has permitted the maintenance of a sustainable equilibrium among pastures, livestock and people for generations (Koocheki and Gliessman, 2005). However, this livestock production system, which does not permit a place for intensive forage production, has limited possibilities for improvement (Pagot, 1992). Notably, milk hygiene practices among herding communities are usually poor. In Uganda, the Karamojong are a classic example. Due to hygienic consideration, milk supplies from the Karamoja sub-region are currently excluded from records on national milk production. As is the case in many countries, raw milk safety is a major public health concern in Uganda. Any improvement in the quality of milk contributes to the insurance of public health while at the same time having positive economic consequences (Grimaud et al., 2007b).

Dominant informal sector in milk marketing

The Dairy Development Authority (DDA), a statutory body under the Ministry of Agriculture, Animal Industry and Fisheries (MAAIF), is responsible for regulating and developing the dairy industry. DDA implements this mandate through enforcement of quality standards and implementation of regulations such as registration, inspection of premises and factories and issuing of licenses for milk collection centres, outlets, small-scale processors, importers, factories, coolers and freezers, Storage facility, milk road tankers, and input suppliers. However, Uganda's dairy industry has too many players along the entire value chain (DDA, 2010). There are millions of small scale dairy farming households, tens of thousands of scattered middlemen trading in milk, and a handful of dairy processors. The informal sector handles about 80% of the traded milk. This informal segment, comprising mostly scattered middlemen and often beyond the reach of DDA operations, is driven by price and not by quality. This poses a big challenge regarding ensuring quality of the milk that reaches the low income segment of the consumers.

STRATEGIES FOR IMPROVING MILK PRODUCTION AND MARKETING

Developments in breeding, nutrition and animal health will continue to contribute to increasing dairy production (Thornton, 2010). Here the following factors are discussed.

Selective crossbreed utilisation

The use of conventional livestock breeding techniques has been largely responsible for the increases in yield of livestock products that have been observed over recent decades (Leakey, 2009). Of the conventional techniques, selection among breeds or crosses with the most appropriate breed or breed cross followed by selection within the resultant population has extensively been exploited to improve livestock production, especially in developed countries (Simm, 1998; Simm et al., 2004). While most of the gains from selective breeding have occurred in developed countries, there are considerable opportunities to do the same in developing countries. Crossbreeding can result in rapid productivity improvements, but new breeds and crosses need to be appropriate for the environment and to fit within production systems that may be characterized by limited resources and other constraints (Simm et al., 2004 Thornton, 2010).

The potential to produce milk in hotter climates, previously not given due attention, can no longer be ignored. Previous studies in the tropics have shown that at various heat intensities above 27°C, half Friesian-Zebu cattle produce more milk when compared to the three-quarter cross during the stage of maximum lactation, despite the higher genetic potential of the latter (Bligh, 1976; Igono and Aliu, 1982). Therefore crossing European breeds, such as the small sized Jersey, with local Zebu cattle that have proven to be more adaptable to local conditions is an appropriate dairy cattle breeding strategy for Uganda. This cross would produce relatively small animals with low feed requirements but high productivity. However, as pointed out by FAO (2007), institutional and policy frameworks that encourage the sustainable use of traditional breeds and *in situ* conservation need to be implemented so as to avoid losing the genetic merit of these local breeds.

Feed resources development

In tropical countries, modern agronomic practices, such as selection of forage species, fertilisation, and irrigation, levels of productivity comparable to the best obtained in temperate countries (Pagot, 1992). In Uganda, a thorough agronomic assessment of grasses and legumes could be conducted in the different agro-ecological zones. Multilocational trials would test various forage legumes, grasses and browses for their adaptability and performance on different soil types and in various environments. These trials would be supported by pot experiments to determine soil nutrient status and pinpoint any trace-element deficiencies (Reynolds, 1981; Kayastha, 1982). Forage grasses and legumes found suitable in multilocational trials could then be recommended for planting (Servoz, 1983). Dairy farmers should be supported to cultivate fodder species of high value on one to two acres of land per dairy cow through provision of planting materials. Examples of such fodder species include grasses such as *Pannisetum purpereum*, *P. maximum*, and *Bracharia mulato*; legumes such *Pueraria phaseolois*, *Chemaecrista rutondifolia*, *Arachis pintoi*, and *Centrosema arienarium*; and trees such as *Sesbania sesban*, *Calliandra calothyrsus*, *Leucaena leucocephala*, and *Leucaena diversfolia*. Trees can be interplanted with legumes.

Disease prevention and control

Through risk analysis, the Ministry of Agriculture, Animal Industry, and Fisheries (MAAIF) could identify areas prone to disease outbreaks and carry out routine strategic vaccinations in those locations. MAAIF also could also enter into contracts with vaccine manufactures to keep vaccine stocks readily available for delivery when disease outbreaks occur (RoU, 2009). Better still, government can establish a national veterinary drugs centre with sufficient resources to procure, store and

avail existing vaccines and drugs of priority endemic diseases in the country. This would be the equivalent of the now largely successful strategy of the National Medical Stores for the health system in Uganda. Rehabilitation and construction of community dip tanks as a key intervention in the control of ticks and tick-borne diseases has been recommended, especially in areas where pastoral or communal grazing is still being practised (Ekou, 2013).

Support for pastoral production systems

Unhygienic handling practices in traditional milk production and in the informal milk trade represent serious obstacles for the introduction of modern dairy processing and marketing. The influence of pooling of different milk batches along the collection and marketing chain exacerbates the problem.

The successful adaptation of pastoral subsistence production to the needs of an improved milk production and marketing system will depend, to a large extent, on safeguarding the milk quality at production, during transport, processing and marketing. Optimising milk hygiene under pastoral conditions requires the availability of safe clean water, which is an unrealistic expectation in most situations.

However, the introduction of clean metal containers to producing herds has had a measurably positive effect on raw milk quality (Younan, 2004). Pastoral communities therefore ought to be supported to acquire simple metal containers in addition to continuous education on hygienic milk handling procedures and practices.

Establishment and support for dairy co-operative societies

Co-operative societies are an important forum for bringing together small holder farmers. Farmers should be mobilized to form cooperative groups in various milk producing areas. Government and development partners could support these farmers with startup capital and training.

Farmers can also be linked to the buyers through close and regular market interaction. These farmer groups would collect milk from various farmers to one collection centre where processors come and buy. This can bring many advantages. Because of the organised bulking (selling in large quantities), there is ready market for milk since processors come and take all the collections from one place.

It makes it easier for farmers to acquire milk coolers, generators and other equipment from processors at friendly terms to ensure quality standardisation and increased production of milk (EADD, 2013). Farmers can even receive other dairy services like farmer education

On improved breeding, disease control, pasture improvement and milk production enhancement from the dairy society.

Conflict of Interests

The author have not declared any conflict of interests.

REFERENCES

Aliguma L Nyoro JK (2004). Regoverning markets. In: Ministry of Agriculture, Animal Industry and Fisheries (ed). Scoping study on dairy products, fresh fruits and vegetables. Entebbe, Uganda.

Bligh, J (1976). Temperature regulation in cattle. In: Johnson HD (ed), Progress in Biometeorology, Swets and Zeitlinger, Amsterdam.

Bahiigwa GBA (1999). Household Food Security in Uganda: An Empirical Analysis. Research Series Kampala: Economic Policy Research Center. P. 25.

Balikowa D (2011). Dairy development in Uganda: A review of Uganda's dairy industry. Ministry of Agriculture, Animal industry and Fisheries, Republic of Uganda and Food and Agricultural Organisation of the United Nations and Dairy Development Authority. Research Report.

Bell-Sakyil L, Koneye EBM, Dogbeyo O, Walker AR (2004). Incidence and prevalence of tick-borne haemoparasites in domestic ruminants in Ghana. Vet. Parasitol. 124:25-42.

Bruinsma J (2003). World agriculture: towards 2015/2030, an FAO perspective. Rome, Italy: Earthscan, FAO.

Chase LE (2006). Climate Change Impacts on Dairy Cattle. Available at: www.climateandfarming.org/pdfs/FactSheets/III.3Cattle.pdf

DDA (2004). Annual Report 2003/2004, Dairy Development Authority, Kampala, Uganda.

DDA (2010). Annual Report 2009/2010, Dairy Development Authority, Kampala, Uganda.

Delgado C, Rosegrant M, Steinfeld H, Ehui S Courbois C (1999). Livestock to 2020', IFPRI Brief. Washington, DC: International Food Policy Research Institute.

EADD (2013). The start, collapse and revival of Uganda's dairy industry. East Africa Agribusiness summit and awards, 10th-11th December, 2013. Kampala Serena Hotel: Kampala.

Ekou J (2013). Eradicating extreme poverty among the rural poor in Uganda through poultry and cattle improvement programmes - A Review. J. Dev. Agric. Econ. 5(11):444-449.

Elepu G (2006).Value Chain Analysis for the Dairy Sub-sector in Uganda. A Final Report. Uganda Agribusiness Development Component, ASPS/DANIDA.

FAO (2007). Global plan of action for animal genetic resources and the Interlaken Declaration. Int. technical conf. on animal genetic resources for food and agriculture, Interlaken, Switzerland, 3–7 September 2007, Rome, Italy: FAO.

FAO (2010). Status of and Prospects for Smallholder Milk Production – A Global Perspective, by T. Hemme J. Otte. Rome.

Faye B, Alary V (2001). Les enjeux des productions animales dans les pays du Sud. INRA Prod. Anim.14:3-13.

Grillet N, Grimaud P, Loiseau G, Wesuta M, Faye B (2005). Qualité sanitaire du lait cru tout au long de la filière dans le district de Mbarara et la ville de Kampala en Ouganda. Rev. Elev. Med. Vet. Trop. 4:245-255.

Grimaud P, Faye B, Mugarura L, Muhoozi E, Bellinguez A (2004). Identification of research activities for the dairy sector development in Uganda: Systemic and participatory approaches. Uganda J. Agric. Sci. 9:879-884.

Grimaud P, Mpairwe D, Chalimbaud J, Messad S, Faye B (2007a). The place of Sanga cattle in dairy production in Uganda. Trop. Anim. Health Prod. 39:217-227.

Grimaud P, Sserunjogi ML, Grillet N (2007b). An evaluation of milk quality in Uganda: value chain assessment and recommendations. Afr. J. Food Agric. Nutr. Develop. 7:5.

Hemme T (2007). IFCN Dairy Report, International Farm Comparison Network, Dairy Research Centre, Kiel, Germany.

IPCC (2007). Climate Change 2007: Impacts, adaptation and vulnerability. Summary for policy makers. Available at: http://www.ipcc.ch/publications_and_data/ar4/wg2/en/spm.html (Accessed 20/10/2013).

Igono MO, Aliu YO (1982). Environmental Profile and Milk Production of Friesian-Zebu Crosses in Nigerian Guinea Savanna. Int. J. Biometeor. 26(2):115-120.

Jain AK, Muladno M (2009). Selection criteria and breeding objectives in improvement of productivity of cattle and buffaloes In Selection and Breeding of Cattle in Asia: Strategies and Criteria for Improved Breeding. IAEA-TECDOC-1620.

Kayastha AK (1982). Technical report on pot trials carried out on soils from Tunguu, Maweni and Kizimbani. Report prepared for the Government of Zanzibar by FAO/UNDP. Working Paper. FAO/UNDP Project URT/81/017.

Koocheki A, Gliessman R (2005). Pastoral Nomadism: A Sustainable System for Grazing Land Management in Arid Areas. J. Develop. Agric. 25:4.

Leakey R (2009). Impacts of AKST (Agricultural Knowledge Science and Technology) on development and sustainability goals. In Agriculture at a crossroads (eds McIntyre BD, Herren HR, Wakhungu J, Watson RT), Washington, DC: Island Press. pp. 145–253.

MAAIF (2010). MAAIF statistical abstract 2010. Ministry of Agriculture, Animal Industry and Fisheries (MAAIF), Entebbe, Republic of Uganda.

MAAIF,UBOS (2009). National Livestock Census Report 2008. Ministry of Agriculture, Animal Industry and Fisheries (MAAIF), Entebbe /Uganda Bureau of Statistics, Kampala, Uganda.

Mbabazi P (2005). Supply Chain and Liberalization of the Milk Industry in Uganda. Available at: https://www.africanbookscollective.com/books/supply-chain-and...(Accessed 23/12/2013).

Mpairwe D (2005). Undernutrition in dairy ruminants and intervention options for coping with feed scarcity in smallholder production systems in Uganda. In: Ayantunde AA., Fernández-Rivera S and G McCrabb (Eds). Coping with feed scarcity in smallholder livestock systems in developing countries. Animal Sciences Group, UR, Wageningen, The Netherlands, University of Reading, Reading, UK, Swiss Federal Institute of Technology, Zurich, Switzerland and International Livestock Research Institute, Nairobi, Kenya.

Norval RAI, Perry BD, Young AS (1992). The Epidemiology of Theileriosis in Africa. Academic Press Ltd., London.

Ocaido M, Muwazi RT, Opuda JA (2009). Economic impact of ticks and tick-borne diseases on cattle production systems around Lake Mburo National Park in South Western Uganda. Trop. Anim. Health Prod. 41(5):731-739.

Pagot J (1992). Animal production in the tropics and subtropics. First edition. The Macmillan press Ltd, London and Basingstoke.

Reynolds SG (1981). Pasture improvement activities. Technical Report. FAO/UNDP Project URT/ 78/028.61.

RoU (2004). Increasing incomes through exports: a plan for zonal agricultural production, agroprocessing and marketing. Republic of Uganda. Available at: http://www.ugandaexportsonline.com/strategies/zoning_plan.pdf.(Accessed 08/10/2013).

RoU (2009).Value for money audit report on the prevention and control of livestock diseases by the Department of Livestock health and Entomology in the Ministry of Agriculture Animal industry and Fisheries. Office of the Auditor General, Kampala, Uganda.

Servoz HM (1983). Consultancy in pasture agronomy. Report prepared for the Government of Zanzibar. FAO/UNDP Project URT/81/017.

Simm G (1998). Genetic improvement of cattle and sheep. Wallingford, UK: CABI Publishing.

Simm G, Bünger L, Villanueva B, Hill WG (2004). Limits to yield of farm species: genetic improvement of livestock. In Yields of farmed species: constraints and opportunities in the 21st century (eds R. Sylvester-Bradley & J. Wiseman), Nottingham, UK: Nottingham University Press. pp. 123–141.

Thornton PK (2010). Livestock production: recent trends, future prospects. Philosophical Phil. Trans. R. Soc. B. 365:2853–2867.

Wozemba D, Nsanja R (2008). Report on Dairy Investment Opportunities in Uganda: Dairy sector analysis, SNV. Richard, York and Associates (Uganda).

Younan M (2004). Milk hygiene and udder health, In Farah Z, Fischer A (ed), Milk and meat from the camel. ETH, Zurich, Switzerland, pp. 67–76.

23

Physiological and biochemical alterations during germination and storage of habanero pepper seeds

Franciele Caixeta, Édila Vilela De Resende Von Pinho, Renato Mendes Guimarães, Pedro Henrique Andrade Rezende Pereira and Hugo Cesar Rodrigues Moreira Catão

Department of Agriculture, Federal University of Lavras, C.P. 3037, 37200-000, Lavras, MG, Brazil.

The objective of this study was to evaluate physiological and biochemical alterations during the development and storage of habanero pepper seeds with a view toward determining the time of harvest. Seeds were manually extracted from the fruit at three stages of development: E1 (fruit with first signs of yellowing), E2 (mature fruit) and E3 (mature fruit submitted to seven days of rest). After drying, seeds with 8% water content were stored at 10°C for 0, 4 and 8 months, and their quality evaluated by means of germination and vigor tests. Activities of the enzymes α-amylase, endo-β-mannanase, esterase, Superoxide Dismutase (SOD), malate dehydrogenase (MDH) and alcohol dehydrogenase (ADH) were evaluated during germination at 0, 48, 96 and 144 h after seeding. A randomized block design was used in a 3 × 3 factorial design (stages of development × storage) with 4 replications. Lower germination and vigor values were observed for the E1 stage seeds at all storage periods. In recently stored seeds, greater germination and vigor values were observed for the E3 stage seeds. Dormancy was observed principally in recently stored seeds and this was overcome at four months of storage. In summary, the physiological tests and activity of the enzymes evaluated indicated that the habanero pepper should be harvested at the E3 stage for a higher seed quality.

Key words: *Capsicum chinense*, seed maturation, seed quality.

INTRODUCTION

The demand for high quality seeds has grown substantially in recent years, which requires that seed companies adopt advanced technologies during production, processing and storage processes. The adoption of new seed technologies requires knowledge of the factors that influence physiological quality on the one hand and on the physiological and biochemical alterations that occur during the germination and storage processes on the other hand. Studies related to maturation and harvest of seeds is important since seeds reach their maximum quality in the field. Such knowledge is necessary for seed producers to determine the ideal time for harvest that would minimize seed quality deterioration caused by a prolonged period in the field, and increase seed production, by preventing a too early harvest since this may result in a large proportion of immature seeds (Vidigal et al., 2009). In species with fleshy fruit, it has been observed that seeds maintained for a certain period of time in the fruit after harvest complete the maturation process, reaching maximum levels of germination and vigor (Barbedo et al., 1994; Vidigal et al., 2006; Dias et al., 2006). Therefore, post-harvest storage of the fruit before seed extraction may be advantageous because it allows early harvest, and avoid exposure to possible unfavorable conditions that may deteriorate seed quality (Barbedo et al., 1994;

Table 1. Mean values of maximum (Tx), minimum (Tn), and mean (Tm) temperatures, relative humidity (UR), total rainfall (Pt) and insolation (I) of the months corresponding to the crop development period of the 2007 harvest.

Months	Tx (°C)	Tn (°C)	Tm (°C)	UR (%)	Pt (mm)	I (h)
January	27.6	18.7	21.9	85.7	554.7	3.1
February	28.9	18.1	22.6	73.1	151.3	7.5
March	30.9	18.1	23.6	66.5	35.4	9.0
April	28.1	17.2	21.8	72.4	35.6	7.3
May	25.7	12.8	18.1	70.6	30.4	7.5
June	25.8	11.1	17.3	66.3	5.9	8.8
July	25.4	11.1	17.1	66.8	17.6	7.7

Source: Agrometeorology Sector of the Engineering Department – UFLA.

Dias et al., 2006).

Deterioration is a process determined by a series of physiological, biochemical, physical and cytological alterations that occur in a progressive manner, leading to lower quality and culminating in death of the seed (Freitas, 2009). The main alterations related to the deterioration process are degradation and inactivation of enzymes, reduction of respiratory activity and loss of integrity of cellular membranes (Copeland and McDonald, 2001). The speed of this deterioration process is influenced by storage conditions. During seed storage, it is necessary to maintain adequate temperature and humidity conditions in the attempt to preserve quality. In this research, physiological and biochemical alterations during development of yellow habanero pepper (*Capsicum chinense*) seeds were evaluated, with a view toward determination of the best time for harvest. In addition, enzymatic alterations were evaluated during the germination process of the seeds processed at different stages of maturity and storage.

MATERIALS AND METHODS

The research was conducted in the experimental area and in the Central Seed Laboratory (Laboratório Central de Sementes) of the Agriculture Department (Departamento de Agricultura) of the Federal University of Lavras (Universidade Federal de Lavras - UFLA), in Lavras, MG. The city located in the Southern Region of Minas Gerais, latitude 21° 14' S and longitude 40° 17' W and at 918.8 m altitude. The annual average temperature is 19.4°C and rainfall is distributed principally from October to April, with annual amounts of 1529.7 mm. In a first stage of research, yellow habanero pepper (*C. chinense*) seedlings were formed for installation of the experiment in the field. Seeds were sown in "styrofoam" trays with 72 cells containing the commercial substrate Plantmax® - hortaliças and 5 ml of 2000 ppm solution of ammonium sulfate per cell. Transplanting of the seedlings was performed 45 days after seeding to an experimental area of the olericulture sector of the Agriculture Department in an area with a Dark Red Latosol/Oxisol and clayey texture; prepared conventionally. Tests were installed in a randomized complete block design with four replications. Each plot consisted of 2 rows of 5 m length with 5 plants per meter and spacing of 1.5 m between rows. Plant cultivation was performed in accordance with Filgueira

(2005). Temperature and relative air humidity data during plant development are presented in Table 1. Seeds were manually extracted from many fruits at three stages of development: E1 (fruit with first signs of yellowing), E2 (mature fruit) and E3 (mature fruit submitted to seven days of rest). Then the seeds were dried in a laboratory oven with air circulation at 35°C until reaching 8% water content. The seeds corresponding to each stage of development of the fruit were packed in airtight plastic packages and stored in a walk-in cooler at 10°C and 50% relative humidity for periods of 0, 4 and 8 months after drying. At the end of each storage period, seed quality was evaluated by means of germination tests; emergence tests (Brazil, 2009); emergence rate index (Maguire, 1962); electrical conductivity (Vidigal et al., 2008) and accelerated aging (Bhering et al., 2006). Furthermore, the activity of the enzymes esterase, superoxide dismutase (SOD), malate dehydrogenase (MDH), alcohol dehydrogenase (ADH) and endo-β-mannanase (Downie et al., 1994) was evaluated.

The activities of the enzymes α-amylase (Alfenas et al., 1991), endo-β-mannanase, esterase, MDH and ADH were evaluated during the germination process of the seeds at 0, 48, 96 and 144 h after seeding. The seeds were ground in the presence of PVP (polyvinylpyrrolidone) and liquid nitrogen in a mortar over ice and afterwards stored at a temperature of -86°C. The experimental design used was randomized complete blocks in a 3 × 3 factorial design, with the factors being: stage of development of the fruit (E1, E2 and E3) and storage periods (0, 4 and 8 months). Analysis of variance was performed for all tests using the statistical program Sisvar (Ferreira, 2000). For comparison among the means, the Scott-Knott test was used at the 5% probability level.

RESULTS

It is observed from the results in Table 2, the interaction between the factors stage of development of the fruit (E1, E2 and E3) and storage periods (0, 4 and 8 months) was significant by F test for all the variables analyzed. Lower germination values were observed for the habanero pepper seeds processed in the first stage of development (E1), in all storage periods (Table 2). At 0 and 8 months of storage, there was no difference between the germination values of the seeds processed at the E2 and E3 stages of development, whereas at 4 months of storage, greater germination values were observed for E3 stage seeds. An increase in germination was observed for E2 and E3 stage seeds at the 4th month of storage

Table 2. Percentage of germinated seedlings (%), percentage of seedling emergence (%), emergence rate index and vigor obtained by the accelerated aging test and electrical conductivity of habanero pepper seeds gathered at different stages of development throughout the storage period.

Storage periods			
Stages	0	4	8
Germination			
E1	1[Ba]	3[Ca]	7[Ba]
E2	25[Ac]	41[Bb]	58[Aa]
E3	32[Ab]	50[Aa]	53[Aa]
Emergence			
E1	5[Cb]	35[Ba]	30[Ba]
E2	60[Bb]	85[Aa]	84[Aa]
E3	75[Ab]	86[Aa]	87[Aa]
Emergence rate index			
E1	0[Cb]	5[Ba]	5[Ca]
E2	8[Bc]	23[Ab]	29[Ba]
E3	13[Ac]	25[Ab]	34[Aa]
Accelerated aging			
E1	1[Cb]	3[Cb]	23[Ca]
E2	20[Bc]	68[Ab]	87[Ba]
E3	43[Ab]	47[Bb]	95[Aa]
Electrical conductivity			
E1	825[Cb]	580[Bb]	634[Ca]
E2	750[Bc]	535[Ab]	463[Ba]
E3	654[Ac]	511[Ab]	406[Aa]

(1) Means followed by the same capital letter in the column and small letter in the row do not differ among themselves by the Scott-Knott test at the 5% probability level.

(Table 2). These values were maintained in E3 stage seeds but increased in E2 stage seeds after eight months of storage. As for E1 stage seeds percentage of germinated seedlings were very low and did not differ significantly during storage. The results obtained in the emergence test and emergence rate index, under greenhouse conditions, are presented in Table 2. Lower plantlet emergence values were observed in seeds processed in the E1 development stage at all storage periods. At 4 months of storage, there was an increase in the percentage of seedling emergence from seeds processed in different stages of development (Table 2). The results of the emergence rate index indicated less vigor in the E1 stage seeds and greater vigor values in the E2 and E3 stage seeds stored for 4 months. However, at 8 months, there was vigor increase for E3 and E2 seeds and maintenance for E1 stage seeds. The results of the accelerated aging test (Table 2) showed greater means of seed vigor in E1, E2, and E3 stage seeds stored for 8 months. At the beginning of storage,

the seeds processed in the E3 stage were more vigorous for other stages. Regardless of the storage period, less vigor was observed in habanero seeds extracted at the E1 stage. By the electrical conductivity test (Table 2), greater vigor was observed in E3 seeds that were either recently stored, or stored for 8 months. There was no significant difference in the conductivity values observed in E2 and E3 stages seeds stored for 4 months. In the same way, greater leaching was also observed at 8 months of storage (Table 2). Regardless of the stage of development of habanero pepper seeds, at the end of 8 months of storage, there were lower means of electrical conductivity.

Considering the electrophoretic analyses, the enzymatic profile of esterase (Figure 1) +showed a greater activity of this enzyme for E1 stage seeds at the three storage periods, with this being attributed to greater immaturity of these seeds but also to seed quality deterioration throughout the storage period. Regarding enzyme SOD (Figure 1), there was an increase in its

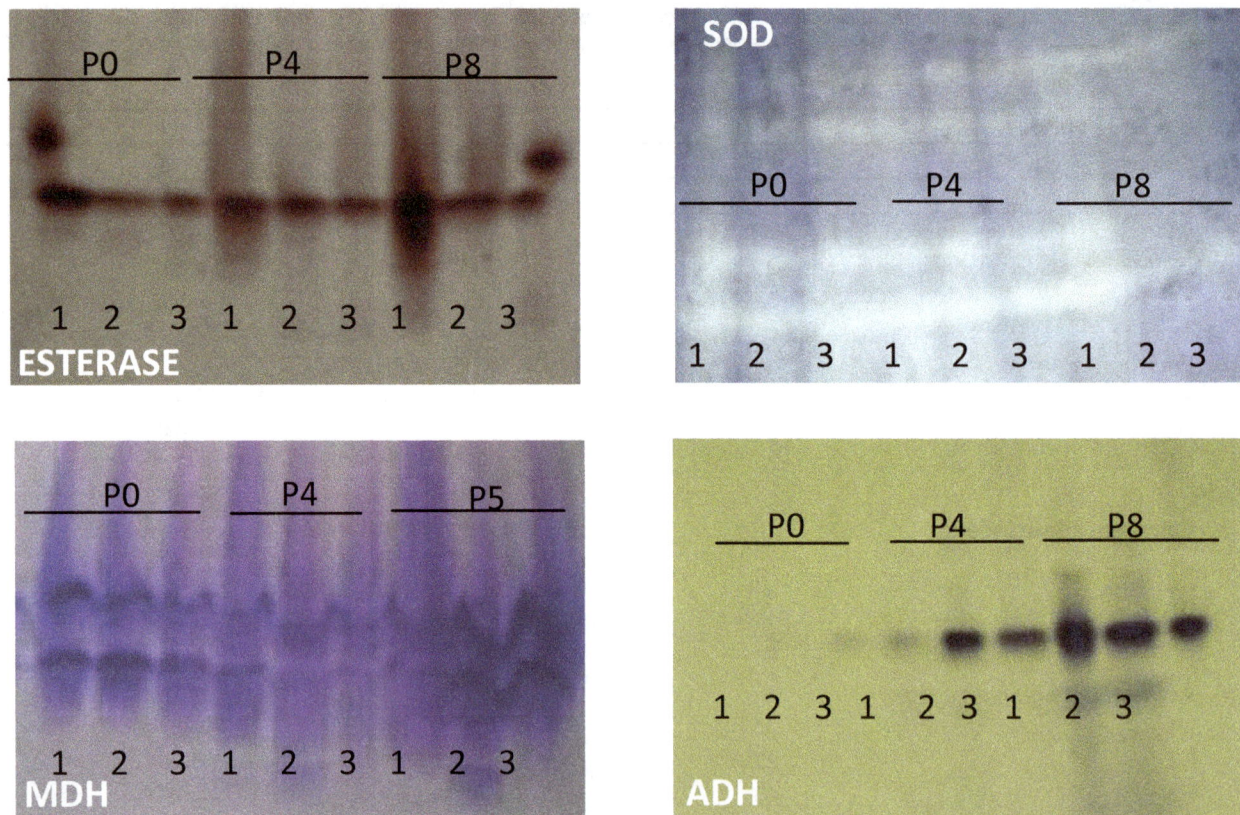

Figure 1. Esterase enzyme profiles, SOD, MDH and ADH of habanero pepper seeds processed in the E1 (1), E2 (2) and E3 (3) stages in the three storage periods 0 (P0), 4 (P4) and 8 (P8) months of storage.

activity at 4 months of storage for all the three stages of development. At 8 months of storage, less activity was observed in E2 and E3 seeds, while in E1 stage seeds there was an increase in the activity of this enzyme. The enzyme MDH (Figura 1) had increased activity in all three stages of development at 8 months of storage. Regarding the enzyme ADH (Figure 1), greater activity in E2 stage seeds was observed at 4 and 8 months. The profile of the enzyme endo-β-mannanase showed an increase in activity (Figure 2) for seeds processed in the most advanced stages of development. Moreover, less activity of this enzyme was observed in recently stored seeds regardless of the maturity stage. In the present research, greater germination and vigor values were observed after storage of the seeds in E2 and E3 stage, leading to the supposition of breaking of dormancy of the seeds during storage (Table 2). Considering the electrophoretic analyses of the enzymatic patterns during germination, there was no difference in the activity of the enzymes in the dry seeds in relation to those soaked for 48 h in all the enzymatic patterns, with the exception of that observed for the enzyme endo-β-mannanase (Figure 3). Variations in the MDH enzyme patterns in habanero pepper seeds (Figure 4) were verified, with less activity in the period of 144 h of soaking in seeds processed in the

different stages of development and different storage times. In relation to the patterns observed for the enzyme ADH, it may be observed that with the advance of the soaking period, the activity of the enzyme ADH diminished in the three storage periods (Figure 4). It was also verified that activity of the enzyme increased throughout the storage period in seeds soaked for 48 h. Like MDH, there was variation in the activity of the enzyme ADH in seeds processed at different stages of maturity, in terms of the storage period and of the soaking period. In habanero seeds (Figure 4), greater activity of the ADH enzyme was observed in E2 and E3 seeds after 48 h of soaking at 0 and 8 months of storage, whereas at 4 months of storage, greater activity was observed in seeds of all three stages of development after 48 h of soaking.

Regarding the enzyme α-amylase, Figure 4 showed that there was variation in the activity of this enzyme in terms of the stage of development at which the seeds were processed, storage period and duration of soaking. The activity of the enzyme α-amylase may become evident through the clearer bands in a bluish background, where the starch was hydrolyzed. Greater activity of the enzyme α-amylase was observed at 4 months of storage compared to 0 and 8 months of storage for all stages of

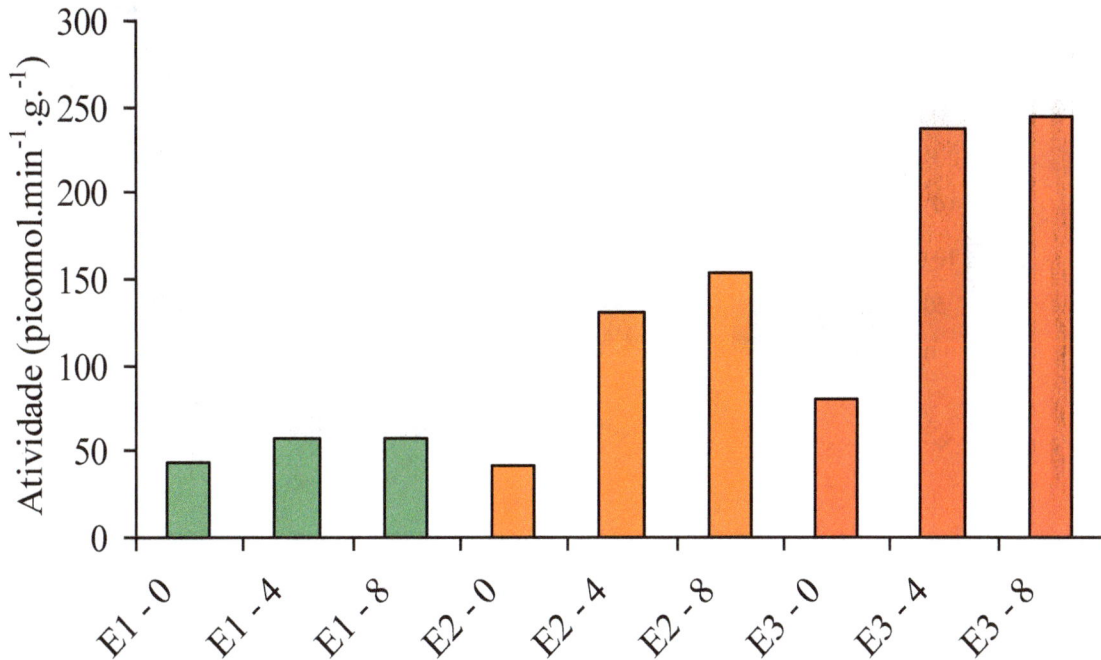

Figure 2. Activity of the enzyme endo-β-mannanase in habanero pepper seeds processed in the E1, E2 and E3 development stages in three storage periods of 0, 4 and 8 months of storage.

seed development. E1 seeds presented greater activity of this enzyme than E2 and E3 seeds at the different soaking times and storage periods, except for recently stored seeds soaked for 48 h and those stored for 8 months and soaked for 96 h.

DISCUSSION

Barbedo et al. (1994) also verified that better quality eggplant seeds were obtained from fruit harvested 50 days after anthesis and submitted to 15 days of post harvest storage. According to Sanchez et al. (1993), green pepper seeds should remain in the mature fruit (50 days after anthesis) after harvest from 7 to 14 days so that maximum germination potential is reached. According to Nascimento et al. (2006), immature fruit, of green color, generally produces seeds with low vigor and germinating power or even poorly formed seeds. The low germination percentage at the beginning of storage may be related to the presence of dormancy in the seeds which was broken throughout the storage period. These results corroborate those found by Bosland and Votava (1999), in which dormancy was observed in recently gathered seeds of species of the genus *Capsicum* (Table 2). Thus, it may be inferred that in recently stored seeds and in those stored for 4 months, aerobic respiration is greater at the beginning of the germination process. It may be observed that the greatest percentage of germination is after 4 months of storage. According to

Nascimento et al. (2006), the sowing of pepper seeds recently extracted from the fruit may represent a risk for obtaining uniform stands, contributing to increased seed expenses. These seeds are induced dormancy to preserve the perpetuation of the species. Since this seed peppers should be stored before being sown. The increase in seedling test emergency seed processed in different stages of development in relation to germination emphasizes that in spite of the reports regarding the occurrence of dormancy in pepper seeds (Bosland and Votava, 1999) (Table 2). This dormancy can be broken down by microorganisms in the substrate, when the seeds of determined cultivars are extracted from completely mature fruit and seeded thereafter (Bolsland and Votava, 1999).

Randle and Honma (1981) verified in work with different cultivars of the genus Capsicum, that the genotype and the age of the fruit influence the intensity of dormancy of the seeds. The authors reported that seeds extracted from mature fruit with days of rest before seed extraction germinate more rapidly, younger fruits being more prompt to increased seed dormancy. According to Barbedo et al. (1994), by the emergence rate index, it is possible to detect small existing differences in the physiological quality of cucumber seeds extracted from fruits harvested 15 to 45 days after anthesis (DAA) and without storage. Valdes and Gray (1998), upon harvesting tomato fruits of differing maturity stages and without post harvest storage, observed that the mean germination time of the seeds differed significantly among

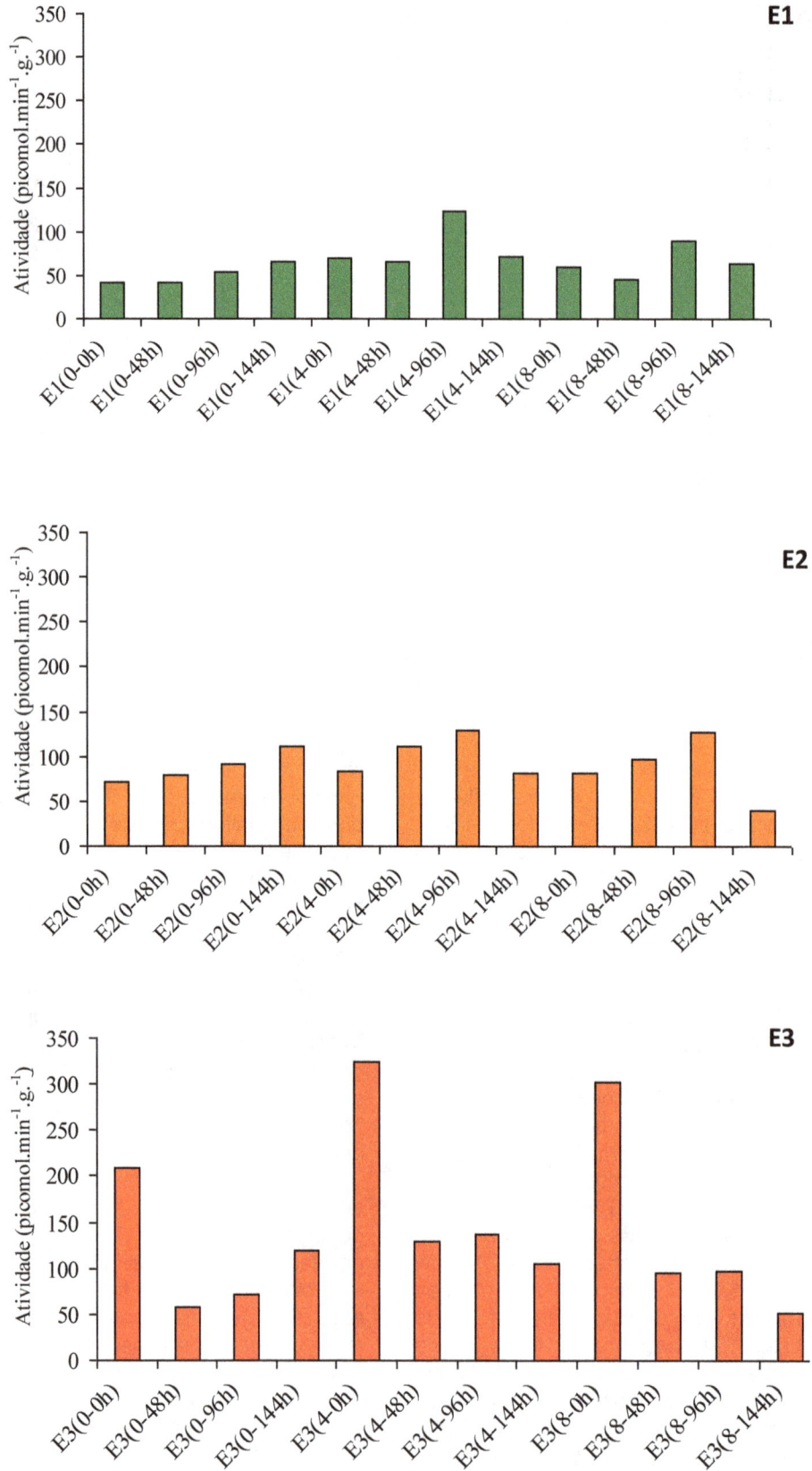

Figure 3. Activity of the enzyme endo-β-mannanase in habanero pepper seeds during 0, 48, 96 and 144 h of germination in the E1, E2 and E3 stage in the three storage periods of 0, 4 and 8 months of storage.

maturity stages, greater germination time being observed in the less mature seeds. Our findings on electrical conductivity were similar to those observed by Vidigal et al. (2006) and Dias et al. (2006) for tomato seeds extracted from fruits with different stages of maturity submitted to post-harvest storage (Table 2). Greater leaching of exudates was observed in immature seeds consistent with less structuring of the organelle and cellular membrane system. Nevertheless, the high conductivity value observed for E1 seeds suggests destructuring of the system of membranes, probably because of their immaturity (Albuquerque et al., 2009), and this fact is also reinforced by the results of the other physiological tests.

As observed by Bhering et al. (2006), the accelerated aging test was efficient to evaluate the effect of pepper seeds, checking statistical difference between the different treatments (Table 2). Esterase is an enzyme that participates in membrane hydrolysis of esters. This fact shows greater lipid peroxidation since this enzyme is involved in ester hydrolysis reactions, being directly connected to lipid metabolism (Santos et al., 2004). Many of these lipids are constituents of membranes, whose degradation increases with deterioration (Figure 1). The enzyme endo-β-mannanase being involved in the degradation of the endosperm in seed germination. In lettuce and coffee seeds, this enzyme is considered as key in the germination process, being involved in mannanase degradation at the time of germination, resulting in weakening of the cell walls of the endosperm (Silva et al., 2004; Veiga, 2005) (Figure 2). Vidigal et al. (2009) observed a small increase in the activity of SOD in chili peppers obtained from fruits harvested 50 DAA and stored for 6 days, greater physiological quality as evaluated by the germination and vigor tests (Figure 1). Enzyme ADH activity reported here was also found by Vidigal et al. (2009) (Figure 1). This enzyme is related to anaerobic respiration, promoting reduction of the acetaldehyde to ethanol. Acetaldehyde accelerates seed deterioration (Buchanan et al., 2005). With the increase of ADH activity, the seeds are more protected against the deleterious action of this compound, which is greater when compared to that of ethanol. In the 8th month of storage, through the fact of the seeds being in more advanced stages of deterioration, there is greater respiratory intensity and consequently greater demand of activity from the enzyme MDH (Figure 1). MDH enzyme patterns varied according to storage and soaking periods and seed maturity stages. In research undertaken by Taiz and Zeiger (2004), no difference in MDH activity was observed in seeds during the maturation process. Nevertheless, the authors reported that the reserve organs in development need greater energy supply and, therefore, respiratory activity in these plant tissues is more intense (Figure 4). These results with MDH and ADH may be associated with the germination and vigor data, in which the seeds processed in the E2 and E3

stages have better quality than the seeds of the E1 stage (Table 2).

According to Nedel et al. (1996), within a group of enzymes, the α and β - amylases are involved in the main starch degradation system. Development of the amylase activity constitutes an important event and may be detected at the beginning of germination with its main role being the making of substrates for the plantlet nutrition until it becomes photosynthetically self-sufficient. A large number of types of seed dormancy arise from blocking the action of α-amylase. The α-amylase present in the dormant seeds is found in small quantities. The activity of this enzyme increases to the extent that the dormancy of rice seeds is overcome during the storage period (Vieira et al., 2008).

In this study, the presence of dormant seeds was observed, principally at the E1 stage of development and in recently stored seeds (Table 2). In these seeds, high activity of α-amylase was observed, which confirms the importance of this enzyme in the germination process of pepper seeds. In E1, E2 and E3 stage seeds and recently stored seeds, there was an increase in the activity of the enzyme endo-β-mannanase (Figure 3) to the extent that the soaking period of the seeds was increased during the germination process. The greatest activity of this enzyme was observed at 144 h of soaking, which coincided with the occurrence of root protrusion. Seeds of the three stages of development stored for 4 and 8 months germinated 96 h after soaking. These data may be correlated with the greater activity of this enzyme in seeds submitted to these treatments. Greater activity of the endo-β-mannanase, in absolute values, was verified in the E2 and E3 stage seeds at all the storage and soaking periods during the germination process. The lowest germination and vigor values were observed in E1 seeds and in recently stored seeds. In these seeds, the activity of the enzyme endo-β-mannanase was, the lowest, indicating the importance of this enzyme in the germination of pepper seeds. The greatest level of activity of this enzyme was observed in seeds stored for 4 months and soaked for 96 h, and this was true for all the three stages of development (Figure 3). The results of the tests used for evaluation of physiological quality, showed an increase in the germination and vigor values for 4 month-stored seeds (Table 2). Based on these results, the presence of dormancy in recently harvested seeds is inferred. This dormancy has probably been overcome in the 4th month of storage as shown by a greater activity of the enzyme endo-β-mannanase.

Conclusions

By means of the physiological tests and the activity of the enzymes evaluated, it was observed that habanero pepper seeds should be extracted from E3 stage fruits, which ensures production of better quality seeds. Seed

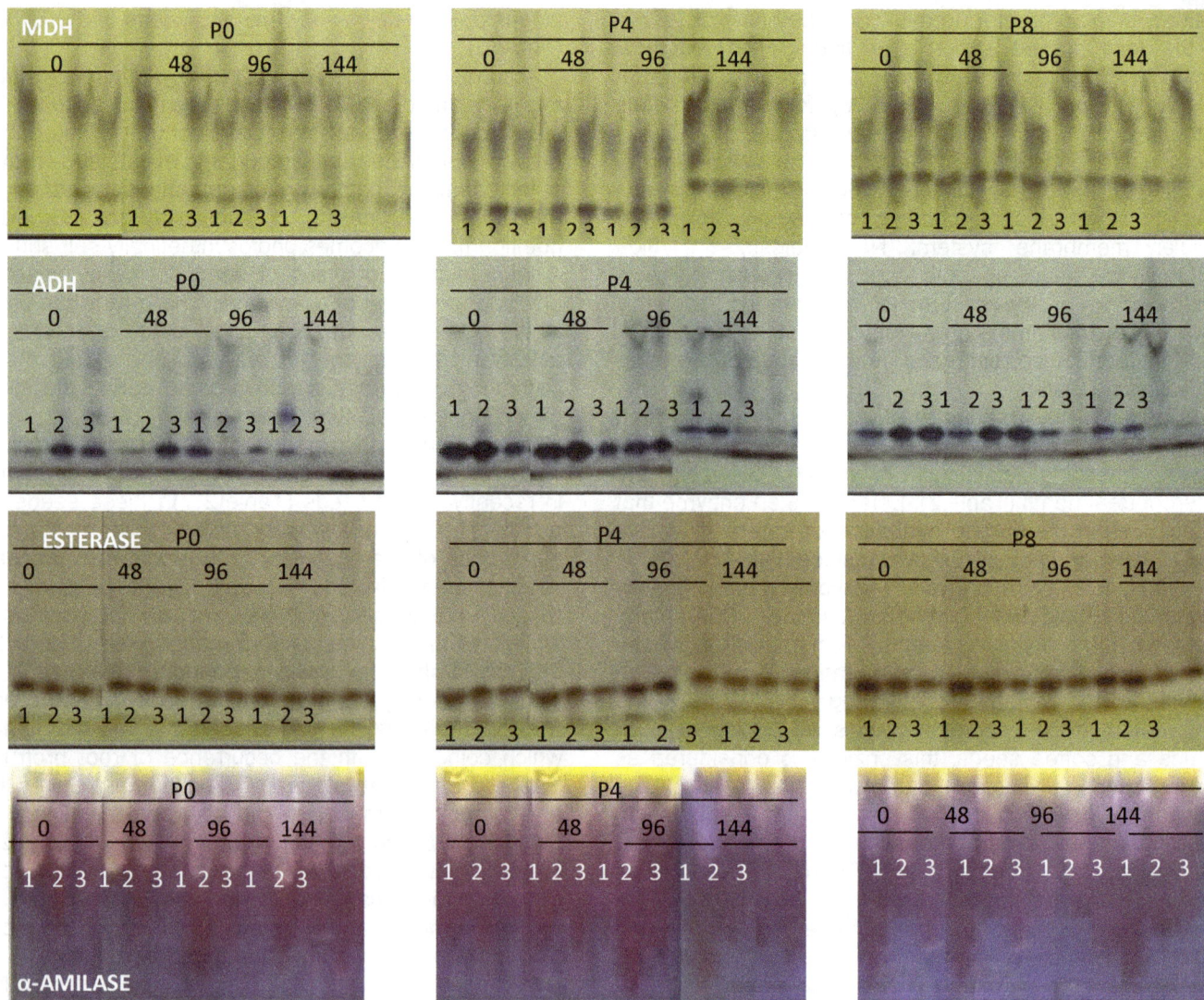

Figure 4. Electrophoretic patterns of the enzyme MDH, ADH, esterase and α-amylase observed in habanero pepper seeds during germination: 0, 48, 96 and 144 h in the E1 (1), E2 (2) and E3 (3) stages with 0 (P0), 4 (P4) and 8 (P8) months of storage.

extraction from fruits harvested at E1 stage must be avoided because of lower physiological quality due to seed dormancy and immaturity. Seed physiological maturity in the species studied does not coincide with the maximum germination and vigor values due to the incidence of dormancy.

REFERENCES

Albuquerque KS, Guimarães RM, Gomes LAA, Vieira AR, Jácome MF (2009). Condicionamento osmótico e giberelina na qualidade fisiológica de sementes de pimentão colhidas em diferentes estádios de maturação. Rev. Bras. Sementes 31(4):100-109.

Alfenas AC, Petres I, Bruce W, Passados GC (1991). Eletroforese de proteínas e isoenzimas de fungos e essências florestais. Viçosa: UFV, P. 242.

Barbedo ASC, Zanin ACW, Barbedo CJ, Nakagawa J (1994). Efeitos da idade e do período de repouso pós-colheita dos frutos sobre a qualidade de sementes de berinjela. Horticultura Brasileira, Brasília, 12 (1):18-21.

Bhering MC, Dias DCFS, Vidigal DDS, Naveira, DDSP (2006). Teste de envelhecimento acelerado em sementes de pimenta. Rev. Bras. Sementes 28(3):64-71.

Bosland PW, Votava EJ (1999). Peppers: vegetable and spice capsicums. Wallingford. Crop Prod. Sci. Hortic. 12:204.

Brazil (2009). Ministério da Agricultura, Pecuária e Abastecimento. Regras para análise de sementes. Brasília, P. 395.

Buchanan BB, Gruissem W, Jones RL (2005). Biochemistry and molecular biology of plants. Rockville: Am. Soc. Plant Physiol. P. 1367.

Copeland LO, McDonald MB (2001). Principles of seed science and technology. 4 ed. New York: Chapman & Hall, P. 467.

Dias DCFS, Ribeiro FP, Dias LAS, Silva DH, Vidigal DS (2006). Tomato seed quality in relation to fruit maturation and post-harvest storage. Seed Sci. Technol. Zürich 34(3):691-699.

Downie B, Hillhorst HWM, Bewley JD (1994). A new assay for quantifying endo-β-mananase activity using Congo Red dye.

Phytochemistry, Oxford, 36(4):829-835.

Ferreira DF (2000). Análises estatísticas por meio do SISVAR para Windows® versão 4.0. In: Reunião anual da região brasileira da sociedade internacional de biometria, 45, 2000, São Carlos, SP. Programas e Resumos... São Carlos: UFSCar, P. 235.

Filgueira FA (2005). Novo manual de olericultura: Agrotecnologia moderna na produção e comercialização de hortaliças. 2 ed. Viçosa: Editora UFV, P. 412.

Freitas RA (2009). Deterioração e armazenamento de sementes de hortaliças. In: Nascimento WM (Ed.). Tecnologia de sementes de hortaliças. Brasília, DF: Embrapa hortaliças pp. 155-184.

Maguire JD (1962). Speed of germination and relation evaluation for seedling emergence vigor. Crop Sci. Madison 2:176-177.

Nascimento WM, Dias DCFS, Freitas AF (2006). Produção de sementes de pimentas. Informe Agropecuário, Belo Horizonte, 27(235):30-39.

Nedel JL, Assis FN, Carmona PS (1996). A planta de arroz: morfologia e fisiologia. Pelotas: UFPEL, 56 p.

Randle WM, Honna S (1981). Dormancy in peppers. Scientia Horticulturae, Alexandria, 14:19-25.

Sanchez VM, Sundstrom GN, McClure GN, Lang NS (1993). Fruit maturity, storage and postharvest maturation treatments affect bell pepper (Capsicum annuum L.) seed quality. Sci. Hortic. Alexandria 54(3):191-201.

Santos CMR, Menezes NL, Vilela FA (2004). Alterações fisiológicas e bioquímicas em sementes de feijão envelhecidas artificialmente. Rev. Bras. Sementes Bras. 26(1):110-119.

Silva EAA, Toorop PE, Aelst AC, Hilhorst HWM (2004). Abscisic acid controls embryo growth potential and endosperm cap weakening during coffee (Coffea arabica cv. Rubi) seed germination. Planta Berlin 220(2):251-261.

Taiz L, Zeiger E (2004). Fisiologia vegetal. 3 ed. Porto Alegre: Artmed, pp. 106-325.

Valdes VM, Gray D (1998). The influence of stage of fruit maturation on seed quality in tomato (Lycopersicon lycopersicum (L.) Karsten). Seed Sci. Technol. Zürich 26(2):309-318.

Veiga AD (2005). Armazenabilidade de sementes de cafeeiro em diferentes estádios de maturação e submetidas à diferentes métodos de secagem. 60 f. Dissertação (Mestrado em Fitotecnia) – Universidade Federal de Lavras, Lavras.

Vidigal DS, Dias DCFS, Naveira DSP, Rocha FB, Bhering MC (2009). Alterações fisiológicas e enzimáticas durante a maturação de sementes de pimenta (Capsicum annuum L.). Rev. Bras. Sementes 31(2):129-136.

Vidigal DS, Dias DCFS, Pinho EVRV Dias LAS (2006). Qualidade fisiológica de sementes de tomate em função da idade e do armazenamento pós-colheita dos frutos. Rev. Bras. Sementes 28(3):87-93.

Vidigal DS, Lima JS, Behring MC, Dias DCFS (2008). Teste de condutividade elétrica para sementes de pimenta. Rev. Bras. Sementes 30(1):168-174.

Vieira AR, Oliveira JA, Guimarães RM, Pinho EVRV, Pereira CE, Clemente ACS (2008). Marcador isoenzimático de dormência em sementes de arroz. Rev. Bras. Sementes 30(1):81-89.

24

Optimization of irradiation and storage temperature for delaying ripening process and maintaining quality of Alphonso mango fruit (*Mangifera indica* L.)

M. K. Yadav[1] and N. L. Patel [2]

[1]Department of Horticulture, N. M. College of Agriculture, Navsari Agricultural University, Dandi Road, Navsari-396450, India.
[2]ASPEE College of Horticulture and Forestry, Navsari Agricultural University, Dandi Road, Navsari -396450, India.

Alphonso mango fruit has high nutritional values, pleasant flavor, delicious taste as well as beautiful appearance and hence is known as the king of mango verities. The experiment was arranged from the 2008 and 2010 with 16 treatment combinations of irradiation dose (that is, 0.00, 0.20, 0.40, and 0.60 kGy) and stored at different storage temperatures viz., ambient at 27 ± 2°C and 60 to 70% RH, 9°C and 90% RH, 12°C and 90% RH, and Control atmospheric storage (12°C, O_2 2%, CO_2 3% and RH 90%). The fruits were exposed to gamma radiation from the source of ^{60}Co. The two years collective data indicated that, the significantly minimum percent reduction in physiological loss in weight, reduced ripening percent, increased marketability of fruits, maximum total soluble solids, total and reducing sugars, and ascorbic acid content and minimum acidity were noted in 0.40 kGy gamma rays irradiated fruits stored at 12°C as compared to the other irradiated or unirradiated fruits stored at ambient condition and other storage environment. Suggestions were made for maximizing maintained physiological changes and quality by use of irradiation and adequate storage facilities for hygiene produce.

Key words: Alphonso mango, irradiation, marketability, ripening, quality, storage temperature.

INTRODUCTION

Asia accounts for 77% of global mango production and the Americas and Africa account for 13 and 19%, respectively (Pereira et al., 2010). India is the global leader in mango production (Tharanathan et al., 2006).The significant mission of any post harvest skill is to be raising the method by which decline of produce is controlled as much as possible during the stage between collect and consumption. Mango (*Mangifera indica* L., family Anacardiacae) is a tropical fruit and classified as climacteric fruit and ripens rapidly after harvest. Mango is generally harvested when physiologically mature and is allow ripening under suitable conditions of temperature and humidity. Therefore, if freshly harvested fruit is allowed to ripen at normal ambient conditions, ripening processes increase rapidly within few days, and quality point of view, it is not good. Mango is susceptible to chilling injury and an optimum temperature of 12 to 13°C is generally recommended (Gomez-Lim, 1993, Yimyong et al., 2011).

Irradiation is a physical process for the treatment of foods akin to conventional process like heating or freezing. It prevents food poisoning, reduces wastage to contamination, and at the same time preserves quality (Mahindru, 2009). However, issues related to quarantine and quality are the major stumbling blocks to trade, both national and international (Yadav et al., 2010). Therefore, the new knowledge is critical because it is important to maintain a balance between the optimum doses required to achieve safety and the minimum change in the sweetness of the fruit. In view of the aforementioned fact, it becomes quite clear that, investigation for mango fruit is very important for not only increase the soluble solids but also to control the conversion of starch into sugars for long time. The loss in sweetness of fruits is likely to reduce the marketability and quality of fruit drastically. Alphonso mangoes from India have captured sizeable Indian market and have very good export potential, but the protocol for their irradiation and post harvest storage yet needed to be standardized. In this paper the results of studies for standardization protocol of irradiation and storage are presented and discussed.

MATERIALS AND METHODS

Fruits and irradiation treatment

The experiment was set from 2008 to 2010 at Department of horticulture, N. M. College of Agriculture, Navsari Agricultural University, Navsari, Gujarat. Export grade mangoes of cv. Alphonso were harvested from the University orchard. The selected mangoes from class I as per the quality parameters specified and described in "post harvest manual for mangoes" published by Agricultural Production and Export Development Authority (Anonymous, 2007). These fruits sorted by uniformity in size, maturity, and freedom from defects. The fruits were kept in plastic crates with cushioned material and transported to cold storage of Post Harvest Technology Unit, Navsari Agricultural University, Navsari (Gujarat) India. Than after, fruits were again sorted to remove those with spotty and having bad appearance. The individual fruit weight was from 250 to 350 g. The selected fruits were washed with chlorine water and after drying, the fruits were packed in corrugated Fiber board boxes cushioned (CFB) with tissue paper. The dimension of CFB box was 370 × 275 × 90 mm and gross weight of box with fruits was 3.0 kg. One box having nine fruits for each treatment and each treatment replicated thrice as per experimental design. The packed boxes kept in cold storage at 12°C for 8 h for pre-cooling treatment. The time gap between harvesting and pre-cooling was not more than 6 h.

After pre-cooling, fruits were transported to irradiation treatment in air conditioned vehicle. It was carried out at ISOMED plant, Board of Radiation and Isotope Technology, Bhabha Atomic Research Centre, Mumbai (India). The fruits were exposed to gamma radiation for different doses from the source radio isotope ^{60}Co with energy 1.33 MeV. There were four irradiation doses that is, I_1 -0.00 kGy (Unirradiated), I_2 -0.20 kGy, I_3 -0.40 kGy, and I_4 -0.60 kGy. The time gap from pre-cooling to irradiation was not more than 9 h. After irradiation, fruits immediately transported to cold storage of university in air conditioned vehicle.

Storage conditions

The boxes were kept in storage at different temperature as per s torage temperature treatments viz., ambient at 27 ± 2°C and 65 ± 5% relative humidity (S_1), 9°C and 90% relative humidity (S_2), 12°C

and 90% relative humidity (S_1) and Control atmospheric storage at 12°C, O_2 2%, CO_2 3%, and 90% relative humidity (S_1). Post harvest biochemical changes of these fruits were studied by measuring the total soluble solids, sugars, acidity and ascorbic acid content of fruits.

Measurement protocols

Determination of physiological parameters

Physiological loss in weight (PLW) (%): Four fruits from each treatment were weighted on 1st day of treatment and subsequently their weight was recorded from 4 to 6 day interval up to the end of shelf life. The PLW was expressed in percentage and calculated as follows;

$$PLW \% = \frac{W_1 - W_2}{W_1} \times 100$$

where, W_1 = initial weight and W_2 = final weight (Shankar et al., 2009).

Ripening percent: Ripening was measured by the 10 number of fruits having change in colour from greenish to yellow and soft in texture were counted from the 4th day of storage to the 6th day intervals up to the eating ripeness and expressed in percentage over total number of fruits taken for study.

Marketable fruits percent: The number of good quality and visibly sound fruits that can be marketed were counted and expressed as percentage over the total number of fruits at prescribed interval up to 90% fruits has marketability.

Quality parameters

Total soluble solids were tested by using a digital hand refractometer PAL-1 (Atago, Japan). Sugar's percentage was determined by titrimetric method of Lane and Eynon described by Rangana (1986). The method is based on the principle that, invert sugar or reducing sugar reduced the copper in the Fehling's solution to red insoluble cuprous oxide. Non-reducing sugars were calculated by subtracting reducing sugar from total sugars. Method for titrable acidity by Rangana (1986) was adopted for estimation of titrable acidity. Ascorbic acid (mg/100 g) determination by the 2, 6-dichloroindophenol titrimetric method described by Rangana (1986) was adopted for estimation of the ascorbic acid content of fruits.

Statistical analysis

Two years thrice replicated data obtained from the experiment was analyzed using ANOVA for completely randomizes deign with factorial concept. Significance differences among treatments were compared using the Fisher's analysis of variance at the 5% probability level, technique as described by Panse and Sukhatme (1967). The data were subjected to appropriate transformation (arcsine) to meet the assumptions of normality.

RESULTS AND DISCUSSION

Physiological loss in weight

The data indicated that, the physiological loss in weight of fruits increased with the advancement of storage period and significantly influenced by irradiation and storage temperature. It was evident from the Table 1 that, the shelf life of fruits exposed with 0.40 and 0.60 kGy irradiation and stored at 9°C was extended more than 34 days. The minimum reduction in PLW was recorded in the fruits exposed with 0.40 kGy irradiation and stored at 9°C (I_3S_2) that is, 5.50% at 34th day, 4.45% at 28th day, 3.23% at 22th day, 2.35% at 16th day, 1.43% at 10th day, and 0.53% at 4th day of storage. The physiological loss in weight of fruits was possibly on account of loss of moisture through transpiration and utilization of some reserve food materials in the process of respiration (Mayer et al., 1960).

The physiological loss in weight of mango fruit was significantly influenced by the various exposed dose of gamma rays and different storage temperature. The irradiation significantly reduced physiological loss in weight during storage period over control which might be attributed to reduction in utilization of reserve food material in the process of respiration (Purohit et al., 2004). The delay in respiration rate as a result of irradiation was also reported by Singh and Pal (2009) in guava (Psidium guajava L.). Similar findings were also observed by Prasadini et al. (2008) and by El-Salhy et al. (2006) in mango.

Similarly, in the different storage conditions, the highest physiological loss in weight was observed in fruits subjected to ambient temperature and this was largely due to water loss through lenticles of fruits, which permit free water vapor movement (Salahddin and Kedar, 2006). Lower physiological loss in weight was noted in temperatures which might be due to lesser water vapour deficit compared to ambient condition and the low temperature which had slowed down the metabolic activities like respiration and transpiration (Mane and Patel, 2010). The observation accordance with the results in mango (Waskar and Masalkar, 1997), in banana (Nagaraju and Reddy, 1995), and in guava (Gutierrez et al,. 2002). The significantly minimum reduction in physiological loss in weight of mango fruits subjected to irradiation and stored at various temperatures that is, at 9 and 12°C and in CA (12°C) might be due to the mutual complementary effect of irradiation and low temperature.

Ripening percent

Irradiated fruits significantly delayed the ripening process over unirradiated fruits irrespective of storage condition (Table 2) and not fully ripe up to 34th day of storage at 9°C. Rest of the treatments had more ripening and the other was discarded due to the lost of their shelf life. The fruits exposed to gamma rays (0.20 and 0.40 kGy) and stored at 9°C were showed at 86.21 and 84.23% ripening, respectively at 9°C (S_3) on 34 days of storage. Rest of the treatments had high ripening or discarded due to complete of their shelf life. Ripening percentage is a

physiological process which designates the period from harvest until the fruits attain the stage of maximum consumer acceptability. The unirradiated mangoes had early ripeness whereas, gamma rays exposed mangoes that had a significantly delayed in ripening. The possible mechanisms that have been postulated include:

a) Irradiations results in decreased sensitivity to ripening action of ethylene.
b) Alteration in carbohydrates metabolism by regulating certain key enzymes, which interfere with production of ATP which is required for various synthetic processes during ripening (Udipi and Ghurge, 2010). Same findings were noted by Farzana (2005) in mango and by Aina et al. (1999) in banana. The decrease of ripening percent and increase in days for ripening at low temperature may be due to desirable inhibition of enzymatic activities leading to reduction in the respiration and ethylene production. These results were supported by Mann and Singh (1975) in mango and by Deka et al. (2006) in banana. The minimum and delayed ripening in fruits due to exposed to gamma rays and storage temperature at 9 and 12°C and in CA (12°C) storage compared to fruits unirradiated and kept at ambient temperature in present study might be due to the joint balancing effect of irradiation and low temperature.

Marketable fruits percent

During storage, few treatments had 100% values for marketability and few had 0.00% marketability due to induction of senescence (Table 3). Irradiation significantly influenced the marketable fruit compared to unirradiated fruits at all conditions of the storage.

The highest marketable fruit (96.46%) was recorded in fruits exposed to 0.40 kGy gamma irradiation and kept at 12°C storage (I_3S_3) at 34 day of storage, and the rest of treatments had lower marketability or discarded due to the end of their shelf life. The marketable fruit was significantly influenced by various doses of gamma irradiation and storage temperatures. The possible reasons might be that, irradiation maintained water content in the fruit and low temperature coupled with high humidity in cold storage maintained the health of the fruits. These results were in conformity with the findings of El-Salhy et al. (2006) with respect to irradiation and Mane and Patel (2010) with respect to low temperature in mango.

Total soluble solids

The data revealed that, total soluble solids in fruits were significantly affected by irradiation, storage temperature, and their interaction. It was evident from the data presented in Table 4 that significantly, the maximum total

Table 1. Optimization of irradiation and storage temperature for maintaining physiological loss in weight of Alphonso mango.

Physiological loss in weight days after storage (%)

4 days after storage

Source	I_1	I_2	I_3	I_4	Mean
S_1	3.54	2.82	2.68	2.79	2.95
S_2	0.89	0.73	0.53	0.78	0.73
S_3	0.92	0.80	0.66	0.88	0.81
S_4	2.73	2.34	1.74	2.40	2.30
Mean	2.02	1.67	1.40	1.71	

Source	I	S	I X S
S. Em ±	0.002	0.004	0.004
CD ($P \leq 0.05$)	0.005	0.016	0.011

10 days after storage

Source	I_1	I_2	I_3	I_4	Mean
S_1	12.21	7.47	7.18	7.26	8.53
S_2	2.36	1.87	1.43	2.05	1.93
S_3	2.56	2.00	1.65	2.23	2.11
S_4	7.12	3.52	3.39	3.62	4.41
Mean	6.06	3.72	3.41	3.79	

Source	I	S	I X S
S. Em ±	0.003	0.003	0.006
CD ($P \leq 0.05$)	0.009	0.009	0.017

16 days after storage

Source	I_1	I_2	I_3	I_4	Mean
S_1	0.00 (1.65)	10.31 (18.72)	10.83 (19.20)	11.50 (19.81)	8.16 (14.84)
S_2	3.19 (10.28)	2.97 (9.91)	2.35 (8.82)	3.10 (10.13)	4.30 (9.79)
S_3	3.88 (11.34)	3.10 (10.13)	2.58 (9.23)	3.38 (10.58)	3.24 (10.32)
S_4	10.11 (18.53)	5.28 (13.27)	5.23 (13.20)	5.48 (13.52)	6.54 (14.63)
Mean	4.30 (10.45)	5.42 (13.01)	5.25 (12.62)	5.87 (13.52)	

Source	I	S	I X S
S. Em ±	0.005	0.005	0.011
CD ($P \leq 0.05$)	0.016	0.016	0.031

22 days after storage

Source	I_1	I_2	I_3	I_4	Mean
S_1	0.00 (1.65)	13.49 (21.54)	12.66 (20.84)	13.83 (21.83)	9.10 (16.46)
S_2	4.66 (12.45)	4.24 (11.87)	3.23 (10.35)	4.37 (12.05)	4.12 (11.68)
S_3	4.83 (12.68)	4.38 (12.07)	3.43 (10.67)	4.63 (12.42)	4.32 (11.96)
S_4	13.33 (21.40)	7.33 (15.69)	6.96 (15.28)	8.00 (16.42)	8.91 (17.20)
Mean	5.70 (12.05)	7.36 (15.29)	6.61 (14.53)	7.42 (15.43)	

Source	I	S	I X S
S. Em ±	0.004	0.004	0.009
CD ($P \leq 0.05$)	0.013	0.013	0.026

28 days after storage

Source	I_1	I_2	I_3	I_4	Mean
S_1	0.00 (1.65)	0.00 (1.65)	0.00 (1.65)	0.00 (1.65)	0.00 (1.65)
S_2	6.43 (14.68)	5.47 (13.52)	4.45 (12.16)	5.78 (13.90)	5.53 (13.57)
S_3	6.95 (15.27)	5.70 (13.80)	4.62 (12.41)	5.99 (14.15)	5.82 (13.91)
S_4	0.00 (1.65)	9.31 (17.75)	9.21 (17.65)	9.51 (17.95)	7.01 (13.75)
Mean	3.350 (8.31)	5.12 (11.78)	4.57 (10.97)	5.32 (11.92)	

Source	I	S	I X S
S. Em ±	0.004	0.005	0.009
CD ($P \leq 0.05$)	0.013	0.014	0.026

34 days after storage

Source	I_1	I_2	I_3	I_4	Mean
S_1	0.00 (1.65)	0.00 (1.65)	0.00 (1.65)	0.00 (1.65)	0.00 (1.65)
S_2	7.74 (16.36)	7.00 (15.33)	5.50 (13.56)	7.10 (15.44)	6.84 (15.17)
S_3	0.00 (1.65)	7.23 (15.59)	5.72 (13.83)	7.52 (15.91)	5.12 (11.74)
S_4	0.00 (1.65)	0.00 (1.65)	0.00 (1.65)	0.00 (1.65)	0.00 (1.65)
Mean	1.94 (5.33)	3.56 (8.55)	2.81 (7.67)	3.66 (8.66)	

Source	I	S	I X S
S. Em ±	0.003	0.003	0.006
CD ($P \leq 0.05$)	0.009	0.010	0.019

Figure in parenthesis indicates ARC SINE transformed value. Where, I= Irradiation, S= Storage temperature.

soluble solids (17.69%) were recorded in fruits exposed to treatment I_3 (0.40 kGy) followed by treatment I_2 (0.20 kGy). The minimum total soluble solids (16.60%) were observed in treatment I_1(0.00 kGy). The higher total soluble solids in medium and lower dose irradiated fruits indicating

Table 2. Optimization of irradiation and storage temperature for maintaining ripening of Alphonso mango.

Ripening days after storage (%)

Source	4 — I₁	I₂	I₃	I₄	Mean	10 — I₁	I₂	I₃	I₄	Mean	16 — I₁	I₂	I₃	I₄	Mean
S₁	0.00	0.00	0.00	0.00	0.00	93.96 (75.73)	0.00 (1.65)	0.00 (1.65)	0.00 (1.65)	23.49 (20.17)	0.00* (1.65)	71.68 (57.82)	69.95 (56.73)	72.96 (58.64)	53.57 (43.7)
S₂	0.00 (1.65)	0.00 (1.65)	0.00 (1.65)	0.00 (1.65)	0.00 (1.65)	0.00 (1.65)	0.00 (1.65)	0.00 (1.65)	0.00 (1.65)	0.00 (1.65)	0.00 (1.65)	0.00 (1.65)	0.00 (1.65)	0.00 (1.65)	0.00 (1.65)
S₃	0.00 (1.65)	0.00 (1.65)	0.00 (1.65)	0.00 (1.65)	0.00 (1.65)	0.00 (1.65)	0.00 (1.65)	0.00 (1.65)	0.00 (1.65)	0.00 (1.65)	0.00 (1.65)	0.00 (1.65)	0.00 (1.65)	0.00 (1.65)	0.00 (1.65)
S₄	0.00 (1.65)	0.00 (1.65)	0.00 (1.65)	0.00 (1.65)	0.00 (1.65)	0.00 (1.65)	0.00 (1.65)	0.00 (1.65)	0.00 (1.65)	0.00 (1.65)	0.00 (1.65)	0.00 (1.65)	0.00 (1.65)	0.00 (1.65)	0.00 (1.65)
Mean	8.36 (10.12)	0.00 (1.65)	0.00 (1.65)	0.00 (1.65)		23.49 (20.17)	0.00 (1.65)	0.00 (1.65)	0.00 (1.65)		0.00 (1.65)	17.95 (15.69)	17.49 (15.42)	18.12 (15.90)	
Source	I			S	IXS	I			S	IXS	I			S	IXS
S. Em ±	0.003			0.003	0.004	0.007			0.007	0.002	0.010			0.010	0.019
CD (P≤0.05)	0.010			0.010	0.014	0.021			0.021	0.007	0.028			0.029	0.055

Source	22 — I₁	I₂	I₃	I₄	Mean	28 — I₁	I₂	I₃	I₄	Mean	34 — I₁	I₂	I₃	I₄	Mean
S₁	0.00* (1.65)	0.00* (1.65)	0.00* (1.65)	0.00* (1.65)	0.00 (1.65)	0.00* (1.65)	0.00* (1.65)	0.00* (1.65)	0.008 (1.65)	0.00 (1.65)	0.00* (1.65)	0.00* (1.65)	0.00* (1.65)	0.00* (1.65)	0.00 (1.65)
S₂	0.00 (1.65)	0.00 (1.65)	0.00 (1.65)	0.00 (1.65)	0.00 (1.65)	65.38 (53.94)	29.98 (33.18)	28.39 (32.19)	47.16 (43.35)	42.73 (40.66)	0.00* (1.65)	86.21 (68.29)	84.23 (66.65)	97.81 (81.56)	66.91 (54.47)
S₃	0.00 (1.65)	0.00 (1.65)	0.00 (1.65)	0.00 (1.65)	0.00 (1.65)	74.75 (59.81)	33.16 (35.14)	28.98 (32.55)	57.47 (49.56)	48.59 (44.27)	0.00* (1.65)	97.08 (80.25)	96.40 (79.08)	98.16 (82.31)	72.91 (60.75)
S₄	77.30 (61.52)	69.53 (56.47)	65.00 (53.71)	73.26 (58.84)	71.27 (57.63)	0.00* (1.65)	0.00* (1.65)	0.00* (1.65)	0.00* (1.65)	0.00 (1.65)	0.00* (1.65)	0.00* (1.65)	0.00* (1.65)	0.00* (1.65)	0.00 (1.65)
Mean	19.33 (16.62)	17.38 (15.36)	16.25 (14.67)	18.32 (15.95)		35.03 (29.26)	15.79 (17.91)	14.34 (17.01)	26.16 (24.06)		0.00 (1.65)	45.82 (37.90)	45.16 (37.23)	48.99 (41.74)	
Source	I			S	IXS	I			S	IXS	I			S	IXS
S. Em ±	0.01			0.02	0.02	0.09			0.04	0.02	0.03			0.03	0.06
CD (P≤0.05)	0.03			0.06	0.06	0.27			0.14	0.05	0.09			0.09	0.17

Figure in parenthesis indicates ARC SINE transformed value   2 * indicate fruits completely discarded Where, I = irradiation, S = storage temperature.

the induction of ripening process due to irradiation (Sudto et al. 2005). These results were in accordance with the findings of El-Salhy et al. (2006) in mango, Wall (2007) in banana, Singh and Pal (2007) in guava, and Silva et al.(2010) in Caja (*Spondias sp.*) fruit. Under various storage

Table 3. Optimization of irradiation and storage temperature for maintaining marketing of Alphonso mango.

Marketable fruits days after storage (%)

Source	16					10					4				
	I_1	I_2	I_3	I_4	Mean	I_1	I_2	I_3	I_4	Mean	I_1	I_2	I_3	I_4	Mean
S_1	0.00 (1.65)	100 (88.31)	100 (88.31)	100 (88.31)	75.00 (60.90)	97.66 (81.19)	100 (88.31)	100 (88.31)	100 (88.31)	99.41 (86.53)	0.00 (1.65)	0.00 (1.65)	0.00 (1.65)	0.00 (1.65)	8.36 (1.65)
S_2	100 (88.31)	100 (88.31)	100 (88.31)	100 (88.31)	100 (88.31)	100 (88.31)	100 (88.31)	100 (88.31)	100 (88.31)	100 (88.31)	0.00 (1.65)	0.00 (1.65)	0.00 (1.65)	0.00 (1.65)	0.00 (1.65)
S_3	100 (88.31)	100 (88.31)	100 (88.31)	100 (88.31)	100 (88.31)	100 (88.31)	100 (88.31)	100 (88.31)	100 (88.31)	100 (88.31)	0.00 (1.65)	0.00 (1.65)	0.00 (1.65)	0.00 (1.65)	0.00 (1.65)
S_4	100 (88.31)	100 (88.31)	100 (88.31)	100 (88.31)	100 (88.31)	100 (88.31)	100 (88.31)	100 (88.31)	100 (88.31)	100 (88.31)	0.00 (1.65)	0.00 (1.65)	0.00 (1.65)	0.00 (1.65)	0.00 (1.65)
Mean	75.00 (60.90)	100 (88.31)	100 (88.31)	100 (88.31)		99.41 (86.53)	100 (88.31)	100 (88.31)	100 (88.31)		8.36 (10.12)	0.00 (1.65)	0.00 (1.65)	0.00 (1.65)	
Source	I			S	I X S	I			S	I X S	I			S	I X S
S. Em ±	0.02			0.02	0.03	0.03			0.03	0.07	0.003			0.003	0.004
CD ($P \leq 0.05$)	0.05			0.05	0.10	0.10			0.10	0.20	0.010			0.010	0.014

Source	34					28					22				
	I_1	I_2	I_3	I_4	Mean	I_1	I_2	I_3	I_4	Mean	I_1	I_2	I_3	I_4	Mean
S_1	0.00 (1.65)	0.00 (1.65)	0.00 (1.65)	0.00 (1.65)	0.00 (1.65)	0.00 (1.65)	0.00 (1.65)	0.00 (1.65)	0.00 (1.65)	0.00 (1.65)	0.00 (1.65)	88.80 (70.43)	89.33 (70.91)	85.96 (67.97)	66.02 (52.74)
S_2	0.00 (1.65)	70.45 (57.05)	74.22 (59.46)	69.11 (56.21)	53.45 (43.59)	84.54 (66.82)	95.94 (78.36)	100 (88.31)	94.00 (75.80)	93.62 (77.32)	100 (88.31)	100 (88.31)	100 (88.31)	100 (88.31)	100 (88.31)
S_3	74.16 (59.42)	94.13 (75.94)	96.46 (79.12)	92.00 (73.54)	89.19 (72.00)	98.44 (82.83)	100 (88.31)	100 (88.31)	100 (88.31)	99.61 (86.94)	100 (88.31)	100 (88.31)	100 (88.31)	100 (88.31)	100 (88.31)
S_4	0.00 (1.65)	0.00 (1.65)	0.00 (1.65)	0.00 (1.65)	0.00 (1.65)	59.41 (50.41)	70.89 (57.32)	84.23 (66.57)	69.27 (56.31)	70.95 (57.65)	94.04 (78.85)	100 (88.31)	100 (88.31)	96.17 (78.70)	97.55 (82.79)
Mean	18.54 (16.10)	41.15 (34.07)	42.67 (35.47)	40.28 (33.26)		60.60 (50.43)	66.71 (56.41)	71.06 (61.21)	65.82 (55.52)		73.51 (63.53)	97.20 (83.84)	97.33 (83.56)	95.53 (80.82)	
Source	I			S	I X S	I			S	I X S	I			S	I X S
S. Em ±	0.02			0.02	0.03	0.03			0.03	0.07	0.003			0.003	0.004
CD ($P \leq 0.05$)	0.05			0.05	0.10	0.10			0.10	0.20	0.010			0.010	0.014

Figure in parenthesis indicates ARC SINE transformed value, I = irradiation, S= storage temperature.

conditions, the maximum total soluble solids (18.04%) were recorded by fruits stored under treatment S_3 (12°C) followed by treatment S_4 (CA at 12°C). The minimum total soluble solids (16.62%) were observed under treatment S_1 (9C). The total soluble solids in fruits at ripening were

Table 4. Optimization of irradiation and storage temperature for maintaining quality of Alphonso mango.

TSS (%)

Source	I_1	I_2	I_3	I_4	Mean
S_1	15.86	17.13	17.36	16.14	16.62
S_2	16.12	17.27	17.36	16.57	16.83
S_3	17.62	18.27	18.46	17.82	18.04
S_4	16.83	17.44	17.57	17.34	17.30
Mean	16.60	17.53	17.69	16.96	
Source	I	S	I X S		
S. Em ±	0.009	0.009	0.017		
CD ($P \leq 0.05$)	0.026	0.025	0.049		

Total sugars (%)

Source	I_1	I_2	I_3	I_4	Mean
S_1	13.12	13.76	13.87	13.48	13.56
S_2	13.52	14.15	14.47	13.81	13.99
S_3	14.21	14.97	15.12	14.36	14.67
S_4	13.68	14.05	14.61	13.95	14.07
Mean	13.63	14.24	14.52	13.90	
Source	I	S	I X S		
S. Em ±	0.007	0.007	0.014		
CD ($P \leq 0.05$)	0.021	0.021	0.041		

Reducing sugars (%)

Source	I_1	I_2	I_3	I_4	Mean
S_1	3.45	3.93	4.12	3.67	3.79
S_2	3.92	4.39	4.68	4.17	4.29
S_3	4.58	5.13	5.20	4.88	4.95
S_4	4.00	4.52	4.97	4.42	4.48
Mean	3.99	4.49	4.74	4.291	
Source	I	S	I X S		
S. Em ±	0.004	0.004	0.009		
CD ($P \leq 0.05$)	0.013	0.013	0.026		

Non-reducing sugars (%)

Source	I_1	I_2	I_3	I_4	Mean
S_1	9.67	9.82	9.74	9.80	9.76
S_2	9.60	9.76	9.79	9.79	9.63
S_3	9.63	9.84	9.91	9.47	9.71
S_4	9.68	9.53	9.64	9.52	9.59
Mean	9.64	9.74	9.77	9.61	
Source	I	S	I X S		
S. Em ±	0.005	0.005	0.011		
CD at 5 %	0.016	0.016	0.031		

Acidity (%)

Source	I_1	I_2	I_3	I_4	Mean
S_1	0.262	0.215	0.193	0.244	0.228
S_2	0.240	0.194	0.182	0.230	0.211
S_3	0.182	0.162	0.145	0.169	0.164
S_4	0.220	0.179	0.171	0.215	0.196
Mean	0.226	0.187	0.172	0.214	
Source	I	S	I X S		
S. Em ±	0.0003	0.0003	0.0006		
CD at 5 %	0.0008	0.0008	0.0017		

Reducing sugars (%)

Source	I_1	I_2	I_3	I_4	Mean
S_1	9.12	9.20	9.45	9.12	9.23
S_2	9.14	9.41	9.53	9.23	9.33
S_3	9.12	10.24	10.58	9.98	9.98
S_4	9.20	9.78	9.87	9.41	9.57
Mean	9.15	9.66	9.86	9.43	
Source	I	S	I X S		
S. Em ±	0.005	0.005	0.010		
CD at 5 %	0.015	0.015	0.029		

Where, I = irradiation, S = storage temperature.

significantly higher in fruits stored at lower temperature storage as compared to minimum at ambient temperature (Table 4). This might be that, the accumulation of total soluble substances due to desired ripening. These findings were also in accordance with the findings of Roy and Joshi (1989) in mango, Plaza et al. (1992) in papaya, and Hussein et al. (1998) in guava. Jointly the maximum total soluble solids (17.36%) were recorded in fruits exposed to gamma rays at 0.40 kGy and stored at 12°C (I_3S_3). The minimum total soluble solids (15.86%) were recorded in unirradiated ambient stored (I_1S_1) fruits at the time of full ripening (Table 4). The maximum total soluble solids was recorded in fruits exposed to various dose of irradiation and stored at 12 and 9°C and in CA (12°C) storage compared to unirradiated fruits stored at ambient temperature in present study might be due to the beneficial effects of irradiation dose and storage temperature.

Sugars (percent)

Effect of irradiation

The data revealed that, total sugar percent of fruits was significantly affected by irradiation, storage temperature and their interaction. It was evident from the data presented in Table 4 that significantly the maximum total sugars (14.52%) were observed in fruits exposed to treatment I_3 (0.40 kGy). The minimum total sugars (13.63%) were observed in treatment I_1 (0.00 kGy). The maximum reducing sugar percent (4.95) was observed in fruits exposed to treatment I_3 (0.40kGy) whereas, minimum reducing sugars (0.00 kGy) were observed in treatment I_1 (0.00 kGy). The maximum non-reducing sugars (9.77%) were observed in fruits exposed to treatment I_4 (0.60 kGy) followed by treatment I_2 (0.20 kGy) compared to minimum (9.61) in unirradiated. The higher rate of increase in sugars content in irradiated fruits might be due to maintained ripening and corresponding greater conversion of starch into sugars. Irradiation might also accelerate the rate of gluconeogenesis (Wall, 2007). Similar findings had been observed by Beyers and Thomas (1979) in mango and Kovacs et al. (1994) in apple.

Effect of storage temperature

It was cleared from the data presented in Table 4 that significantly maximum total sugar (14.61%) was recorded by fruits stored under treatment S_3 (12°C), and minimum (13.56%) were under treatment S_1. The maximum reducing sugar percent (4.95) was recorded in fruits stored under treatment S_3 (12°C). The minimum reducing sugars (3.79%) were observed under ambient temperature (S_1). The maximum non-reducing sugar (9.71%) was recorded in fruits stored under treatment S_3 (12°C) compared to minimum (9.63%) were observed under at 9°C (S_4). The increase in the total and reducing sugars were maintained till the end of shelf life in storage temperature at 12 and 9°C and in CA (12°C) storage might be due to suppression in the respiration rate and enzyme activities and therefore, the conversion of starch into sugars might had been at slower rate and reaching maximum at the end of storage. Same trend of results were noticed by Narayana and Singh (2000) in mango and Purwoko et al. (2002) in banana.

Combined effect of irradiation and storage temperature

Results obtained during experimentation indicating (Table 4) significantly that, the maximum total sugars (15.12%) were recorded in fruits exposed to gamma rays at the dose of 0.40 kGy and stored at 12°C (I_3S_3. The minimum total sugar (13.12%) was recorded in unirradiated ambient stored (I_1S_1) fruits at the time of complete ripening. The maximum reducing sugar (5.20%) was recorded in fruits exposed to gamma rays at the dose of 0.40 kGy and stored at 12°C (I_3S_3). Results obtained during experimentation indicating that, maximum non-reducing sugar (9.91%) was recorded in fruits exposed to gamma rays at the dose of 0.40 kGy and stored at 12°C (I_3S_3) whereas, minimum non-reducing sugar (9.67%) was recorded in unirradiated ambient (I_1S_1) stored fruits at the time of full ripening. The maximum total and reducing sugar were recorded in fruits exposed to various doses of irradiation and storage temperature of 12 and 9°C and in CA (12°C) storage compared to unirradiated fruits stored at ambient temperature in present study which might be due to the beneficial effects of irradiation dose and storage temperature. The total and reducing sugars increased during storage but the non-reducing sugars did not exhibit same pattern during storage, since it represented a product of subtraction of reducing sugars from total sugars.

Titrable acidity

The data revealed that, acidity fruits was significantly affected by irradiation, storage temperature and their interaction. It was evident from the Table 4 that, significantly minimum acidity (0.172%) was observed in fruits exposed to treatment I_3 (0.40 kGy) compared to maximum (0.226%) was observed in treatment I_1 (0.00 kGy). The reduction in acidity by irradiation reflects a possible decrease in organic acids (Wall, 2007). These results were in accordance with the findings of Upadhyay (1992) and El-Salhy et al. (2006) in mango and Sornsrivichai at el. (1990) in apple. Under storage conditions it is cleared from the Table 4 that, significantly the minimum acidity (0.164%) was recorded in fruits stored under treatment S_3 (12°C). The maximum acidity (0.228%) was observed under treatment S_1 (ambient temperature). The lower acidity at low temperature might be due to utilization of acids in the process of respiration during ripening and reduced supply of sugars (Mane and Patel 2010). Same findings noted by Plaza et al. (1992) in papaya and Singh and Pal (2007) in guava. Combined results obtained during experimentation indicating that, significantly the minimum acidity (0.145%) was recorded from fruits exposed to gamma rays at the dose of 0.40 kGy and stored at 12°C (I_3S_3) whereas, maximum acidity (0.262%) was recorded in unirradiated ambient stored (I_1S_1) fruits at the time of full ripening. The minimum acidity was recorded in fruits exposed to various dose of irradiation and stored at 12 and 9°C and in CA (12°C) storage as compared to unirradiated fruits stored at ambient temperature in present study might be due to the beneficial effects of irradiation dose and storage temperature.

Ascorbic acid

Significantly the maximum ascorbic acid (9.86 mg/100g pulp) was observed in fruits exposed to 0.40 kGy (I_3), and the minimum ascorbic acid (9.15 mg/100 g pulp) was observed in treatment I_1 (0.00 kGy) (Table 4). The higher ascorbic acid due to irradiation was in accordance with the findings of Dhakar et al. (1966) in mango and Bhushan and Thomas (1990) in apple. Significantly the maximum ascorbic acid (9.98mg/100g pulp) was recorded in fruits stored under treatment S_3 (12°C) as compared to minimum (9.23 mg/100 g pulp) was observed under treatment S_1. Also, significantly the maximum ascorbic acid was recorded in fruits stored at 12 and 9°C temperature and in CA (12°C) storage whereas, minimum ascorbic was recorded under ambient temperature stored fruits (Table 4). Same findings was noted by Ray and Joshi (1989) in mango and Plaza et al. (1992) in papaya.

RESULTS

Results obtained (Table 4) during the experimentation indicates that, significantly the maximum ascorbic acid (10.58 mg/100 g pulp) was recorded in fruits exposed to gamma rays at 0.40 kGy and stored at 12°C (I_3S_3) whereas, minimum ascorbic acid (9.12 mg/100 g pulp) was recorded in unirradiated ambient stored (I_1S_1) fruits at the time of complete ripening. The fruits of Alphonso mango subjected to 0.40 kGy gamma rays irradiation subsequently stored at 9°C delayed the ripening process which maintained lower percentage of physiological loss in weight and ripening percentage, higher percentage of marketable fruits, and increase the shelf life for longer period. The data also indicated that, the maximum total soluble solids, total, and reducing sugars, ascorbic acid, and minimum acidity were noted in 0.40 kGy gamma rays irradiated fruits were stored at 12°C as compared to unirradiated fruits stored at ambient condition at ripening stage.

ACKNOWLEDGEMENTS

Thanks are due to Dr. R. Chander, Dr. L. N. Bandi, Dr. A. Shrivastva, and Shri Jyotis, ISOMED (Board of Radiation and Isotope Technology) Sir Bhabha Atomic Research Centre, Mumbai(India) for providing necessary facility for irradiation.

REFERENCES

Aina JO, Adesiji OF, Ferris SRB (1999). Effect of gamma irradiation on post harvest ripening of plantain fruit (*Musa paradisiaca* L.). J Sci Food. Agric. 79(5):653-656.

Anonymous (2007). Post harvest manual for mangoes. APEDA, Ministry of Agriculture, Govt of India.

Beyers M, Thomas AC (1979). Gamma irradiation of subtropical fruits 4. Changes in certain nutrients present in mangoes, papayas and litchis during canning, freezing and gamma irradiation. J Agric Food Chem. 27(1):48-51.

Bhushan B, Thomas P (1990). Quality of apples following gamma irradiation and cold storage. Int. J. Food Sci. Nut. 49(66):485-492.

Deka BC, Choudhury A, Bhattacharyya KH, Begum, Neog M (2006). Postharvest treatment for shelf life extension of banana under different storage environments. Acta Hort. 2:841-849.

Dhakar SD, Savagraon KA, Srirangarajan AN, Sreenivasan A (1966). Irradiation of mangoes. I. Radiation- induced delay in ripening of Alphonso mangoes. J Food Sci. 31(6):863-869.

El-Salhy FTA, Khafagy SAA, Haggay LF (2006). The changes that occur in mango fruits treated by irradiation and hot water during cold storage. J. Appl .Res. 2(11):864-868.

Farzana P (2005). Post harvest technology of mango fruits, its development, physiology, pathology and marketing in Pakistan. Digital VerlagGmbH Pub Germany.

Gomez-Lim MR (1993). Mango fruit ripening: Physiological and

Molecular biology. Acta Hort. 341:484-496.

Gutierrez AO, Nieto AD, Martinez D, Dominguez AMT, Delgadillo S, Qutierrez AJG (2002). Low temperature plastic film, maturity stage and shelf life of guava fruits. Revista Chapingo Serie Hort. 8(2):283-301.

Hussein AM, Sabrou MB, Zaghloul AE (1998). Post harvest physical and biochemical changes of common and late types of seedy guava fruits during storage. Alexandria J. Agric. Res. 43(3):187-204.

Kovacs E, Djediro GA, Sass P (1994). Metabolism of source in apples (*Malus domestica* Borkh) as a function of ripeness, cultivar, radiation dose and storage time. Acta Hort. 60:235-242.

Mahindru SN (2009) Food preservation and irradiation. APH publishing corporation, New Delhi.

Mane SR, Patel BN (2010). Effect of maturity indices, postharvest treatments and storage temperatures on physiological changes of mango (*Mangifera indica* L) cv Kesar. In: National seminar on precision farming in horticulture during, CoHF, Jhalawar, India, P. 291.

Mann SS, Singh RN (1975). Studies on cold storage of mango fruits (*Mangifera indica* L.) cv. Langra. Indian J. Hort. 32(1):7-14.

Mayer BS, Anderson DS, Bhing RH (1960). Introduction to plant physiology. D Van Nastrand Co Ltd, London.

Nagaraju CG, Reddy TV (1995). Deferral of banana fruit ripening by cool chamber storage. Adv. Hort. Sci. 9(4):162-166.

Narayana CK, Singh BP (2000). Effect of chilling temperature on ripening behavior of mango. Haryana J. Hort Sci. 29(3-4):168-170.

Panse VG, Sukhatme PV (1967). Statistical Methods for Agricultural workers. ICAR, New Delhi.

Pereira T, Tijskens LMM, Vanoli M, Rizzolo A, Eccherzerbini P, Torricelli A, Spinelli L, Filgueiras H (2010). Assessing the harvest maturity of Brazilian mangoes. In EW Hevett et al. (Eds) Proc IS on Post Harvest Pacifica 2009. Acta Hort, P. 880.

Plaza J, L-de la Alique R, Calvo L, Moure J (1992). Control of modifications in quality of the papaya fruit during cold storage in polyethylene bags. In Rodriguez J H (eds) Proceeding of a conference held in Granada, Spain, pp. 19-27.

Prasadini PP, Khan MA, Reddy PG (2008). Effect of irradiation on shelf life and microbiological quality of mangoes (*Mangifera indica* L.). J. Res. ANGRU. 36(4):14-23.

Purohit AK, Rawat TS, Kumar A (2004). Shelf life and quality of ber fruit cv. Umran in response to post harvest application of ultra violate radiation and paclobutrazole. Pl Foods for Human Nut. 58(3):1-7.

Purwoko BS, Susanto S, Kodir KA, Novita T, Harran S (2002). Studies on the physiology of polyamines and ethylene during ripening of banana and papaya fruits. Acta Hort. 575(2):651-657.

Rangana S (1986). Manual of analysis of fruits and vegetable products. Tata McGraw Hill Publishing Co Ltd New Delhi.

Roy SK, Joshi GD (1989). An approach to integrated post harvest handling of mango. Acta Hort. 231:469-661.

Salahddin ME, Kedar AA (1980). Post harvest physiology and storage behaviour of pomegranate fruits. Sci. Hort. 24:287-298.

Shankar V, Veeragavathatham D, Kannan M (2009). Effect of organic farming practices on post harvest storage life and organoleptic quality of yellow onion (*Allium cepa* L.). Indian J. Agric. Sci. 79(8):608-614.

Silva JM, Correia LCSA, DeMoura NP, Maciel MIS, Villar HP (2010). Use of technology radiation as a method of reducing the microorganism and conservation postharvest of caja during storage. In 10th International Working Conference on Stored Product Protection, Cidade University, Brazil, pp. 573-577.

Singh SP, Pal PK (2007). Post harvest fruit fly disinfestations strategies in rainy season guava crop. Acta Hort. 375:591-596.

Singh SP, Pal RK (2009). Ionizing radiation treatment to improve postharvest life and maintain quality of fresh guava fruit. Radiation Phy. Chem. 78:135-140.

Singh SP, Pal RK (2009). Ionizing radiation treatment to improve postharvest life and maintain quality of fresh guava fruit. Radiation Phy. Chem. 78:135-140.

Sornsrivichai J, Jampanil R, Gomolmanee S, Tuntawiroon O, Boonthan K (1990). Post harvest colouration improvement of Anna apple by white fluorescent light. Acta Hort. 279:501-509.

Sudto T, Uthariaratanakij A, Jitareerat R, Photchanachai S, Vongcheeree S (2005). Effect of gamma irradiation on ripening process of Morn-Thong durian. In International symposium on new frontier irradiated food and non-food products, Bangkok, Thailand.

Tharanathan RN,Yashoda HM, Prabha TN (2006). Mango (*Mangifera indica* L) the king of fruits – An overview. Food Rev Int. 22:95-123.

Udipi SA, Ghurge PS (2010). Applications of food irradiation. In: Udipi SA, Ghugre PS (ed) Food irradiation, Agrotech Publishing Academy, Udaipur, pp. 40-71.

Upadhyay PI (1992). Effect of gamma irradiation and hot water treatment on the shelf life of mango (*Mangifera indica* L) var Red. In Asian Inst Tech, Bangkok.

Wall MM (2007). Post harvest quality and ripening of Dwarf Brazillian bananas (*Musa Sp*) after x-ray irradiation quarantine treatment. HortSci. 42(1):130-134.

Waskar DP, Masalkar SD (1997). Effect of hydrocooling and bavistin dip on the shelf life and quality of mango during storage under various environments. Acta Hort. 455:687-695.

Yadav MK, Patel NL, Hazarika Ankita, Chaudhary PM (2010). Physiological changes in Kesar mango as influenced by treatment and storage. The Hort. J. 23(1):16-17.

Yimyong S, Datsenka TU, Handa AK, Sereypheap K (2011). Hot water treatment delays ripening associated metabolic shift in 'Okrong' mango fruit during storage. J. Am. Soc. Hort Sci. 136(6):441-451.

25

Trypsin (serine protease) inhibitors in peanut genotypes aiming for control of stored grain pests

Patrícia de Lima Martins[1], Roseane Cavalcanti dos Santos[2], João Luis da Silva Filho[2], Fábia Suelly Lima Pinto[2] and Liziane Maria de Lima[1,2]

[1]Universidade Estadual da Paraíba – UEPB, Pós-graduação em Ciências Agrárias, Rua Baraúnas n. 351, CEP: 58429-500 Bairro Universitário, Campina Grande, PB, Brazil.
[2]Empresa Brasileira de Pesquisa Agropecuária, Embrapa Algodão. Rua Oswaldo Cruz n. 1143, CEP: 58428-095, Centenário, Campina Grande, PB, Brazil.

Peanut seeds from different genotypes were evaluated for activities of trypsin inhibitors (serine protease), based on *in vitro* and *in vivo* assays, aiming to further selection of genitors in breeding programs for tolerance of stored grain pests. The *in vitro* assays were based on inhibition of insect digestive enzymes and also on thermal and pH stabilities of seed protein, while the *in vivo* assays were performed with insects *Alphitobius diaperinus*, *Tribolium castaneum*, *Tenebrio molitor* and *Spodoptera frugiperda*. Seed inhibitors of all genotypes inhibited bovine trypsin at 70 to 94%. The seed extract of BRS Havana inhibited *T. castaneum* and *T. molitor* up to 80% while the extract of BRS 151 L7 inhibited *A. diapennus* at nearly 20%. The seed inhibitors of both cultivars were stable at 80°C and also at different pH values. The two peanut genotypes are recommended as promising parents for breeding program aiming to selecting lines with tolerance to *Tenebrio* and *Alphitobius* insects.

Key words: Trypsin inhibitor, *Arachis hypogaea*, lepidoptera, coleoptera.

INTRODUCTION

Peanut is an important oleaginous known for its broad environmental adaptation. One of the major bottlenecks in the management is the post harvest phase mainly storage of grains. In this stage, the problems with storage pests are recurrent especially those caused by weevils: *Tribolium castaneum* (Horbst, 1797) (Coleoptera: Tenebrionidae), *Alphitobius diaperinus* (Panzer, 1797) (Coleoptera: Tenebrionidae), *Plodia interpunctella* (Hübner, 1813) (Lepidoptera: Pyralidae), *Tenebrio molitor* (Linnaeus, 1758) (Coleoptera: Tenebrionidae), *Corcyra* *cephalonica* (Stainton, 1865) (Lepidoptera: Pyralidae) and *Tenebroides mauritanicus* (Linnaeus, 1758) (Coleoptera: Ostomidae). Depending on the level of infestation, the control of weevils can become unfeasible due to high costs with chemical treatment.

Genetic resistance to insect-pests is a desired goal by the most plant breeder. However, the acquisition of this trait is a big challenge due to heavy interactions between genetic and environmental factors. Thus, other natural strategies should be researched to detect tolerant

Table 1. Main traits of peanut genotypes used in this study.

Genotype	G	SP	BT	GH	C	SS	O	CS	100S
BRS Perola Branca	C	H	Vi	R	110-115	L	50-52	White	55
176 AM	L	F	V	U	90-95	M	46-48	Red	43
173 AM	L	F	V	U	90-95	M	46-48	Tan	51
Florunner	C	H	Vi	R	120-130	L	49-51	Tan	50
IAC Caiapó	C	H	Vi	R	120-130	L	48-50	Tan	51
186 AM	L	F	V	U	90-95	L	46-48	Red	46
BRS Havana	C	F	V	U	88-90	M	43-45	Tan	50
L7 bege	L	F	V	U	88-90	L	45-47	Tan	61
175 AM	L	F	V	U	90-95	M	46-48	Tan	47
LGoPE-06	L	H	Vi	R	120-130	EL	50-52	Tan	72
BRS 151 L7	C	F	V	U	85-87	L	46-48	Red	67

G: Genealogy, C- Cultivar, L- Top line; SP- Subspecies: F- fastigiata, H- hypogaea; BT- Botanic type: V- Valencia, Vi- Virginia; GH- Growth habit: R- Runner. U- Upright; C- Cycle (days); SS- Seed size: M- Medium, L- Large, EL- Extra large; O- Oil content (%); SC- Seed colour and 100S- Average of 100 seed weight.

genotypes.

Focusing on stored grain pests, it is known that several plant species produce proteins with insecticidal property, such as inhibitor of proteases (IPs) which play a key role in plant defense against various orders of insects (Marinho et al., 2008; Pereira et al., 2007). The insecticidal activity of IPs is due to inhibition of proteolytic enzymes in the midgut of insects, leading to malnutrition, delay in larvae development and even death (Mosolov and Valueva, 2008). Among the groups of proteolytic enzymes affected by IPs, serine proteinases are the most investigated (Oliveira et al., 2005; Habib and Fazili, 2007).

Peanut seeds have different levels of IPs but information on using this trait for selecting genotypes tolerant to pests of stored grain is limited (Suzuki et al., 1987; Norioka et al., 1981). Considering the wide genetic base between intraspecific accessions of A. hypogaea, it is possible to identify promising materials for further use in hybridization works aiming subsequent selection of top lines with different level of tolerance to these pests.

The present research aimed to estimating the inhibitory activity of trypsin (serine protease) in seeds of different peanut genotypes for further recommendation of parents in breeding programs to tolerance to store grain pests.

MATERIALS AND METHODS

Extraction of proteins and determination of antitryptic activity

The peanut seeds used in this study were collected in January 2011 and January 2012 in Barbalha, CE (07 ° 18'18 "S, 39 ° 18'07" W, 414 m), semiarid region of northeastern Brazil. The study begin when the seeds were 8% moisture. The main traits of peanut genotypes are found in Table 1. Total crude protein of each genotype was extracted using methodology described in Bland and Lax (2000). The proteins were quantified by Bradford method (Bradford, 1976) at 595 nm using bovine serum albumin (BSA) asanalytical standard.

A previously described methodology was used to determine the antitryptic activity based on the following steps: a) bovine trypsin assay with seed total crude extract; b) bovine trypsin assay with partial purified seed trypsin inhibitor; c) insect digestive enzyme preparations assay with seed total crude extract; d) insect digestive enzyme preparations assay with partial purified seed trypsin inhibitor. A summary of the methodology is described below (Kakade et al., 1969).

In a microtube, 1.5 ml was performed following reaction: 5 µl of bovine trypsin (1 µg/µl) or 5 µl of insect digestive enzyme preparations (1 µg/µl), 20 µl of seed total crude extract or partial purified seed trypsin inhibitor (5 µg), 125 µl of 50 mM Tris-HCl buffer pH 8.5. The reaction was pre-incubated at 37°C for 20 min and 200 µl azocasein (1.5%, m/v) was added and again incubated at the same temperature and period. The reaction was discontinued with 300 µl of 20% trichloroacetic acid. After 5 min at room temperature, samples were centrifuged at 12.000 x g for 10 min. An aliquot of 250 µl of supernatant was collected and added to 250 µl of 2 mM NaOH; the reading was performed in a spectrophotometer (Femto, model 700S) at 440 nm. All assays were performed with three repetitions. Reagents from Sigma Aldrich (USA) were used in this assay.

Intestinal extract assays

Twenty third-instar larvae of A. diaperinus, T. molitor, T. castaneum and Spodoptera frugiperda (Lepidoptera: Noctuidae) were fed on artificial diet and then dissected for collecting the guts. S. frugiperda was included because it is an important crop pest and is susceptible to protease inhibitors of soybean (Glycine max L.) and bean (Vigna radiata L. Wilczek) seeds (Brioschi et al., 2007; Paulillo et al., 2000).

Dissected tissues were immediately immersed in 50 µl of 50 mM sodium phosphate buffer pH 7.5 and maintained at 4°C. After macerated, the homogenates were centrifuged at 12.000 x g for 20 min at 4°C. The supernatants were collected for further use in enzymatic assays for antitryptic activity assays with the seed total crude extract (Terra et al., 1977).

Hatching bioassay with A. diaperinus

Bioassays were carried out in order to estimate the hatching rate of

Table 2. Inhibitory activity of the seed total crude extract of peanut genotypes with bovine trypsin.

Genotype	Inhibition rate (%)
175 AM	94.2[a]
Florunner	92.6[a]
176 AM	92.2[a]
BRS Havana	90.3[ab]
186 AM	89.7[b]
173 AM	89.7[b]
IAC Caiapó	88.2[bc]
L7 Bege	79.3[bc]
BRS Pérola Branca	78.5[c]
LGoPE-06	78.0[c]
BRS 151 L7	70.7[d]

Means followed by the same letter do not differ significantly by the Tukey test (p≤0.05). Variance analysis to Inhibition rate: Mean square of treatment: 292.31, Standard error: 0.32, Freedom degree: 10, F test: 929.24**, Average: 85.60, Coefficient of variation: 0.66. **significant by the Friedman test (p≤0.01).

A. diaperinus fed on peanuts seeds (50 g). The seeds of each genotype were infested with 20 adults sexed and placed in plastic pots (8.0 cm of height x 11.0 cm in diameter). The pots were stored at room temperature for 90 days. Corn bran was used in control treatment. The bioassays were completely randomized with four replications. The number of larvae was registered at 53 and 90 days and hatching rates were estimated (Azevedo et al., 2010). The data were analyzed by the Friedman test (p ≤ 0.05) and the means were compared by Student Newman Keuls test (p ≤ 0.05).

Stability of seed trypsin inhibitors

This assay was performed in order to study the stability of seed-protease inhibitors at different temperature and pHs. The crude extracts were previously fractionated by sequential precipitation with ammonium sulfate at saturation ranges of 0 to 30, 30 to 60 and 60 to 90%. The 60 to 90% fraction showed the highest antitryptic activity (data not shown), and therefore was chosen as the peanut seed trypsin inhibitor fraction for stability assays. The procedures were based on that described by Gomes et al. (2005). To investigate thermal stability, the 60 to 90% fraction (1 ml, 1 µg/µl) was incubated for 30 min at 40, 60, 80 and 100°C, and thereafter cooled to 4°C. For characterization of the pH stability, the same volume of the 60 to 90% fraction was dialyzed for 16 h using the following buffers: 50 mM sodium phosphate, pHs of 5 to 8 and 50 mM Tris-HCl, pHs of 9 to 11. Then, the samples were incubated at 37°C and again dialyzed for 4 h in 50 mM Tris-HCl, pH 8.5. An aliquot of 5 µl of each sample was used for assessment of the remaining antitryptic activity with bovine trypsin. All assays were carried out in five replications. Data were analyzed by the Friedman test (p ≤ 0.05) and the means were compared by Tukey test (p ≤ 0.05).

RESULTS

Determination of antitryptic activity

Bovine trypsin was inhibited in all peanut samples

analyzed, whose means ranged from 70 to 94% (Table 2). The genotypes 175 AM, Florunner, 176 AM and BRS Havana showed the same statistical classification with average of inhibition rate of 92%.

Based on the results in Table 2, all genotypes with inhibition rate at least 90% were selected for intestinal inhibition assays with *T. castaneum*, *T. molitor*, *S. frugiperda* and *A. diaperinus*. The genotype BRS 151 L7, with the lowest inhibition rate was also chosen considering that many have different response as to the types of protease inhibitors likely present in the seeds. As seen in Figure 1, the percentages of inhibitory activity of seed crude protein extract from different genotypes were almost uniform for *T. castaneum* (between 70 and 81%) and *S. frugiperda* (45 to 48%). However, different responses were obtained for *T. molitor* (35 and 83.6%) and *A. diaperinus* (8 to 19%).

The extract obtained from BRS Havana was very promising to inhibit *T. castaneum* and *T. molitor*, with inhibition rate up to 80%. For *S. frugiperda*, the extract of all genotypes showed average inhibition rate of 47% while for *A. diaperinus* BRS 151 L7 showed the highest inhibition rate close to 20% (Figure 1).

Feeding bioassay with *A. diaperinus*

In order to validate the information contained in Figure 1, a hatching bioassay was carried out using sexed adults of *A. diaperinus* fed on peanut seeds. This species was chosen due to high incidence in Brazilian grain storages. It was observed that larvae hatching were negatively affected by feed supplied and incubation period (Table 3). At 90 days, the number of larvae fed on peanut seeds was 33% less than at 53 days. As this average involved

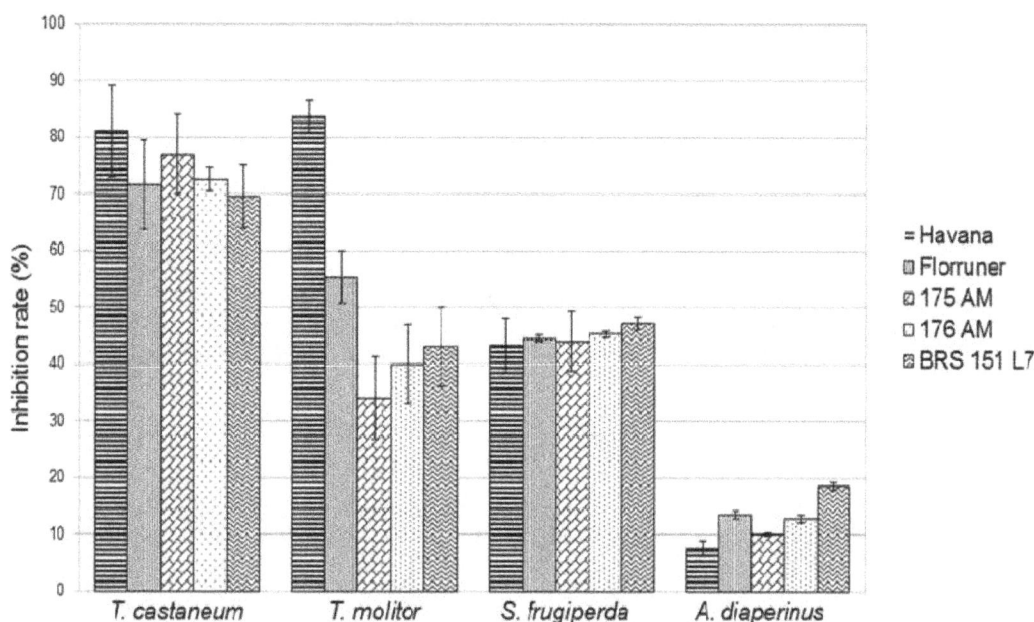

Figure 1. Inhibitory activity of seed total crude protein extracts from the peanut genotypes with insect digestive enzyme preparations of *T. castaneum*, *T. molitor*, *S. frugiperda* and *A. diaperinus*. Data are the mean values ± SD of three biological replicates.

Table 3. Number of hatched larvae of *A. diaperinus* fed on peanut seeds and corn bran.

Treatment	Number of larvae	DRC (%)
Period of incubation (days)		
53	9.00[a]	-
90	6.08[b]	33
Diet		
BRS 151 L7	5.52[bc]	51
Florunner	5.89[b]	47
175 AM	6.34[b]	43
176 AM	6.51[b]	42
BRS Havana	9.39[ab]	16
Corn bran (control)	11.17[a]	-

DRC- difference in hatching rate. Means with the same letter do not differ significantly by the Tukey test (p ≤ 0.05). Means were transformed into ⊠x for statistical analysis. Variance analysis to period of incubation: Standard error: 91.16, Freedom degree: 1, F test: 24.41**, Average: 7.54; Variance analysis to Diet - Standard error: 40.91, Freedom degree: 5, F test: 10.77**; Average: 7.47. Coefficient of variation: 22.47%. ** significant by the Friedman test (p ≤ 0.01).

most treatments with peanut seeds, it was suggested that this reduction may be associated with the greater period of insect feeding. This can be evident by the number of hatched larvae in different treatments. It was also verified that insects fed for 90 days on seeds from BRS 151 L7, Florunner, 175 and 176 AM had hatching rate reduced around 46% compared to control (corn bran). Among these genotypes, however, BRS 151 L7 revealed greater inhibitory rate (51%), which was in agreement with data recorded in Figure 1.

Stability assay of peanut seed trypsin inhibitors

The partial purified seed trypsin inhibitors from peanut genotypes showed thermostability in an interval of 40 to 100°C, with inhibition rate up to 60%, highlighting 175 AM who kept the inhibitory rate above 85% at all temperatures (Figure 2). An exception was verified to Florunner that had inhibitory activity substantially decreased at 100°C. As to pH stability of protease inhibitors, it was verified that the inhibitors from most

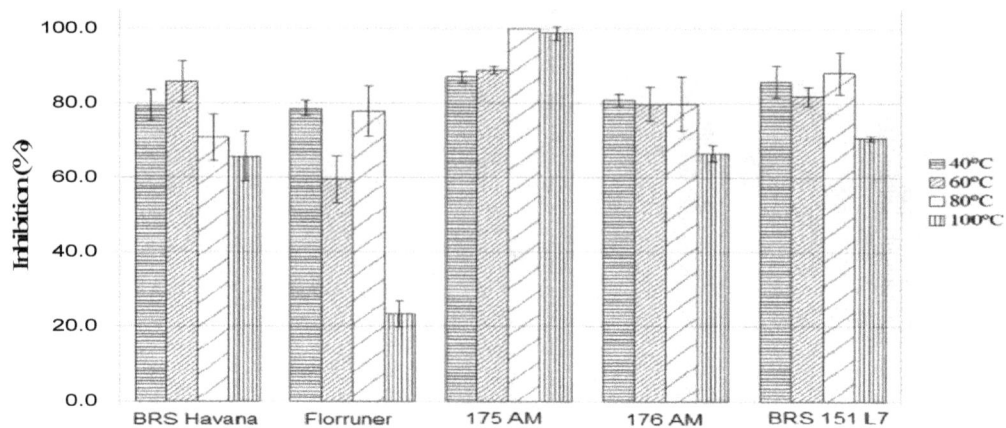

Figure 2. Thermal stability of peanut seed trypsin inhibitors from 60 to 90% fraction pre-incubated at different temperatures and tested with bovine trypsin. Data are the mean values ± SD of three biological replicates.

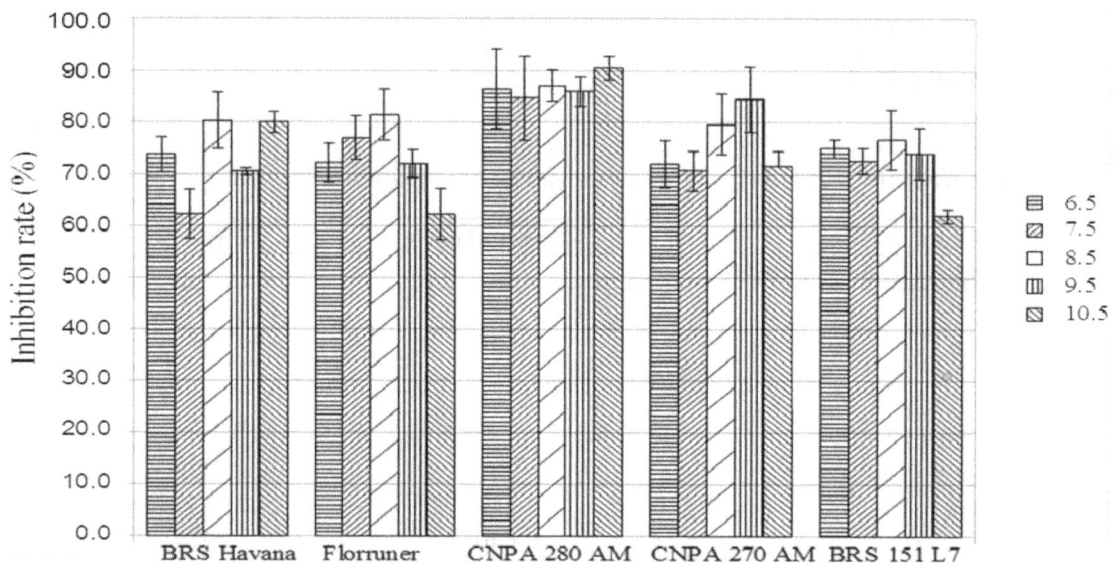

Figure 3. pH stability of peanut seed trypsin inhibitors from 60 to 90% fraction pre-incubated at different pHs and tested with bovine trypsin. Data are the mean values ± SD of three biological replicates.

genotypes maintained inhibition rate up to 70% at pH ranging between 6.5 and 9.5. Above this value, a small reduction in the rate was seen in extracts from Florunner and BRS 151 L7. Even so, it surpassed 60% which is still a reasonable rate of inhibition (Figure 3). The top line 175 AM showed large pH stability keeping the inhibition rate up to 80% at all pH range evaluated.

DISCUSSION

Based on high inhibition rates obtained in antitryptic activity assays using different peanut genotypes, we concluded that all materials showed great potential for further use in breeding programs aiming to tolerance to stored grain pests. Considering the rates obtained (70 to 94%, Table 2), it is suggested that proteases inhibitors are abundant in the seeds and this trait must be quantitative with high heritability. Such inference is based on studies developed by Dam and Baldwin (2003) and Kollipara et al. (1996), involving genetic inheritance of genes encoding to high expression of IPs in *Glycine tomentella* and *Nicotiana attenuata*. According to these authors, the inheritance is dominant and highly heritable. These findings are relevant when considering the perspective of transferring IP genes by conventional hybridization.

Despite relevance of these data, *in vitro*-inhibition assays are not enough to define promising genotypes for a breeding program aiming to tolerance to stored grain pests. The bioassays with insects are essential to complement the information, mainly because a high inhibition rate to a given insect cannot be the same to another, due to particular differences in the families of intestinal proteases as well as to variations in pHs of insect guts. These are what really influence on binding and expressiveness of IPs to target proteins associated to insect digestion (Linser et al., 2009; Vinokurov et al., 2006; Dow, 1992).

The results shown in Tables 2 and 3 demonstrated the assertive in inclusion of cv. BRS 151 L7 in additional assays planned in this work. Although, this cultivar showed the lowest inhibition rate *in vitro* assay (70%, Table 2), it revealed expressive reduction in hatching rate (51%, Table 3), by using *A. diaperinus. T. castaneum* and *T. molitor*, the cv. BRS Havana was the most promising genotype due to its high inhibition rate (about 80%) in bioassays with intestinal homogenate (Figure 1). This is an expressive result comparing inhibition rate of others leguminous species. With *Crotalaria pallida*, Gomes et al. (2005) found 74% inhibition in intestinal homogenate of *Callosobruchus maculatus*. In further works, it would be interesting to test the extract of BRS Havana also with this coleoptera.

The inhibition rates obtained to *S. frugiperda* were reasonable but should be taken with caution in a breeding program since larvae are able to overcome the deleterious effects of IPs, possibly due to wide ability to activating new trypsin-like enzymes, which are less sensitive to inhibitors produced by plants (Paulillo et al., 2000; Xavier et al., 2005). This is a natural defense process also verified in several classes of insects.

In relation to thermal and pH stabilities of peanut seed trypsin inhibitors, these traits are quite important for further selection of genotypes because all seeds can be facing thermal variations in post-harvest processes. The instability of the trypsin inhibitors may be a negative factor for selection in a breeding program, even though genotype shows satisfactory results in inhibition assays. In this work, the results of pH and thermal stabilities of the partial purified trypsin inhibitors of the peanut genotypes were very expressive. The inhibitory potential of the seed extracts remained between pH 6.5 to 9.5 (Figure 3) indicating that peanut seed-IPs can affect a range of insect pests, especially Coleoptera and Lepidoptera, which intestinal lumen-pH varies from neutral to alkaline, acid to neutral or acid to alkaline, depending on the species (Linser et al., 2009; Vinokurov et al., 2006; Dow, 1992). The thermostability of the proteins were also satisfactory, since all extracts kept the inhibition above 60°C, excepting to Florunner that did not tolerate temperatures at 100°C. Plants rich in IPs and that hold the stability of inhibition in this range of pH

and temperature are excellent sources genetic resources for plant defense by conventional hybridization or by transgenesis.

For food crops, such as peanuts, this information becomes more relevant due to several commercial products that are processed from the grains. Since many cereals are consumed boiled, losses in the activity of this compound may lead to the inactivation of its function. Serquiz (2012) tested the thermal and pH stabilities of trypsin inhibitor (TI) from peanut candy and verified that peanut-TI was fairly resistant, keeping the total inhibitory activity over trypsin when heated to 80°C and reducing only 8% when tested at 100°C. Gomes et al. (2005) evaluated the effect of trypsin inhibitor from *Crotalaria pallida* seeds on *Callosobruchus maculates*, *Ceratitis capitata* and obtained good results of inhibition from fresh seeds. However, when the extract was heated at 100°C for 30 min, the inhibition activity was reduced to 50%.

The results presented in this work can contribute greatly to the planning of a peanut breeding program aiming to tolerance to stored grain pests. The both earliness BRS Havana and BRS 151 L7, developed by Embrapa were promising candidates for this proposal. Additionally, the segregating arising from this crossing would also offer a range of variability for grain and oil, based on the characters presented in Table 1. The top line 175 AM was also a promising material based on the results obtained in inhibition assays against *T. castaneum* (Figure 1) and feeding assay with *A. diaperinus* (Table 3). Furthermore, the thermal and pH stabilities also contribute to explain its selection. The combination of this genotype, that is a Florunner-descendant (Gomes et al., 2007), with BRS Havana and BRS 151 L7 would generate large variability populations due to their broad genetic base which involves ancestors of fastigiata, hypogaea and vulgaris subspecies (Duarte et al., 2013; Gomes et al., 2007). The perspective in obtaining rich-IPs lines would naturally increase considering the dominant inheritance that controls the trait and also the percentage of inhibition shown in Table 2. Before beginning the breeding procedures, however, it is recommended to validate the information in natural storage conditions.

Conflict of Interests

The authors have not declared any conflict of interests.

ACKNOWLEDGEMENTS

The authors are thankful to the National Research Network on Agricultural Biodiversity and Sustainability (Rede Nacional de Pesquisa em Agrobiodiversidade e Sustentabilidade Agropecuária - REPENSA (MCT/ CNPq/ MEC/ CAPES/ CT AGRO/ CT IDRO/ FAPS/ EMBRAPA) for the financial support and grants.

REFERENCES

Azevedo AIB, Lira AS, Cunha LC, Almeida FAC, Almeida RP (2010). Bioatividade do óleo de nim sobre Alphitobius diaperinus (Coleoptera: Tenebrionidae) em sementes de amendoim. Rev Bras Eng Agríc Ambient. 14:309-313. http://dx.doi.org/10.1590/S1415-43662010000300011

Bland JM, Lax A (2000). Isolation and characterization of a peanut maturity-associated protein. J. Agric. Food Chem. 48:3275-3279. http://dx.doi.org/10.1021/jf000307h

Bradford M (1976). A rapid and sensitive method for the quantitation of microgram quantities of protein utilizing the principle of protein-dye binding. Anal. Biochem. 72:248-254. http://dx.doi.org/10.1016/0003-2697(76)90527-3

Brioschi D, Nadalini LD, Bengtson MH, Sogayar MC, Moura DS, Silva-Filho MC (2007). General up regulation of Spodoptera frugiperda trypsins and chymotrypsins allows its adaptation to soybean proteinase inhibitor. Insect. Biochem. Mole. Biol. 37:1283-1290. http://dx.doi.org/10.1016/j.ibmb.2007.07.016

Dam NM, Baldwin IT (2003). Heritability of a quantitative and qualitative protease inhibitor polymorphism in Nicotiana attenuata. Plant. Biol. 5:179-185. http://dx.doi.org/10.1055/s-2003-40719,

Dow JAT (1992). pH gradientes in lepidopteran midgut. J. Exp. Biol. 172:355-375.

Duarte EAA, Melo Filho PA, Santos RC (2013). Características agronômicas e índice de colheita de diferentes genótipos de amendoim submetidos a estresse hídrico. Rev. Bras. Eng. Agríc. Ambient. 17:843-847. http://dx.doi.org/10.1590/S1415-43662013000800007,

Gomes CE, Barbosa AE, Macedo LL, Pitanga JC, Moura FT, Oliveira AS, Moura RM, Queiroz AF, Macedo FP, Andrade LB, Vidal MS, Sales MP (2005). Effect of trypsin inhibitor from Crotalaria pallida seeds on Callosobruchus maculatus (cowpea weevil) and Ceratitis capitata (fruit fly). Plant. Physiol. Biochem. 43:1095-1102. http://dx.doi.org/10.1016/j.plaphy.2005.11.004

Gomes LR, Santos RC, Filho CJA, Melo PAF (2007). Adaptabilidade e estabilidade fenotípica de genótipos de amendoim de porte ereto. Pesq Agropec. Bras. 42:985-989. http://dx.doi.org/10.1590/S0100-204X2007000700010

Habib H, Fazili KM (2007). Plant protease inhibitors: a defense strategy in plants. Biotechnol. Mol. Biol. Rev. 2:068-085.

Kakade ML, Simons N, Liemer IE (1969). An evaluation of natural vc. Synthetic substrates for measuring the anti tryptic activity of soybean samples. Cereal. Chem. 46:518-526. http://dx.doi.org/10.1590/S0100-67622008000600018

Linser PJ, Smith KE, Seron TJ, Oviedo MN (2009). Carbonic anhydrases and anion transport in mosquito midgut pH regulation. J Exp Biol 212:1662-1671. http://jeb.biologists.org/content/212/11/1662.full.pdf

Marinho JS, Oliveira MGA, Guedes RNC, Pallini A, Oliveira CL (2008). Inibidores de proteases de hospedeiros nativos e exóticos e sua ação em intestinos de lagartas de Thyrinteina leucoceraea. Rev Árvore 32:1125-1132. http://dx.doi.org/10.1590/S0100-67622008000600018

Mosolov W, Valueva TA (2008). Proteinase inhibitors in plant biotechnology: a review. Appl Biochem Microbiol 44:261-269. http://www.academicjournals.org/article/article1380100578_Habeeb%20and%20Khalid.pdf

Norioka S, Omichi K, Ikenaka T (1981). Purification and characterization of protease inhibitors from peanuts (Arachis hypogaea). J Biochem 91:1427-1434. https://www.jstage.jst.go.jp/article/biochemistry1922/91/4/91_4_1427/_pdf

Oliveira MGA, Simone CG, Xavier LP, Guedes RNC (2005). Partial purification and characterization of digestive trypsin-like proteases from the velvet bean caterpillar, Anticarsia gemmatalis. Comp Biochem Phisyol 140:369-380. http://dx.doi.org/10.1016/j.cbpc.2004.10.018

Paulillo LC, Lopes AR, Cristofoletti PT, Parra JR, Terra WR, Silva-Filho MC (2000). Changes in midgut endopeptidase activity of Spodoptera frugiperda (Lepidoptera: Noctuidae) are responsible for adaptation to soybean proteinase inhibitors. J Econ Entomol 93:892-896.

Pereira RA, Valencia-Jiménez A, Magalhães CP, Prates MV, Melo JAT, Lima LM, Sales MP, Nakasu EYT, Silva MCM, Grossi de Sá MF (2007). Effect of a Bowman-Birk proteinase inhibitor from Phaseolus coccineus on Hypothenemus hampei gut proteinases in vitro. J Agric Food Chem 55:10714-10719. http://pubs.acs.org/doi/pdf/10.1021/jf072155x

Suzuki A, Tsunogae Y, Tanaka I, Yamane T, Ashida T, Noriokas S, Hara S, Ikenaka T (1987). The structure of Bowman-Birk type protease inhibitor a-II from peanut (Arachis hypogaea) at 3.3A resolution. J Biochem 101:267-274. https://www.jstage.jst.go.jp/article/biochemistry1922/101/1/101_1_267/_pdf

Terra WR, Ferreira C, Bianchi AG (1977). Action pattern, kinetical properties and electrophoretical studies of an alpha-amylase present in midgut homogenates from Rhynchonsciara americana (Diptera) larvae. Comp Biochem Physiol 56B:201-209.

Vinokurov KS, Elpidina EN, Oppert B, Prabhakar S, Zhuzhikov DP, Dunaevsky YE, Belozersky MA (2006). Diversity of digestive proteinases in Tenebrio molitor (Coleoptera: Tenebrionidae) larvae. Comp Biochem Physiol 145B:126–137. http://dx.doi.org/10.1016/j.cbpb.2006.05.005

Xavier LP, Oliveira MGA, Guedes RNC, Santos AV, Simone SG (2005). Trypsin-like activity of membrane-bound midgut proteases from Anticarsia gemmatalis (Lepidoptera: Noctuidae). Eur J Entomol 102:147-153. http://www.eje.cz/pdfs/eje/2005/02/04.pdf

26

Self-build silos for storage of cereals in African rural villages

Matteo Barbari, Massimo Monti, Giuseppe Rossi, Stefano Simonini and
Francesco Sorbetti Guerri

Department of Agricultural, Food and Forestry Systems(GESAAF) -University of Firenze
Via San Bonaventura, 13 – 50145 Firenze - Italy.

Prototypes of silos for the storage of cereals were designed and built at the Department of Agricultural, Food and Forestry Systems of University of Firenze. The silos were planned specifically to be used in Tanzania, by individual farmers operating in the inland areas of the country. Nevertheless, with a capacity of 1 m³ and 2 m³, they can be considered suitable for several African areas. Except for some details that require local blacksmith workshops, the users can self-build the silos. The possibility to use materials and equipment normally available on site is included among the criteria to take into account during the design phase. For example, corrugated galvanized iron, employed by the local population as cover for houses, can be considered a suitable material, as well as raw-earth, traditionally used in African rural areas. To demonstrate the reliability of the design and the functionality of the adopted solutions, unskilled staff of the Department built the silo in the workshop, using only simple tools available almost anywhere. The paper gives suitable information to replicate the building of this kind of silos, where unskilled labor is available at the family and/or farm level.

Key words: Silos, self-building, raw-earth, rural villages, maize storage, Tanzania.

INTRODUCTION

Especially for the development of the agricultural sector, two main categories of interventions can be schematized in developing countries in support of the economy and social progress (Dixon et al., 2001; Barker, 2007; Hoffmann, 2011).

The first category includes interventions that can transfer on-site facilities and organization having a markedly industrial nature, similar to those present in the countries of origin. In these cases, to be effective and

long lasting, the actions generally need to be set in an industrialized context, capable of providing the new body the contribution of complementary tools, continuity of energy supply and qualified technical and administrative workers, essential for operation. When these conditions do not occur, the intervention is likely to result in an ephemeral attempt, as not infrequently it has happened in the past. These interventions are characterized by a significant financial commitment and by limited needs related to the number and duration of the presence of qualified people provided by donors. Anyhow, this type of intervention has a low impact on the social-economic local environment, at least in the short and medium term,

since it involves only a small number of highly qualified local technicians (HSI, 2011).

The second category comprises the interventions that at first analyze local conditions, in particular taking into consideration the socio-economic situation and the traditional production means and techniques actually available on site. As a result, the introduction of few elements of scientific progress improves the agricultural and livestock productions thank to adoption of intervention strategies by local people, which do not seriously change traditional methods. This type of action is characterized by a limited financial commitment and, mainly in the initial stages, by a need for the presence of technical personnel on site. The social impact of these actions is generally very high since, in addition to technical personnel for the control and dissemination of knowledge, numerous local agricultural operators are directly involved. These interventions can result in extensive agro-livestock chains able to develop the socio-economic conditions of relatively large areas, in gradual but sustained and solid way (Kwa, 2001; Smalley, 2013).

The Department of Agricultural, Food and Forestry Systems of University of Firenze (GESAAF) has operated in various ways in the context of both types of interventions above described, also carrying out trainings in Italy and locally with the implementation of on-site pilot interventions.

The solution proposed in this work can be considered an intervention of the second type. Silos have been planned for the specific needs of a typical family in the area of Itigi, Singida Region, Tanzania that is a typical rural village of Africa, where GESAAF has been called to solve problems related to grains storage. Usually, farmers in this area owns an average of 1 to 2 ha of farmland, which has a typical yield of about 1.000 kg/ha of maize or other grains (FAO, 2013).

In this area, the rainy season starts in early November and ends in late March so the sowing is done in late October and the harvesting in early April (FAO, 2013). After about two weeks of outdoor drying, the maize reaches a water content of around 12 to 13%, and then it can be stored with a water content below 14%, level considered safe for maize storage (Golob, 2009; Wambugu et al., 2009; Yakubu et al., 2010). The crop must be able to be stored both for the next sowing, which occurs 6 months after harvest, and for food use until the next harvest, which occurs after 11 to 12 months.

Currently a significant portion of the grain stored is lost because of poor storage conditions (Proctor, 1994; Coulter and Schneider, 2004; Golob et al., 2009; Yakubu et al., 2010). The stored material can undergo direct attack by animals (insects, small mammals and birds), while the increase of moisture and temperature, in particular in the course of the rainy season, can cause the onset of

mildew and microorganisms, and the germination of the seeds. GESAAF has designed containers able to reduce losses of food. Besides, the storage of maize, the designed silos can store other kind of cereals (wheat, barley) and other seeds (e.g. beans).

The total size of the silos depends on the presence of walls and empty spaces, planned to increase its insulation from the heat. Cereal grain can generate heat, but this process is generally negligible for small amounts, and in any case cannot be avoided unless ventilation is used. In this case, however, other, more severe, drawbacks have to be highlighted such as the increase of humidity (Wambugu et al., 2009).

MATERIALS AND METHODS

Two silos were designed, one of 1 m^3 and another of 2 m^3, constructed in corrugated galvanized iron and insulated with earthen walls, with materials and equipment existing in Itigi, Tanzania. The same equipment and materials are available in many other rural areas in African Countries. The container has been structurally designed according to the relevant Eurocodes (EN 1991-4, 2006; EN 1993-1-3, 2006; EN 1993-4-1, 2007). The drawing of the content must be made on a daily basis without the outside agents penetrate inside (Petrovskij, 1990; Proctor, 1994; Udoh et al., 2000; Coulter and Schneider, 2004), so silos should be impenetrable, airtight and able to maintain a temperature as low and constant as possible. It has to be accessible for cleaning and filling on a yearly basis.

For the construction of the silo the following materials are necessary: corrugated galvanized steel sheet, galvanized smooth sheet, angle section bars 30×30×3 mm, rivets 4 mm nominal diameter, strips of waxed fabric, e.g. truck tarpaulin, which can be formed both from organic or synthetic material, sturdy fabric, sturdy thread to sew the fabric, twine, bitumen, galvanized network, 20x20 mm mesh, clayey soil, coarse gravel, leafy branches, aluminum thin sheets, if available.

The tools needed for the construction are: hand shears (necessary to cut sheets), electric or hand driller and 4 mm diameter drill bits, hand riveter, welder, a small metal container to heat the bitumen, a brush to spread the bitumen, sturdy needle to sew the fabric. The silo consists of the following elements shown in Figure 1.

1. Bottom sheet: round edges allow adapting the container inside the perimeter wall (Figure 1:1).
2. Sidewall of the container cylinder: this part consists of three sheets joined together by rivets; the joints are sealed with outside applied bitumen (Figure 1:2).
3. Angle section bar frame: this structure has the task to strengthen the upper closure of the container, to allow a person to rest over in order to enter into the container (Figure 1:3).
4. Upper closure: the hole allows a person to enter into the container; round edges allow to adapt the container inside the perimeter wall and to eliminate the trouble of alignment with the bottom (Figure 1:4).
5. Lid: it must be affixed and sealed after filling; yearly it must be removed and repositioned (Figure 1:5).
6. Drawing device (Figure 2): it consists of an air lock chamber closed by a sluice valve; the length of the device is due to the need to reach the outside of the external walls (Figure 1:6).
7. Bag: to complete the emptying (Figure 1:7); the rear and the lateral parts of the bag are raised by means of the twines according to the scheme in Figure 3.

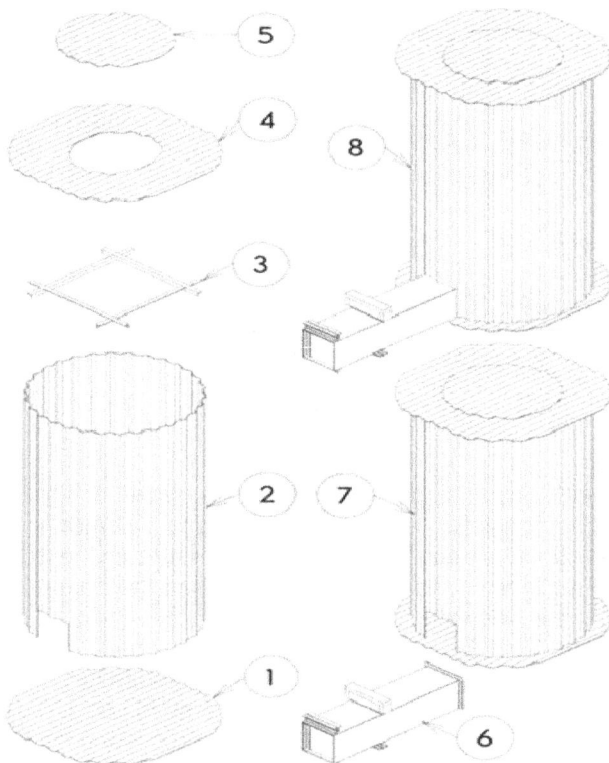

Figure 1. The silo and its components.

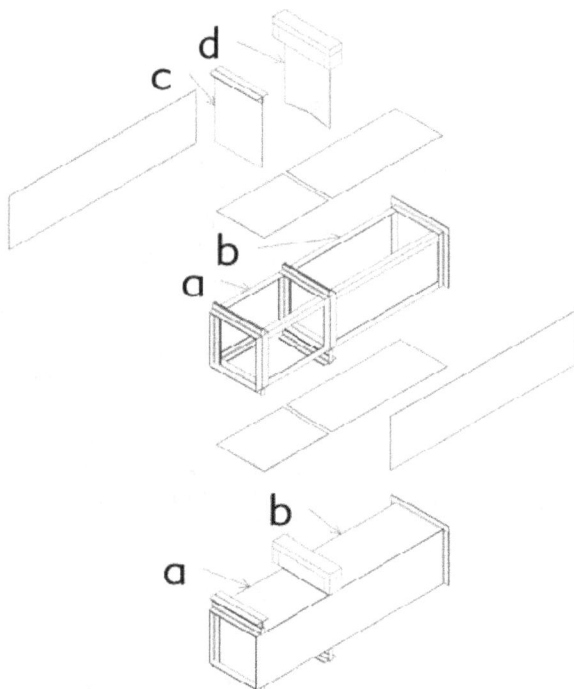

Figure 2. The drawing device and its components: a) drawing chamber b) extension outside of the wall to allow the actuation of the device d; c) outside guillotine for grains drawing; d) inside guillotine (valve) for closure during the drawing.

8. The container assembled is shown in Figure 1: 8.

The main container must be built with corrugated galvanized steel sheet, a thin sheet (0,4 mm) which is normally used by the local population as cover for houses. Also galvanized smooth sheets thicker (1-2 mm) are locally available, but because they are more expensive, they can be used only in small quantities for special applications. Due to the low thickness the sheets are not easily weldable. Therefore, the sheets should be joined each other and to the section bars by means of rivets.

For the construction of the container it is necessary to use a sturdy fabric, which can be formed either from vegetable or synthetic material. To seal the container, hot bitumen of the type used for asphalting roads is used. The bitumen is always coated on strips of waxed fabric, e.g., truck tarpaulin, which are arranged outside of the container, so that the bitumen is never in contact with the edible content.

The construction of raw earth brick walls ensures the thermal insulation of the container. In order to insulate it from the rainwater, from the moisture of the soil and from insects and other animals that burrow in the soil, a frustum of a pyramid constructed of a coarse gravel base supports the whole construction.

To avoid direct sunlight, a thick layer of leafy branches must cover the vertical walls and the roof of the construction. All these materials are available on site. The use of aluminum thin sheets to increase the thermal insulation from radiation heat transfer would be helpful. However, it is not certain that this material is available on site.

A test has been prepared in order to assess the ease of construction of the silo and to verify the functionality of the design solutions adopted. The workshop of the department was specifically arranged to create the same atmospheric conditions and the typical working situations of the Itigi village in Tanzania. In particular, the temperature and the humidity of the environment and the lack of resources and materials were taken into account. In this background, two unskilled workers built a prototype of the smaller silo using only the tools and the materials envisaged in the plan.

RESULTS AND DISCUSSION

The assembling of the silo, except the earth walls, required two days of work, during which several issues related with the construction and the use of materials were highlighted. Main considerations explaining some of the measures and decisions taken to improve the design of silo after the results stressed in the test are reported.

To ensure sealing of the container to external agents during the extraction of the grains, the silo is equipped with a slide valve. As illustrated in Figure 2, the ensiled material is present in part b, which is in direct communication with the interior of the container. Opening the guillotine d, the material enters the part a, which is closed by the guillotine c. Closing the guillotine, the gain to drawing chamber can be accessed and the contents can be drawn without external agents enter in b. The part b must be sufficiently long to allow the operation of the guillotine d, when the silo is inside the thermal insulation walls.

The above-said behavior happens until the ensiled material has reached the angle of repose, which is about 35° for maize and wheat (EN 1991-4, 2006). To allow the complete emptying of the silo, it is necessary that the grain is contained within a fabric bag which, reached the second phase (Figure 3), can be operated by means of

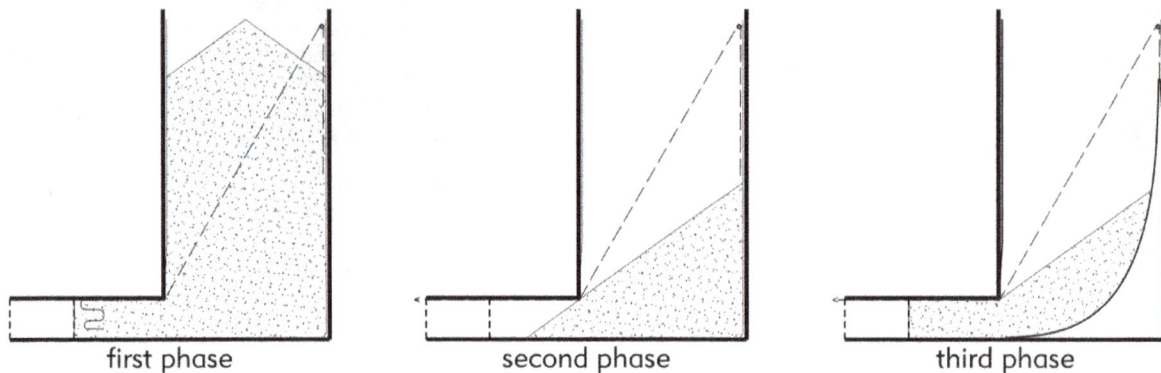

Figure 3. Phases of the container emptying.

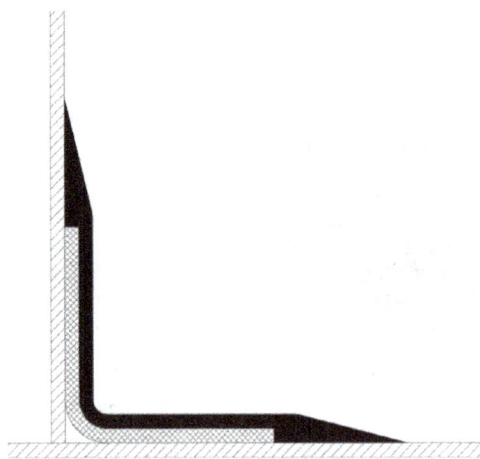

Figure 4. Joints ribbon-bitumen sealing.

Figure 5. The prototype under construction.

small ropes protruding from the outside of the mouth, so as to convey the content to the drawing device.

L-shaped elements, fastened with rivets to the two parts, made the joint between the vertical and horizontal

sheets. By cutting and bending little pieces of sheet, it is also possible to create L-shaped elements. The cylindrical sidewall sustains the pressure of the content, so it slightly stresses the joint between the bottom sheet and the sidewall.

The joint between the vertical and horizontal sheets and between the lid and the upper closure must be sealed affixing a ribbon of waxed fabric on which the bitumen is spread, according to the scheme shown in Figure 4.

In this way, it is possible to seal the spaces corresponding to the corrugations of the sheets. Furthermore, in comparison to the direct sealing of the sheets, this constitutes a safer, less sensitive to the deformations of the support sheets joint. It is also less prone to contamination of the contents.

A prototype of the inside metal container has been made in the workshops of GESAAF by two unskilled people (Figure 5).

The silo should be placed on a foundation consisting of a loose coarse gravel embankment, resting on the ground level. In this way, it is insulated from moisture and meteoric water. Furthermore, it is hindered the access to the internal cavities by soil-borne insects and mammals. On the top of the foundation, a sheet of material impervious to moisture and animals should be placed.

In order to store up the grains in the most appropriate manner, it is very important to minimize the increase in their temperature caused by external factors. For this purpose, the container must be thermally insulated from the external environment. To obtain this goal in the designed silos, it was decided to use the raw earth for the external structure. The raw material is widely available on the same building site and most people may carry out its processing, as it requires no special skill or complex equipment.

Raw earth buildings, if properly executed, have considerable strength and, if properly maintained, can have a virtually unlimited life (Barbari et al., 2014a,b). Maintenance consists essentially in repair by plastering parts that might be damaged by weathering, performed using the same clay mortar. Raw earth has a high

Figure 6. Phases of the silo completion (type A).

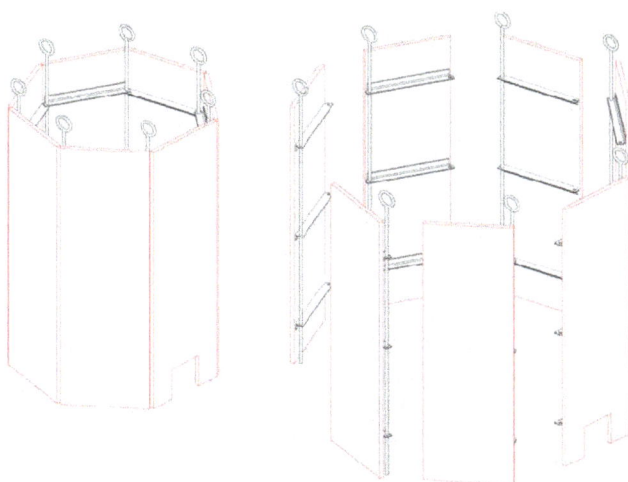

Figure 7. The dismantable internal formwork.

thermal inertia. Therefore, when a raw earth wall is heated on one face, it takes some time so that, while heating the mass of the wall, the temperature rises also on the opposite face. Since the temperature in the external environment has a sinusoidal trend in the 24 h, the temperature on the unheated face has not the time to reach the external value before night cooling, so it remains lower than outside. If the values of the thickness of the wall and of the day/night temperature swing are sufficiently high, the temperature rise of the inner face can be reduced to very small values or also to zero.

The daytime temperature rise of the outer face of the wall can be greatly limited if solar radiation does not

reach it directly. This can be achieved with an effective shading that can be realized thereby charging to the outer surfaces a sufficiently thick layer of plant material, such as grasses, straw, reeds, brushwood and similar as well as with suitably arranged trees.

The temperature of the outer face of the wall can be maintained at a value very similar to outside air temperature realizing a ventilated wall that is built outside a second wall to form a cavity in which air can circulates for the chimney effect. It is possible to place on the outer face of the internal wall a layer of reflective material. In this way, the heat radiation from the outer wall is rejected and the thermal insulation increased.

For these reasons, to achieve the thermal insulation a building made of considerable thick raw earth walls with a significant thermal inertia enclosed the container. An effective thermal lag in heat transfer is achieved (Doat et al., 1991; NZS, 1998; Morton, 2008; Minke, 2009). The roof covering must be removable to allow the annual cleaning and filling. The upper opening is closed with corrugated sheets, on which several layers of unbaked bricks are placed in order to obtain the same effect of thermal inertia characteristics of the vertical walls. On these bricks, galvanized corrugated sheets are placed in turn covered by a thick layer of plant material.

In applying the above principles to the designed silos, two construction methods are proposed, which use: a) unbaked clay bricks (adobe), b) rammed earth monolithic walls.

a) Unbaked clay bricks

Outside the first wall, a second wall is built, forming a cavity wall with natural air circulation. If available, on the outer surface of the first wall a reflective aluminum thin sheet should be applied. Furthermore, the vertical walls and the roof of the construction must be covered with a thick layer of leafy branches to avoid direct sunlight.

This solution requires, for the 1 m^3 capacity silo, the construction of about 800 bricks, of various measures, which must be walled using as mortar the same mixture used for the bricks, but with a different degree of humidity. Finally, the exterior vertical surfaces of the building should be plastered (Figure 6).

b) Rammed earth monolithic walls

The rammed earth construction technique consists in laying raw earth in specially crafted formworks, and ramming it until the desired consistency is obtained. After removing the formwork, the result is a wall of considerable strength and durability.

In order to carry out the construction, two formworks are required: an internal formwork (Figure 7) and an external one (Figure 8). Both formworks must be able to be disassembled for their reuse. In order to favor carrying

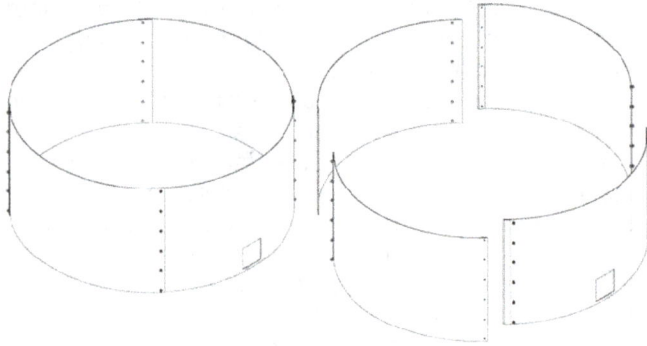

Figure 8. The dismantable external formwork.

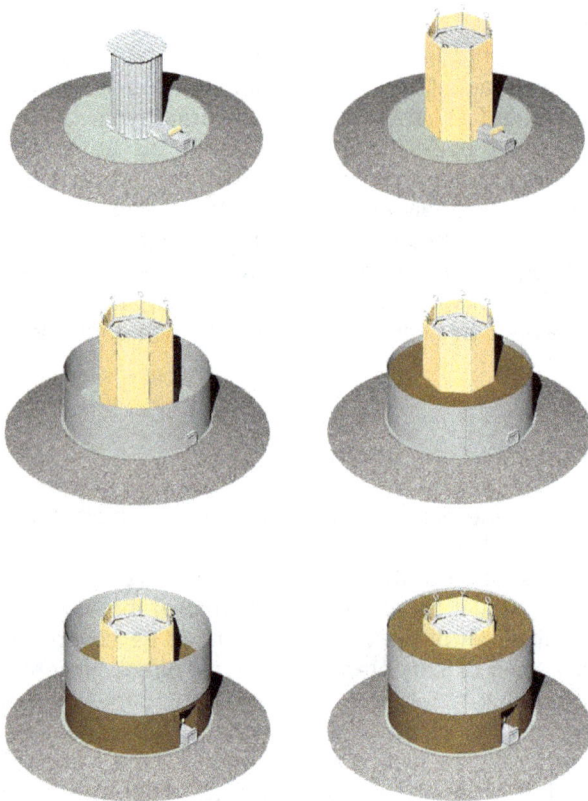

Figure 9. First phases of the silo completion (type B).

Figure 10. Second phases of the silo completion (type B).

and moving, it is necessary to contain the weights of formworks. Therefore, the external formwork has been designed to perform the casting of the wall in two phases, moving it up (Figures 9 and 10).

Conclusions

The proposed silos can be a suitable solution to store cereals and other seeds in African rural villages, thanks to the low cost of realization and to the simplicity of building, turning to unskilled labor of farmer's family.

The container and brick walls of adobe were made in the workshops of GESAAF using only materials and equipment actually available in the area to which the project is intended. Furthermore, the staff who implemented the construction did not have professional skills. This allowed us to verify the actual possibility of implementing the project on site.

Several improvements were introduced in the design after the test phase, related with the construction and the use of materials. In particular, a better system of unloading and a proper way to provide thermal insulation were developed.

Next step of the research will be the construction of the silo in an area of use in order to record and evaluate operating parameters.

Conflict of Interests

The authors have not declared any conflict of interests.

ACKNOWLEDGEMENTS

Numerous information about local conditions have been

provided by the Missionaries of the Precious Blood, which operate at the Hospital of St. Gaspar in Itigi, Singida Region, Tanzania, and by the technicians of the Cooperativa di Legnaia of Firenze, which work on site in support of the Mission.

REFERENCES

Barbari M, Monti M, Rossi G, Simonini S, Sorbetti GF (2014). Proposal for a simple method of structural calculation for ordinary earthen buildings in rural areas. J. Food, Agric. Environ. ISSN 12(2):1459-0263

Barbari M, Monti M, Rossi G, Simonini S, Sorbetti Guerri F (2014). Simple methods and tools to determine the mechanical strength of adobe in rural areas. J. Food, Agric. Environ. ISSN 12(2):1459-0263

Barker D (2007). The rise and predictable fall of globalized industrial agriculture. International Forum on Globalization, San Francisco, CA, USA.

Coulter J, Schneider K (2004). Feasibility study of post-harvest project in Mozambique and Tanzania. Consultancy report. Swiss Agency for Development and Cooperation, Berne, Switzerland.

Dixon J, Gulliver A, Gibbon D (2001). Farming systems and poverty - improving farmers' livelihoods in a changing world. Food and Agriculture Organization and World Bank. Roma and Washington D.C. ISBN 92-5-104627-1.

Doat P, Hays A, Houben H., Matuk S, Vitoux F (1991). Building with earth. France CRATerre. The Mud Village Society, New Delhi, India.

EN 1991-4 (2006). Eurocode 1: Actions on structures - Part 4: Silos and tanks. European Committee for Standardization, Bruxelles, Belgium.

EN 1993-1-3 (2006). Eurocode 3: Design of steel structures - Part 1-3: General rules - Supplementary rules for cold-formed members and sheeting. European Committee for Standardization, Bruxelles, Belgium.

EN 1993-4-1 (2007). Eurocode 3: Design of steel structures - Part 4-1: Silos, tanks, pipelines - Silos. European Committee for Standardization, Bruxelles, Belgium.

FAO (2013). The United Republic of Tanzania - Agriculture sector. FAO Country Profiles. Food and Agriculture Organization (FAO), Rome, Italy.

Golob P, Boxall R, Gallat S (2009). On-farm post-harvest management of food grains. A manual for extension workers with special reference to Africa. Agricultural and Food Engineering Training and Resource Materials (FAO), no. 2. Rural Infrastructure and Agro-Industries Div. Food and Agriculture Organization, Rome, Italy. ISBN 978-92-5-106327-9

Hoffmann U (2011). Assuring food security in developing countries under the challenges of climate change: key trade and development issues of a fundamental transformation of agriculture. United Nations Conference on Trade and Development.

HSI (2011). The impact of industrial farm animal production on food security in the developing world. HSI Report. The Humane Society International. Washington D.C., U.S.

Kwa A (2001). Agriculture in developing Countries: Which way forward? Focus on the Global South. New Delhi, India.

Minke G (2013). Building with earth – Design and technology of a sustainable architecture. Third and revised edition, Basel: Birkhäuser Verlag.http://dx.doi.org/10.1515/9783034608725

Morton T (2008). Earth masonry - Design and construction guidelines. Bracknell: IHS BRE Press.

NZS 4299 (1998). Earth buildings not requiring specific design. Standards New Zealand. Wellington, New Zealand.

Petrovskij L (1990). Saving seed in Senegal. International Ag-Sieve, 3 Rodale Institute, Pennsylvania, USA.

Proctor, D. L. (1994). Grain storage techniques: evolution and trends in developing countries. FAO Agricultural Services Bulletin (FAO), Agricultural Services Div. Food and Agriculture Organization, Rome, Italy. ISBN 92-5-103456-7. P. 109.

Smalley R (2013). Plantations, contract farming and commercial farming areas in Africa: a comparative review. Land and Agricultural Commercialization in Africa. Future Agricultures Consortium. University of Sussex, Brighton, UK.

Udoh JM, Cardwell KF, Ikotun T (2000). Storage structures and aflatoxin content of maize in five agroecological zones of Nigeria. J. Stored Prod. Res. pp. 187-201. http://dx.doi.org/10.1016/S0022-474X(99)00042-9

Wambugu PW, Mathenge PW, Auma EO, van Rheenen HA (2009). Efficacy of traditional maize (Zea mays l.) seed storage methods in western Kenya. AJFAND. Afr. J. Food Agric. Nutr. Develop. 9(4):1110-1129.

Yakubu A, Bern CJ, Coats JR, Bailey TB (2010). Non-chemical on-farm hermetic maize storage in east Africa. 10th International Working Conference on Stored Product Protection, Estoril, Portugal. pp. 338-345.

27

Influence of the storage conditions on moisture and bixin levels in the seeds of *Bixa orellana* L.

Biego G. H. M.[1,2], **Yao K. D.**[1], **Koffi K. M.**[1,3] **Ezoua P.**[1] **and Kouadio L. P.**[2]

[1]Laboratory of Biochemistry and Food Sciences, UFR Biosciences, University of Abidjan-Cocody, 25 BP 313 Abidjan 25, Cote d'Ivoire.
[2]Department of Public Health, Hydrology and Toxicology, Faculty of Pharmaceutical and Biological Sciences, University of Abidjan-Cocody, BP 34 Abidjan, Cote d'Ivoire.
[3]Central Laboratory for Food, Hygiene and Agro-Industry (LCHAI), National Laboratory for Agriculture Development Support (LANADA), Department of Animal Production and Fish Resources, Abidjan, Cote d'Ivoire.

This study was conducted to analyze the changes in the moisture and bixin levels in the seeds of *Bixa orellana* L. under four storage conditions: Storage without tarp (SB), storage over tarp (BD), seeds fully covered with tarp (BE) and storage in a room chilled to 4°C (OB). For 13 weeks, moisture content and bixin levels were monitored. The decreases in bixin after 13 weeks of storage at the different storage modes SB, BD, BE and OB were respectively 28.6, 25.9, 15.6 and 11.0%, while the increments in the moisture content were respectively 63.4, 57.0, 55.7 and 31.3%. No matter the storage mode, the fall of bixin was proportional to the moisture absorption. The storage mode OB brought the smallest loss of bixin and the smallest increment of moisture content, followed by BE. In tropical areas, however, the full covering of seeds with tarp would be more effective both in terms of cost and practicality. Moreover, the establishment of sorption isotherms allowed to suggest an optimal storage of the seeds of *Bixa orellana* L. in a dry enough environment having a relative humidity comprises between 12 and 30%.

Key words: *Bixa orellana* L., bixin levels, storage conditions, moisture, water activity.

INTRODUCTION

Côte d'Ivoire is located in West Africa, in the area between the tropics, at the north of the Gulf of Guinea. Its main export crops are annuity products such as coffee, cocoa, cotton and palm oil. However, the drastic fall in the prices of these raw materials from the nineties has gradually led the Ivorian farmers to shift to new crops, including the *Bixa orellana* L. as indicated in reports by the Ministry of Agriculture and Animal Resources of Côte d'Ivoire and OCDE (MARA, 1999; OCDE, 2002). *B. orellana* L. is mainly grown for its seeds which are covered with a dye composed of 70 to 80% of bixin and 20 to 30% of orelline (Dora and Flores, 1988; Aparnathi et al., 1990). Its chemical composition characterized by the presence of carotenoids (bixin and norabixine), flavonoids and terpenoids (Satyanarayana et al., 2003; Jondiko and Pattenden, 1989) accounts for its therapeutic and food interests (Chengaiah et al., 2010; Antunes et al., 2005; Agner et al., 2005; Fleischera et al., 2003).

Moisture absorption and mold growth remain the major

Table 1. Volume of distilled water added to construct the absorption isotherm.

Receptable number	Water volume added to the seeds (ml)
1	0.0
2	0.3
3	0.6
4	0.9
5	1.2
6	1.5
7	1.8
8	2.1
9	2.4
10	2.7

concerns that producers have to contend (Mara, 1999). The deterioration of seeds during drying and, especially, during storage prior export results in poor marketability products (Lavelli et al., 2007). Needless to say, post-harvest storage conditions of seeds are critical parameters in the conservation of the physicochemical properties especially as regards the content of natural dyes (Fleischera et al., 2003).

Thus, a pilot study was conducted to study the evolution of the amount of dye depending on the storage conditions.

MATERIALS AND METHODS

Sampling

The biological material consisted of annatto seeds (B. orellana L.) from Bondoukou and Tanda, two regions of high production, in the Northeast of Côte d'Ivoire. These seeds were collected during the high season and the early season (January-October) of the year 2009. For each storage mode, three batches of 12 bags of 50 kg each were made and four storage modes were considered: storage without tarp (SB), storage over tarp (BD), seeds fully covered with tarp (BE) and storage in a room chilled to 4°C (OB). The different batches were stored for 13 weeks. Before storage, samples were taken from batches at the rate of one sample (1 kg) per week. Samples were sent to the laboratory for determination of moisture, pigment (bixin) and water activity. Each sample was analyzed in triplicate and a total of 1008 trials were analyzed.

Determination of moisture

The moisture content was determined according to the method described by AOAC (1990). A sample of 5 g of annatto seed (B. orellana L.) was introduced into an aluminum box and then placed in an oven at 105°C till a constant weight was reached. Each test was performed in triplicate. The result was expressed as follows:

Moisture (%) = $100 - [(M1-M0) \times 100/Me]$

M0 is mass of the empty box; M1 is mass of the box containing the sample after drying and cooling and Me is sample mass.

Extraction of pigments from the seeds of *B. orellana* L.

A volume of 50 ml of KOH 5N was then placed in a stirring system from Lab-Line Instruments which alternates phases of agitation and rest for 24 h. The flask was finally adjusted to the mark with distilled water. The resulting solution was used for colorimetric analysis (McCormick, 1996).

Determination of bixin levels

The determination of bixin was performed using the method of McCormick (1996). A colorimeter from Milton Roy Spectronic was calibrated with a solution of 0.075N KOH and an aliquot of 3 ml was introduced into a flask of 100 ml and adjust to the mark with distilled water. Three absorbance measurements were made at a wavelength of 454 nm. The percentage of bixin was calculated using the following formula:

Bixin% = $(A \times V1 \times V2)/(M \times E \times V3)$

A is absorbance; M is mass of sample; E is extinction coefficient of bixin = 3200; V1 is initial volume of dilution (ml); V2 is final volume dilution (ml), and V3 is volume of aliquot (ml).

Construction of sorption isotherms

The relationship between water content and water activity of a product at a constant temperature can be represented by the sorption isotherm that allows expressing the phenomenon of hysteresis. A hygrometer from HygroLabRotronic was used to measure water activity according to the method by McCormick (1995).

Absorption isotherm

A sample of 5 g seed of B. orellana L. was placed in 10 Aw containers void of any trace of water. Increasing volumes of distilled water were added to each container (Table 1). After two minutes of soaking, measurements of water activity and moisture were carried out. A plot of the evolution of water activity (Aw) as a function of moisture content was constructed.

Desorption isotherm

A volume of 500 ml of distilled water were added to 100 g of seeds of B. orellana. After two minutes of soaking, water was removed through a sieve. Then, 5 g of wet seeds were then placed in a series of 10 aluminum crucibles and heated in an oven at 105°C for two hours. After this drying, the crucibles were successively removed from the oven at regular intervals of 10 min. The moisture and water activity were then determined for each pot of the series and a curve of changes in water activity as a function of moisture content was constructed.

Statistical analysis

Data obtained were seized under Excel software then statistically analyzed using Statistica version 7.1, analysis of variance (ANOVA). Averages and standard deviations of the analyzed parameters were classified using Newman-Keuls testand mean values were compared used student test. Differences were considered to be statistically significant when $P < 0.05$ level.

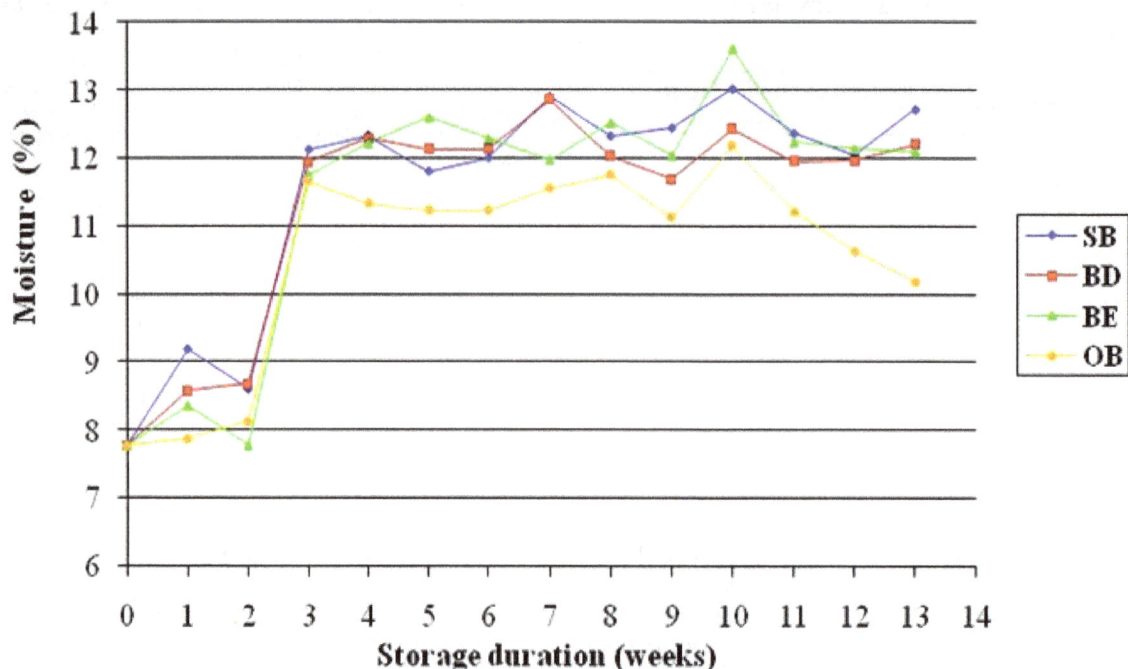

Figure 1. Evolution of moisture content depending on the storage duration (storage without tarp (SB), storage over tarp (BD), seeds fully covered with tarp (BE) and storage in a room chilled to 4°C (OB).

RESULTS

Evolution of moisture and bixin contents during storage

The results of changes in moisture percentage as a function of storage duration are shown in Figure 1. Figure 2 shows the changes in bixin levels in the same conditions. This study shows a decrease in bixin levels and an increase in moisture, whatever the storage method (Figures 1 and 2). The initial rates of bixin and moisture before storage were respectively 2.42, 2.43, 2.66 and 2.79% for the storage modes SB, BD, BE and OB that was a respective loss of bixin of 28.6, 25.9, 15.6 and 11.0%. The final moisture contents were respectively 11.5, 11.3, 11.3 and 10.5% for the storage modes SB, BD, BE and OB that was a respective moisture absorption of 63.4, 57.0, 55.7 and 31.3%. Considering the four storage modes, the significance of the reduction of bixin was as follow: OB<BE<BD<SB. Generally, this decrease is made while humidity increases (Figures 1 and 2). The storage mode that causes the smallest loss of bixin and the smallest increase in moisture content was the storage in room chilled to 4°C (OB), followed by the storage of seeds fully covered with tarp (BE). The storage without tarp is responsible for the most important loss of pigment and the highest increase in moisture. The test of the least significant difference revealed a significant difference at 5% between the averages of loss of bixin for the different storage modes.

Sorption isotherm curves

Figure 3 shows the results of absorption isotherm of *B. orellana* seeds. Water activity and moisture absorption increase in same time during *B. orellana* absorption. This absorption followed hyperbola form. The first was an ascending phase of weak slope with moisture absorption from 5 to 11%, corresponds to a water activity of 0.119 to 0.95; and a second vertical phase from 11 to 40% of moisture absorption corresponding a water activity of 0.97. Desorption isotherm of *B. orellana* seed describes a linear regression and curve present three distinct phases (Figure 4), the first ascending phase with moisture absorption from 10 to 31% corresponds to a water activity between 0.119 and 0.319; the second phase of weak slope shows the water activity from 0.319 to 0.9 and correspond to moisture absorption between 31 to 35%; and the last phase with water activity between 0.9 to 0.97 corresponding a moisture absorption from 35 to 80%.

DISCUSSION

Regarding humidity, the storage mode OB differed significantly from the other storage modes BE, BD and SB. This study clearly shows the influence of seeds moisture on the reduction of bixin. Gloria et al. (1995) showed deterioration in the stability of bixin in the presence or absence of air and light. They note that the half-life of bixin was less than 8 days when storage

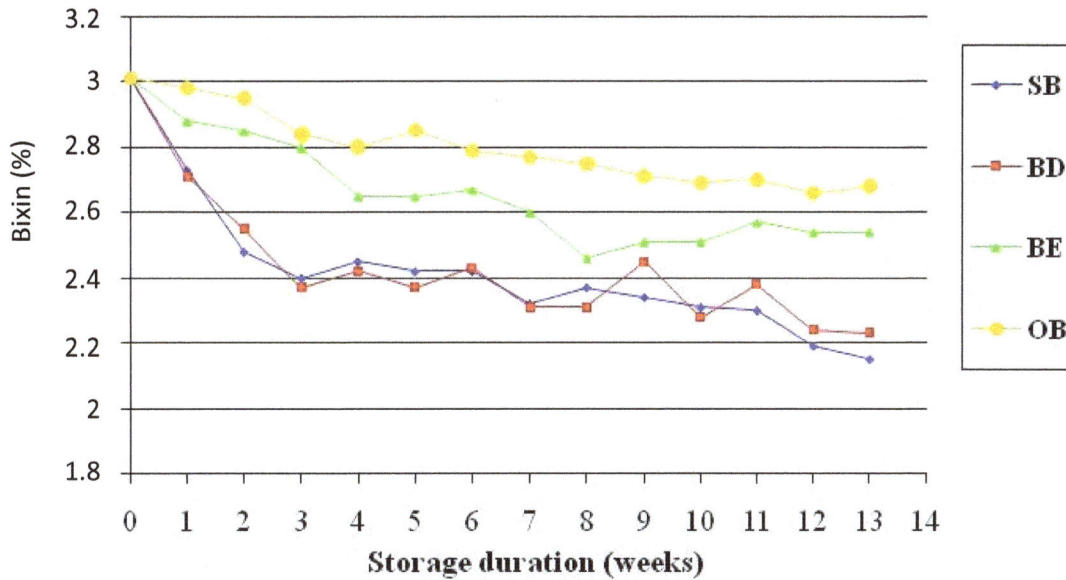

Figure 2. Evolution of bixin rates depending on the duration of storage (storage without tarp (SB), storage over tarp (BD), seeds fully covered with tarp (BE) and storage in a room chilled to 4°C (OB).

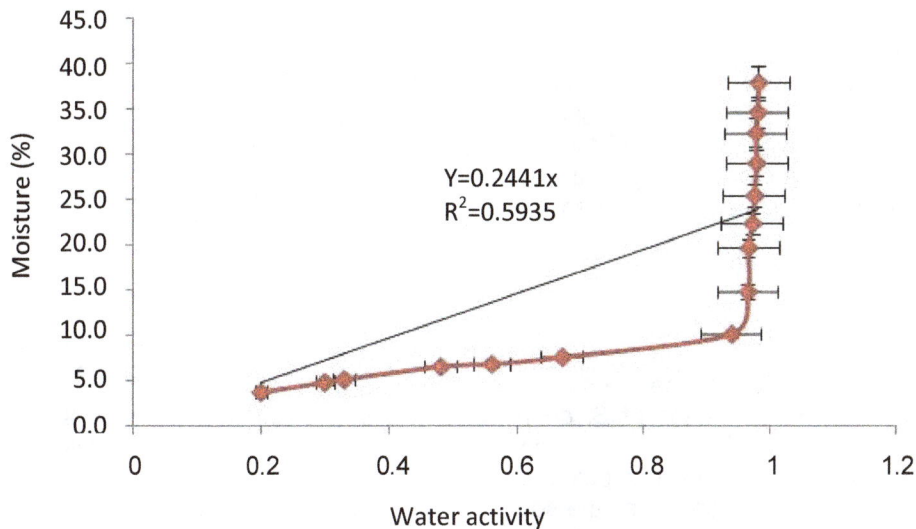

$$Y = 0.2441x$$
$$R^2 = 0.5935$$

Figure 3. Absorption isotherm of *Bixa orellana* L. seeds.

occurs in the presence of light and air, while it was over 55 days in the absence of light and under nitrogen. Several authors have also demonstrated a degradation of bixin and β-carotene of dehydrated carrots during storage (Goldman et al., 1983; Najar et al., 1988; Ribiero et al., 2005; Lavelliet al., 2007). The loss of bixin, activated by light, is caused by its oxidation in the presence of air (Di Mascio et al., 1989; Bradley and Min, 1992; Gloria et al., 1995).

The huge concern of degradation of *B. orellana* and mostly the lack of technical management and definition of quality standards according to permissible humidity of seeds have led producers to abandon its cultivation. Thus, it is of utmost importance for the Ivorian authorities to become more involved in the sector by organizing, training and educating all the stakeholders. Absorption and desorption isotherms were different because of the irreversible phenomena of porosity. The linear regression curves associated with them can predict the value of water activity for seed moistures known. Thus, at the beginning of the storage, the seeds had moisture content of 7.78% and depending on whether it is based on the

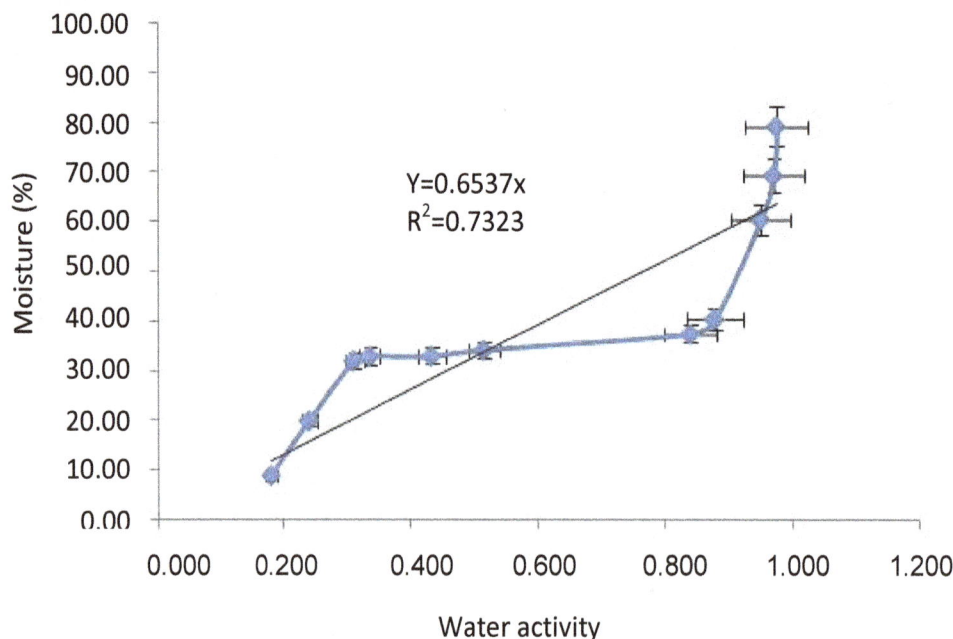

Figure 4. Desorption isotherm of *Bixa orellana* L. seeds.

absorption isotherm or desorption, moisture corresponded to a water activity between 0.119 and 0.319. To avoid moisture absorption, seeds should be stored in an environment where the moisture is in equilibrium with their water activity. We note therefore, the corresponding relative moisture was between 11. 99 and 31.9%.

Conclusion

This study showed a close relationship between changes in bixin and moisture contents of seeds of *B. orellana*. The storage in a room chilled to 4°C appeared to be the most appropriate type of storage for these seeds, but too expensive for Ivorian farmers. They could use the second most appropriate mode storage instead: the storage of seeds fully covered with tarp. Finally, the construction of sorption isotherms allowed to suggest a storage of seeds in a dry environment with humidity between 12 and 30%.

REFERENCES

Agner AR, Bazo AP, Ribeiro LR, Salvadori DM (2005). DNA damage and aberrant crypt foci as putative biomarkers to evaluate the chemopreventive effect of annatto (*Bixaorellana* L.) in rat colon carcinogenesis. Mutat. Res. 582:146-154.

Antunes LM, Pascoal LM, Bianchi M, Dias FL (2005). Evaluation of the clastogenicity and anticlastogenicity of the carotenoid bixin in human lymphocyte cultures. Mutat. Res. 585:113-119.

AOAC (1990). Official methods of analysis.Association of Official Analytical Chemists 13th edition, Arlington, Virginia, USA.

Aparnathi KD, Lata R, Sharma RS (1990). Annatto (*Bixa orellana* L.) its

cultivation, preparation and usage. Int. J. Trop. Agric. 8:80-88.

Bradley DG, Min DB (1992). Singlet oxygen oxidation of foods. Crit. Rev. Food Sci. Nutr. 31:211-236.

Chengaiah B, Rao KM, Kumar KM, Alagusundaram M, Chetty CM (2010). Medicinal importance of natural dyes a review. Int. J. Pharm. Tech. Res. 2:144-154.

Di Mascio P, Kaiser S, Sies HE (1989). Lycopene as the most efficient biological carotenoid singlet oxygen quencher. Arch. Biochem. Biophys. 274:532-538.

Dora R, Flores EM (1988). Le beurre, la margarine, l'huile, le riz, le bonbon et certains fromages à teinte rouge contiennent le pigment de rocou. Species Plantarum 1:422-575.

Fleischera TC, Ameadea EPK, Mensaha MLK, Sawerb IK (2003). Antimicrobial activity of the leaves and seeds of *Bixaorellana*. Fitoterapia 74:136-138.

Gloria MBA, Vale SR, Bobbio PA (1995). Effect of water activity on the stability of bixin in an annatto extract-microcrystalline cellulose model system. Food Chem. 52:389-391.

Goldman M, Horev B, Saguy I (1983). Decolorization of beta-carotene in model systems simulating dehydrated foods. Mechanism and kinetic principle. J. Food Sci. 48:751-757.

Jondiko IJO, Pattenden G (1989). Terpenoids and an apocarotenoid from seeds of *Bixaorellana*. Phytochemistry 28:3159-3162.

Lavelli V, Zanoni B, Zaniboni A (2007). Effect of water activity on carotenoid degradation in dehydrated carrots. Food Chem. 104:1705-1711.

MARA (1999). L'agriculture ivoirienne à l'aube du 21ème siècle. Rapport interministériels des Ministères de l'Agriculture et des Ressources Animales, de l'Environnement et de la forêt, de l'enseignement Supérieur et de la Recherche scientifique. Abidjan, Côte d'Ivoire. P. 243.

McCormick (1995). Determination of water activity.McCormick and Company, Inc. Manual of technical methods and procedures. Baltimore, USA.

McCormick (1996). Extractable color in Annatto *Bixaorellana*, oil soluble and encapsulated Annatto samples. McCormick and Company, Inc. Manual of technical methods and procedures.Baltimore, USA.

Najar SV, Bobbio FO, Bobbio PA (1988). Effects of light, air antioxidants and pro-oxidants on annatto extracts (*Bixa orellana*). Food Chem.

Influence of the storage conditions on moisture and bixin levels in the seeds of Bixa orellana L.

185

29:283-289.

OCDE (2002). L'économie locale de Bondoukou: comptes, acteurs et dynamisme de l'économie locale. Rapport de l'Organisation de Coopération et de Développement Economique OCDE SAH/D521/2001, Paris, France. P. 104.

Ribiero JA, Oliveira DT, Passos ML, Barrozo MAS (2005). The use of nonlinearity to discriminate the equilibrium moisture equations for BixaOrellana seeds. J. Food Eng. 66:63-68.

Satyanarayana A, Rao PGP, Rao DG (2003). Chemistry, processing and toxicology of annatto (Bixa orellana L.). J. Food Sci. Tech. Mys. 40:131-141.

28

Cassava post-harvest processing and storage in Nigeria: A review

Onyenwoke C. A. and Simonyan K. J.

Agricultural and Bioresources Engineering Department, Michael Okpara University of Agriculture, Umudike, P. M. B. 7267, Umuahia, Abia State, Nigeria.

Cassava is an important root crop consumed as a staple food, boiled, baked or often fermented into other foods and beverages all over the world. It is a very good vehicle for addressing some health related problems and also serve as security food. Cassava undergoes postharvest physiological deterioration (PPD) once the tubers are separated from the main plant. PPD is one of the main obstacles currently preventing farmers from exporting fresh cassava abroad thereby generating income from foreign exchange. Cassava can be preserved in various ways such as coating with wax and freezing. Recent development in plant breeding has resulted in cassava that is tolerant to PPD. Genetic manipulation was considered most appropriate to solving the PPD challenge by adding new traits to elite genotypes without altering other desired characteristics. Processing cassava affects the nutritional value of cassava roots through modification and losses in nutrients of high value. The processing methods include peeling, boiling, steaming, slicing, grating, soaking or seeping, fermenting, pounding, roasting, pressing, drying, and milling. The products from cassava are: High Quality Cassava Flour (HQCF), cassava chips, garri, starch, ethanol etc.

Key words: Post-harvest, storage, processing, cassava, high quality cassava flour (hqcf).

INTRODUCTION

Cassava is a major staple crop in Nigeria, as cassava and its product are found in the daily meals of Nigerians. Currently, cassava is undergoing a transition from a mere subsistent crop found on the field of peasants to a commercial crop grown in plantations. This unprecedented expansion on this crop is attributed to its discovery as a cheap source of edible carbohydrate that could be processed into different forms of human delicacies and animal feeds. Cassava is drought-tolerant, staple food crop grown in tropics and subtropical areas.

Cassava is to African peasant farmers what rice is to Asian farmers or wheat and potatoes are to European farmers (El-Sharkawy, 2003).

Cassava (*Manihot esculenta* Crantz) is a perennial woody shrub with an edible root, which grows in tropical and subtropical areas of the world Cassava plays a particularly important role in agriculture in developing countries, especially in sub-Saharan Africa, because it does well on poor soils and with low rainfall, and because it is a perennial crop that can be harvested as required.

Its wide harvesting window allows it to act as a famine reserve and invaluable in managing labour schedules. It offers flexibility to resource-poor farmers because it serves as either subsistence or a cash crop (Stone, 2002).

Furthermore, cassava is the source of raw materials for a number of industrial products such as starch, flour and ethanol. The production of cassava is relatively easy as it is tolerant to the biotic and edaphic encumbrances that hamper the production of other crops. Cassava's roots are used only to store energy, unlike the roots of sweet potato and yam that are reproductive organs. Despite their agronomic advantages, root crops are far more perishable than the other staple food crops. Once out of the ground, some root crops have a shelf life of only few days. Roots as living organs of plants continue to metabolize and respire after harvest. Cassava has a shelf life that is generally accepted to be of the order of 24 to 48 h after harvest (Andrew, 2002). Cassava utilization patterns vary considerably in different parts of the world. In Nigeria, the majority of cassava produced (90%) is used for human food (IITA, 2010). Cassava is very versatile and its derivatives and starch are applicable in many types of products such as foods, confectionery, sweeteners, glues, plywood, textiles, paper, biodegradable products, monosodium glutamate, and drugs. Cassava chips and pellets are used in animal feed and alcohol production. Animal feed and starch production are only minor uses of the crop in Nigeria. Cassava, in its processed form, is a reliable and convenient source of food for tens of millions of rural and urban dwellers in Nigeria (IITA, 2010). The aim of this study is to review the post harvest processing and storage of cassava in Nigeria, in order to improve on the processing and storage equipment for cassava.

Global situation of cassava

Nigeria currently produces about 54 million metric tonnes (MT) per annum (FAO, 2013), making her the highest cassava producer in the world, producing a third more than Brazil and almost double the production capacity of Thailand and Indonesia. However, Nigeria is not an active participant in cassava trade in the international markets because most of her cassava is targeted at the domestic food market. The production methods are primarily subsistence in nature and therefore unable to support industrial level demands (FAO, 2013). More than 248 million tons of cassava was produced worldwide in 2012 of which Africa accounted for 58% (IITA, 2012). In Ghana, Cassava accounts for a daily caloric intake of 30% and is grown by nearly every farming family. The importance of cassava to many Africans is epitomised in the ewe (a language spoken in Ghana, Togo and Benin) name for the plant, agbeli, meaning "there is life" (IITA, 2010). Cassava leaves are important in some countries,

for instant; in Democratic Republic of Congo cassava leaves have greater market value than roots.

In the subtropical region of southern China, cassava is the fifth-largest crop in term of production, after rice, sweet potato, sugar cane and maize. China is also the largest export market for cassava produced in Vietnam and Thailand. Over 60% of cassava production in China is concentrated in a single province, Guangxi, averaging over 7 million tons annually (Frederick, 2008). The world trade in pellets have long been dominated by Thailand, beginning around 1967, a few years after the start of its cassava exports to the European Union (EU). Although Thailand exports cassava chips and pellets to other Asian countries, especially China, where pellets are used both for animal feed and for the production of ethanol, the production and trade in cassava starch has significantly increased in recent years. Cassava starch has product characteristics that are technically superior to those of corn (maize) starch and this sub-sector promises to be a viable new market segment for industrial cassava. Already, in order to meet the global starch demand, large companies specializing in the production of starch and modified starch have invested hugely in Thailand, Brazil and Indonesia. Cassava flour is widely consumed in Brazil and in most of Latin America, as farinha (farinha is important just in Brazil), with various levels of sophistication in its processing from primitive family to large mechanized methods in factories.

THE NIGERIAN CASSAVA INDUSTRY

Over time, cassava has evolved from being a peasant's crop to cash and industrial crop. Cassava in Nigeria is used for two main purposes: 90% as human food and only 5 to 10% as secondary industrial material (used mostly as animal feed). About 10% of Nigeria's industrial demand consists of high quality cassava flour (HQCF) used in biscuits and confectioneries, dextrin pregelled starch for adhesives, starch and hydrolysates for pharmaceuticals products and seasonings, Seventy percent (70%) of cassava processed as human food is gari (Cassava Master Plan, 2006). Other common cassava products for human foods are lafun and fufu/Akpu. Processed products can be classified into primary and secondary products. The former, e.g. gari, fufu, starch, chips, pellets are primary products which are obtained directly from raw cassava roots, while the latter are obtained from further processing of primary products (e.g. glucose syrup, dextrin, and adhesive are obtained from starch).

Cassava production in Nigeria is increasing every year but Nigeria continues to import starch, flour, sweeteners that can be made from cassava (Cassava Master Plan, 2006). This paradox is due to how cassava is produced, marketed, and consumed in Nigeria, in a largely subsistence to semi-commercial manner. To fully exploit cassava's immense potential, especially as a replacement of imported raw materials and as an export commodity, there is a need to change how cassava is grown and traded in the country using a value-chain development approach. Nigerian cassava-based industrial products are just a fraction of imports, and the growth potential is huge (Cassava Master Plan, 2006).

Cassava transformation that builds upon two previous efforts has been embarked upon under the Agricultural Transformation Program of President Goodluck Jonathan and implementation by the Honourable Minister of Agriculture, Dr. Akinkumi Adesina. The cassava transformation seeks to create a new generation of cassava farmers, oriented towards commercial production and farming as a business, and to link them up to reliable demand, either from processors or a guaranteed minimum price scheme of the government. The overarching strategy of the cassava transformation is to turn the cassava sector in Nigeria into a major player in local and international starch, sweeteners, ethanol, HQCF, and dried Chips industries by adopting improved production and processing technologies, and organizing producers and processors into efficient value-added chains. There are three major limitations of increased utilization of cassava roots: poor shelf life, low protein content and their naturally occurring cyanogens (IITA, 2012).

Nutritional value of cassava roots

The nutritional composition of cassava depends on the specific tissue (root or leaf) and on several factors, such as geographic location, variety, age of the plant, and environmental conditions. The roots and leaves, which constitute 50 and 6% of the mature cassava plant, respectively, are the nutritionally valuable parts of cassava (Tewe and Lutaladio, 2004). The nutritional value of cassava roots is important because they are the main part of the plant consumed in developing countries.

Cassava root is an energy-dense food. In this regard, cassava shows very efficient carbohydrate production per hectare. It produces about 250,000 calories/hectare/day (Julie et al., 2009), which ranks it before maize, rice, sorghum, and wheat. The root is a physiological energy reserve with high carbohydrate content, which ranges from 32 to 35% on a fresh weight (FW) basis, and from 80 to 90% on a dry matter (DM) basis (Julie et al., 2009). Eighty percent of the carbohydrates produced is starch (Gil and Buitrago, 2002); 83% is in the form of amyl pectin and 17% is amylose (Rawel and Kroll, 2003). Roots contain small quantities of sucrose, glucose, fructose, and maltose (Tewe and Lutaladio, 2004). Cassava has bitter and sweet varieties. In the latter varieties, up to 17% of the root is sucrose with small amounts of dextrose and fructose (Charles et al., 2005). Raw cassava root has more carbohydrate than potatoes

and less carbohydrate than wheat, rice, yellow corn, and sorghum on a 100-g basis. The fibre content in cassava roots depends on the variety and the age of the root. Usually its content does not exceed 1.5% in fresh root and 4% in root flour (Gil and Buitrago, 2002). The lipid content in cassava roots ranges from 0.1 to 0.3% on a FW basis. This content is relatively low compared to maize and sorghum, but higher than potato and comparable to rice.

Cassava roots have calcium, iron, potassium, magnesium, copper, zinc, and manganese contents comparable to those of many legumes, with the exception of soybeans. The calcium content is relatively high compared to that of other staple crops and ranges between 15 and 35 mg/100 g edible portion. The vitamin C (ascorbic acid) content is also high and between 15 to 45 mg/100 g edible portions (Charles et al., 2004). Cassava roots contain low amounts of the B vitamins, that is, thiamine, riboflavin, and niacin (Table 1), and part of these nutrients is lost during processing. Usually the mineral and vitamin contents are lower in cassava roots than in sorghum and maize (Gil and Buitrago, 2002). The protein, fat, fibre, and minerals are found in larger quantities in the root peel than in the peeled root. However, the carbohydrates, determined by the nitrogen-free extract, are more concentrated in the peeled root (central cylinder or pulp) (Gil and Buitrago, 2002). Thus, cassava roots are rich in calories but low in protein, fat, and some minerals and vitamins. Their nutritional value is, consequently, lower than those of cereals, legumes, and some other root and tuber crops such as potato and yam.

Processing effects on nutritional value

Processing cassava affects the nutritional value of cassava roots through modification and losses in nutrients of high value. Analysis of the nutrient retention for each cassava edible product (Table 2) shows that raw and boiled cassava root keep the majority of high-value nutrients except riboflavin and iron. *Gari* is a common root product that involves grating, fermenting, and roasting. *Gari* and products obtained after retting of cassava root with peel are less efficient than boiled root in keeping nutrients of high value but are better than products obtained after retting of cassava roots. However, the latter is richer in riboflavin than sun-dried flour. *Fufu,* an important staple in Africa, is a mashed cassava root product that is allowed to ferment with *Lactobacillus* bacteria (Sanni et al., 2002). *Medua-me mbong* is a root product that requires only boiling and prolonged washing. However, *medua-me-mbong* has the poorest nutritional value compared to other cassava products with the exception of calcium content (Julie et al., 2009). In contrast to boiled cassava, processed root loss a major part of dry matter, carbohydrates, protein,

Table 1. Proximate, vitamin, and mineral composition of cassava roots and leaves.

Proximate composition	Raw cassava (100 g)	Cassava roots	Cassava leaves
Food energy (kcal)	160	110 - 149	91
Food energy (KJ)	667	526 - 611	209 - 251
Moisture (g)	59.68	45.9 to 85.3	64.8 to 88.6
Dry weight (g)	40.32	29.8 to 39.3	19 to 28.3
Protein (g)	1.36	0.3 to 3.5	1.0 to 10.0
Lipid (g)	0.28	0.03 to 0.5	0.2 to 2.9
Carbohydrate, total (g)	38.06	25.3 to 35.7	7 to 18.3
Dietary fiber (g)	1.8	0.1 to 3.7	0.5 to 10.0
Ashe (g)	0.62	0.4 to 1.7	0.7 to 4.5
Vitamins			
Thiamin (mg)	0.087	0.03 to 0.28	0.06 to 0.31
Riboflavin (mg)	0.048	0.03 to 0.06	0.21 to 0.74
Niacin (mg)	0.854	0.6 to 1.09	1.3 to 2.8
Ascorbic acid (mg)	20.6	14.9 to 50	60 to 370
Vitamin A (µg)	---	5.0 to 35.0	8300 to 11800
Minerals			
Calcium (mg)	16	19 to 176	34 to 708
Phosphorus, total (mg)	27	6 to 152	27 to 211
Ca/P	0.6	1.6 to 5.48	2.5
Iron (mg)	0.27	0.3 to 14.0	0.4 to 8.3
Potassium (%)	---	0.25 (0.72)	0.35 (1.23)
Magnesium (%)	---	0.03 (0.08)	0.12 (0.42)
Copper (ppm)	---	2.00 (6.00)	3.00 (12.0)
Zinc (ppm)	---	14.00 (41.00)	71.0 (249.0)
Sodium (ppm)		76.00 (213.00)	51.0 (177.0)
Manganese (ppm)	---	3.00 (10.00)	72.0 (252.0)

Source: United States Department of Agriculture (USDA) (2009).

Table 2. Nutritional value after processing 100 g of cassava root.

Nutrient	Whole root	Peeled root	Boiled root	Bˆaton or Chikwangue	Gari	Flour (retting and no peel)	Flour (retting and peel)	Washed cooked
Wet root (g)	100	77.0	87.6	38.5	49.2	25.3 to 29.6	27.9 to 34.0	66.8
Dry matter (g)	40.0	32.3	28.3	21.6	29.7	21.3 to 25.6	20.8 to 28.7	19.0
Calories	157	127	112	86	119	85 to 102	83 to 115	76
Protein (g)	1.0	0.48	0.38	0.18	0.37	0.16 to 0.22	0.26 to 0.51	0.16
Fat (g)	0.1	0.1	0.04	0.02	0.2	0.04 to 0.06	0.04 to 0.12	0.03
Carbohydrates(g)	37.9	31.0	27.4	21.2	28.8	20.9 to 25.1	20.3 to 28.1	18.8
Fiber (g)	1. 3	0.6	0.5	0.4	0.6	0.4	0.3 to 0.6	0.3
Ash (g)	0.90	0.57	0.46	0.21	0.34	0.16 to 0.19	0.24 to 0.50	0.06
Calcium (mg)	26	13	12	7	10	6.0 to 8.0	7.0 to 15.0	11
Phosphorus (mg)	47	39	31	13	18	9.0 to 11.0	10.0 to 21.0	7
Iron (mg)	3.5	0.4	0.4	3.1	1.5	0.2 to 0.7	0.8 to 11.9	0.2
Thiamin (µg)	72	31	20	10	18	6.0 to 12.0	13	3
Riboflavin (µg)	34	18	16	21	15	10.0 to 12.0	8.0 to 21.0	6
Niacin (mg)	0.73	0.52	0.41	0.16	0.33	0.11 to 0.18	0.17 to 0.37	0.03
Vitamin C (mg)	33	20	1	1	2	0	0	0

Source: Institute of Food Technologists (2009).

and thus calories. Although raw cassava root contains significant vitamin C, it is very sensitive to heat and easily leaches into water, and therefore almost all of the processing techniques seriously affect its content (Julie et al., 2009). Boiled cassava, gari, and products resulting from retting of cassava root with peel, retain thiamin and niacin better than products obtained after retting of shucked cassava roots, smoked-dried flour, and medua-me-mbong. Riboflavin is well retained in boiled cassava, gari, and smoked-dried cassava flour obtained after retting of cassava root with peel in contrast, the losses of vitamin B2 (riboflavin) (Julie et al., 2009).

POSTHARVEST HANDLING AND STORAGE

Cassava is harvested by hand by raising the lower part of the stem and pulling the roots out of the ground, then removing them from the base of the plant. The upper parts of the stems with the leaves are plucked off before harvest. Cassava undergoes postharvest physiological deterioration (PPD) once the tubers are separated from the main plant. The tubers, when damaged, normally respond with a healing mechanism. However, the same mechanism, which involves coumaric acids, initiates about 15 min after damage, and fails to switch off in harvested tubers (Sánchez et al., 2010). It continues until the entire tuber is oxidized and blackened within two to three days after harvest, rendering it unpalatable and useless. PPD is one of the main obstacles currently preventing farmers from exporting cassava abroad and generating foreign exchange income. Post-harvest strategies include the development of effective and simple machines and tools that reduce processing time and labour, and production losses. With these machines, losses can be reduced by 50% and labour by 75% (Sánchez et al., 2010). Cassava can be preserved in various ways such as coating in wax or freezing. Plant breeding has resulted in cassava that is tolerant to PPD. (Sánchez et al., 2010) identified four different sources of tolerance to PPD. One comes from Walker's Manihot (M. walkerae) of southern Texas in the United States and Tamaulipas in Mexico. A second source was induced by mutagenic levels of gamma rays, which putatively silenced one of the genes involved in PPD genesis. A third source was a group of high-carotene clones. The antioxidant properties of carotenoids are postulated to protect the roots from PPD (basically an oxidative process). Finally, tolerance was also observed in a waxy-starch (amylase-free) mutant (Sánchez et al., 2010). This tolerance to PPD was thought to be co-segregated with the starch mutation, and is not a pleiotropic effect of the latter (Sánchez et al., 2010).

Two types of post harvest deterioration are recognized: Primary physiological deterioration that involves internal discoloration and is the initial cause of loss of market acceptability and secondary deterioration due to microbial spoilage. The former is thought to be a consequence of tissue damage during harvesting, in most cases it is seen as a blue-black discoloration of the vascular tissue referred to vascular streaking. These initial symptoms are followed by a more general discoloration of starch-bearing tissue (Andrew, 2002).

Processing techniques of cassava root

Fresh cassava roots cannot be stored for long because they rot within 48 h of harvest. They are bulky with about 70% moisture content (Hahn, 1994). Therefore, cassava must be processed into various forms in order to increase the shelf life of the products, facilitate transportation and marketing, reduce cyanide content and improve palatability. The nutritional status of cassava can also be improved through fortification with other protein-rich crops. Processing reduces food losses and stabilizes seasonal fluctuations in the supply of the crop. Traditionally, cassava roots are processed by various methods into numerous products and utilized in various ways according to local customs and preferences. Traditional cassava processing methods in use in Africa probably originated from tropical America, particularly north-eastern Brazil and may have been adapted from indigenous techniques for processing yams (Hahn, 1994). The processing methods include peeling, boiling, steaming, slicing, grating, soaking or seeping, fermenting, pounding, roasting, pressing, drying, and milling as shown in Figure 1.

Storage techniques

The storage of agricultural raw materials is an essential aspect of food processing that ensures that food remains available even in time of scarcity (Osunde and Fadeyibi, 2011). Traditional marketing and storage systems have been adapted to avoid root perishability (Aristizabal and Sánchez, 2007). These adaptations include processing centered in proximity to the areas of production to ensure a daily supply of raw material, processing into storable forms (through sun drying, fermentation, etc.) at the farm level and the common practice of trading of small quantities of roots (Weham, 1995; Westby, 2002). A common way of avoiding root losses due to PPD is to leave the roots unharvested in the soil after the period of optimal root development, until the roots can be immediately consumed, processed or marketed. Cassava roots are known to last in soil up to three years. This strategy has disadvantages because large areas of land are used by the standing crop, unavailable for additional agriculture production. Furthermore, even though the roots may increase in size they become more woody and fibrous, decreasing palatability and increasing the cooking time, respectively, if left longer than the optimal

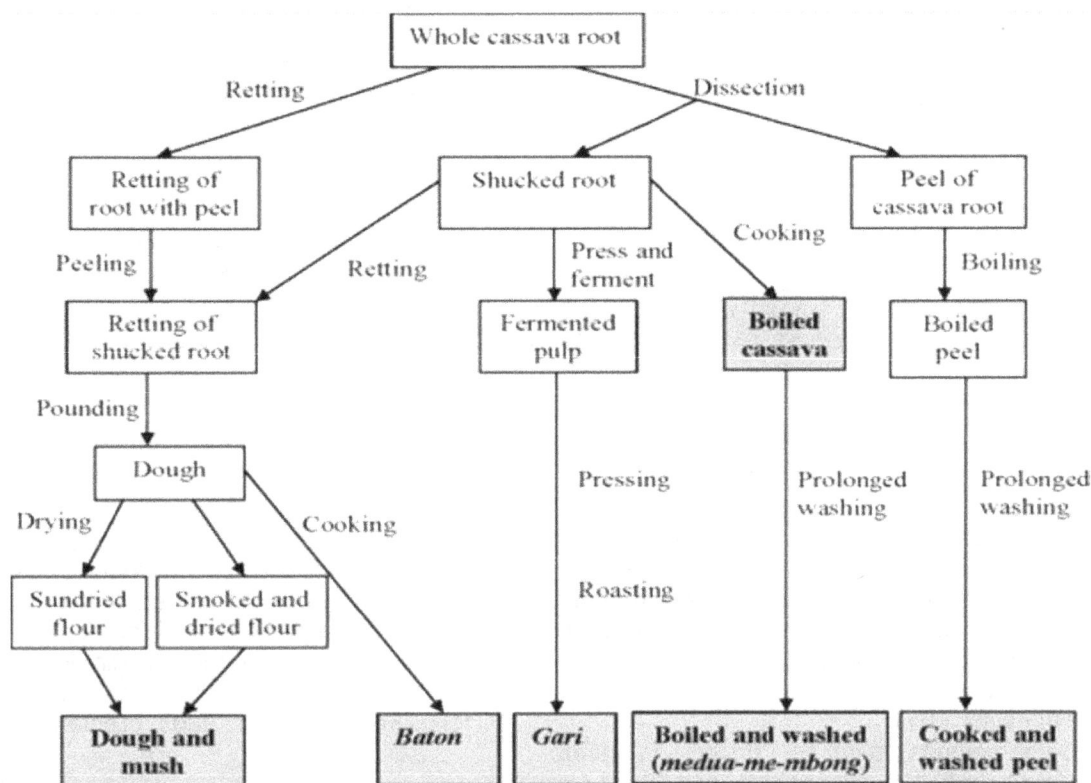

Figure 1. Different processing techniques for whole cassava root. The edible forms of cassava root are shaded in gray adapted of Julie et al. (2009).

harvest time of 10 to 12 months after planting. Another negative effect occurring due to extensive in-field storage of cassava roots is their increased susceptibility to attack by pathogens as well as the reduction of extractable starch (Wenham, 1995; Ravi et al., 1996).

Fresh cassava roots cannot be stored for long because they rot within 24 to 48 h of harvest. They are bulky with about 70% moisture content, and therefore transportation of the tubers to urban markets is difficult and expensive. Good storage depends on the moisture content of the products and temperature and relative humidity of the storage environment. The moisture content of gari for safe storage is belong 12.7% (Osunde and Fadeyibi, 2011), when temperature and relative humidity are above 27°C and 70% respectively, gari goes bad. The type of bag used for packing also affects shelf life depending on the ability of the material to maintain safe product moisture levels. During the last twenty years there have been some developments in improving storage methods capable of extending the shelf life of fresh cassava roots by at least two weeks. These, amongst other advantages, make it possible to market the crop further and give an increased margin to the opportunity of holding stocks of fresh cassava, even for few days, at a processing plant. A joint project between the National Resources Institute, and Centro Internacional de Agricultura Tropical (CIAT) studied alternative storage methods to the traditional

reburial procedures. These included storage in pits, in field clamps and in boxes with moist sawdust. All the storage methods investigated favoured curing conditions in a high humidity and high temperature environment in order to slow down the rates of physiological and microbiological deterioration (Osunde and Fadeyibi, 2011).

However, to be successful they all require careful harvesting and selection of the roots prior to storage, since curing is not effective if root damage is extensive (Crentsil et al., 1995). Storage in boxes lined with moist sawdust or wood shavings involves putting alternative layers of sawdust and cassava roots, starting and finishing with a layer of sawdust. As an alternative to sawdust, wood shavings or any other suitable packing material can be used. However, the packing material must be moist but not wet. Physiological deterioration occurred if the material was too dry and microbial decay accelerated when it was too wet. In Uganda this storage method was tested in combination with the lining of box with plastic (Nahdy and Odong, 1995). The study indicated that 75% of the roots remained healthy after four weeks in store, provided the roots were packed immediately on the day of harvest. With a delay of one day only 50% of the roots were rated as acceptable. This technique has been used for some export markets but the higher transport cost involved because of the box

Table 3. Demand estimates of cassava supply in Nigeria.

Product	Demand estimate (Tons)	Substitution (%)	Potentials (Tons)	Fresh root (Tons)	Utilization estimate (MT)
Flour	7,525,000	30	2,257,500	22,575,000	7,537,479.95
Starch	4,970,000	50	2,485,000	12,425000	1,076,000
Livestock feed	91,243,248	20	18,248,649.6	72,994,598.4	1,614,000
Ethanol	3,600,000	50	1,800,000	14,400,000	538,000
Total				10,765,479.95	

Conversion factor: 1 t of starch = 400 L Ethanol (98% efficiency). Starch: 5: 1 (raw: starch). Livestock: 4:1 (raw: chip). Land required: 356,250 ha@20 t/ha Conversion factor for ethanol: 1:8. Source: Personal Field observations, FAOSTAT (2011) Phillips et al. (2004) and Kormawa et al. (2003).

containers has precluded its use for domestic market (Osunde and Fadeyibi, 2011).

Storage in plastic bags or plastic film wraps appears to be the most practical and promising method of storing cassava roots intended for the urban markets. A number of studies have shown that cassava roots treated with an appropriate fungicide and kept in an airtight plastic bag or a plastic film wrap can be stored for two to three weeks (Osunde and Fadeyibi, 2011).

Some modern methods, such as refrigeration, deep freezing, waxing, controlled atmosphere and chemical treatments, have been suggested for the storage of fresh cassava. Freezing and waxing have been used primarily for export markets in Europe and America, where the customers of African and Latin American origin are prepared to pay high prices. These techniques require specialized equipment and skills and are very capital intensive (Crentsil et al., 1995). A more common modern method of limiting PPD is covering cassava roots with paraffin wax by dipping the root in paraffin wax (at a temperature of 55 to 65°C for a few seconds) after treatment with fungicide. Use of wax has been reported to prolong shelf-life of cassava roots up to 2 months (Ravi et al., 1996; Aristizabal and Sánchez, 2007). Cassava roots can also be stored for 2 weeks between 0 to 4°C without any internal deterioration. The most favourable temperature for storing fresh cassava is 3°C but after 4 weeks microbial infection takes place and will increase with subsequent storage time. However, even after 6.5 months of storage between 0 to 4°C, the part of the root without decay usually is in excellent condition and is suitable for human consumption (Ravi et al., 1996; Oirschotet al., 2000). At temperatures above 4°C roots develop the PPD symptoms more rapidly and have to be discarded after 2 weeks of storage (Ravi et al., 1996). Alternatively, entire roots or more usually pieces of root can be stored frozen under deep-freeze conditions in polyethylene bags and the roots were quite palatable after thawing, although some sponginess was present, and was able to be kept for a further 4 days. This technique is used at a commercial scale in many Latin American countries such as Brazil, Colombia, Costa Rica and Puerto Rico (Ravi et al., 1996).

DEMAND ESTIMATES OF CASSAVA SUPPLY IN NIGERIA

The tolerance of cassava to extreme stress conditions, its low production resource requirements, its biological efficiency in the production of food energy, its availability throughout the year and its stability for farming systems, will make cassava products gain more popularity in Nigeria (Kormawa et al., 2003) (Table 3).

INDUSTRIAL PRODUCTS FROM CASSAVA

Four primary industrial products from cassava stand out as important for Nigeria. These are (a) cassava flour, (b) crude ethanol, (c) native starch, and (d) animal feed/cassava chips and pellets. These products are commonly traded and show the highest potential for growth in demand, and are associated with medium and large scale processing.

In the domestic market, industrial cassava products compete with traditional cassava products, mainly gari. Furthermore, each of the main industrial products (cassava flour, chips for animal feed, chips for food grade ethanol, and cassava starch) faces competition from (a) identical imported products, and (b) substitute products that are either being imported or locally grown. For domestic cassava flour the main competitive product is wheat flour. For cassava chips/pellets it is feed grains. For ethanol it is ethanol from other sources, and for starch it is corn/maize starch. Quite clearly, significantly lowering the cost of raw materials (ex-factory price) would greatly reduce the cost of the final product, making them more competitive. One strategy to achieve this is the vertical integration of commercial farms to each processing plant.

SECONDARY PRODUCTS FROM CASSAVA

Cassava can be processed into various secondary products, including modified cassava starch, glucose syrup, extra neutral alcohol, noodle, bakery and confectionery industries, meat and textile processing. It is also industrially processed as a raw material in the coating of pharmaceutical products, the manufacture of glues and adhesives and oil drilling starch (EFDI-Techno Serve, 2005).

Glucose syrup: is a concentrated aqueous solution of glucose maltose and other nutritive saccharine made from edible starch. Glucose or dextrose sugar is found naturally in sweet fruits such as grapes or honey. It is less sweet than sucrose (cane sugar) and is used in large quantities in fruits, liquors, crystallized fruits, bakery products, pharmaceuticals, and breweries.

Noodles: are a long thin extruded food product made from a mixture of flour, water, and eggs usually cooked in soup or boiling water (Sanni, 2005). At 12.5%, cassava starch/flour forms an integral part of the final product.

Cassava based adhesives: like the cereal starch adhesives, are of three main types:

i) *Liquid starch adhesives* are supplied by the adhesives manufacturer in liquid form usually in plastic or lined metal drums, jerry cans and bottles.
ii) *Pre-gel starch adhesives* are produced in dry flakes and milled to specific particles sizes. They are packed in waterproof lined multi-wall paper bags/sacks and are very suitable for export.
iii) *Dextrin based adhesives* are delivered to consumers in liquid and dry forms depending on specification and requirement. The liquid dextrin adhesives are packed as the liquid starch adhesives, while the dry dextrin adhesives are packed as the milled pre-gel adhesives. Dry dextrin adhesives are very suitable for export as intermediate raw materials used especially in Europe and America by the food and industrial companies.

High Quality Cassava Flour (HQCF)

Nigeria imports over one million tonnes of wheat annually. At 10% substitution of cassava flour in wheat flour and with the current national demand, 300,120,000 metric tonnes of HQCF (assuming the national demand for wheat flour is 1.2 million tonnes), is required (IITA, 2011). 30% of the total wheat can be replaced by cassava flour in bread making, and 100% cassava flour is currently being used in pastries and confectioneries (Onabolu et al., 1998). However, with poor regulation and standardization, some bakeries have complained about problems: including presence of impurities such as sand; odour; shorter product shelf life (e.g. biscuits); brittleness; gradual change of colour (biscuits turning pale); unreliable supply; poor final product quality in cases where the cassava flour had partially fermented. With other domestic uses for cassava flour in snacks, a more realistic estimate for the annual demand of cassava

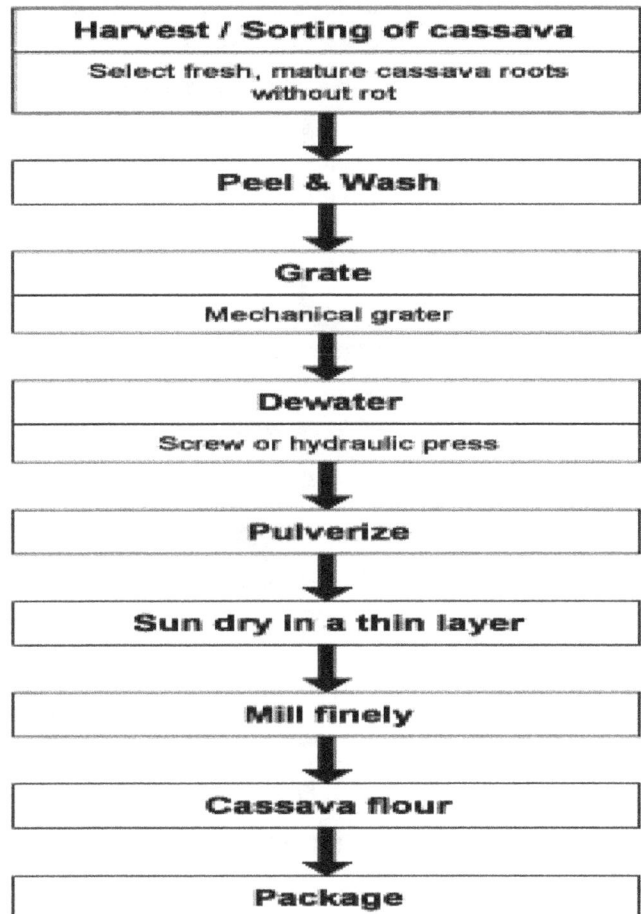

Figure 2. Process flow chart for high quality cassava flour (Cassava Master Plan, 2006).

flour is therefore 250,000 to 300,000 MT (Cassava Master Plan, 2006), a figure impossible for small holders to supply. The process of flour production is described below.

Cassava chips

Cassava chips are dried irregular slices of roots, which vary in size but should not exceed 5 cm in length (CIAT, 2004). The tuberous roots, either peeled or unpeeled, are cut up into chips (cossettes) and dried. Chips from peeled roots are used for human consumption and in animal feed industry and generally store better than flour (IITA, 1990). Chips are the most common form in which dried cassava roots are marketed and most exporting countries produce them. The standard method of processing chips consists of peeling, washing, chipping the cassava roots, and then sun drying the slices. The recovery rate of chips from roots is 20 to 40% depending on the initial dry-matter content of the cassava roots and the final moisture of the chips (Cassava Master Plan, 2006) (Figure 2).

In Nigeria, cassava chips were processed into animal

Cassava flour

Water and alpha-amylase enzyme

↓

Liquification

(90-95 °C, pH 4-4.5) 400rpm

↓

Saccharification

(55-65 °C, pH 4-4.5)
Glucosidase enzymes

↓

Cooling

(30-33 °C)

↓

Fermenter

(Yeast added) (Carbon dioxide out)

↓

Distillation

Feed recovery

↓

Ethanol

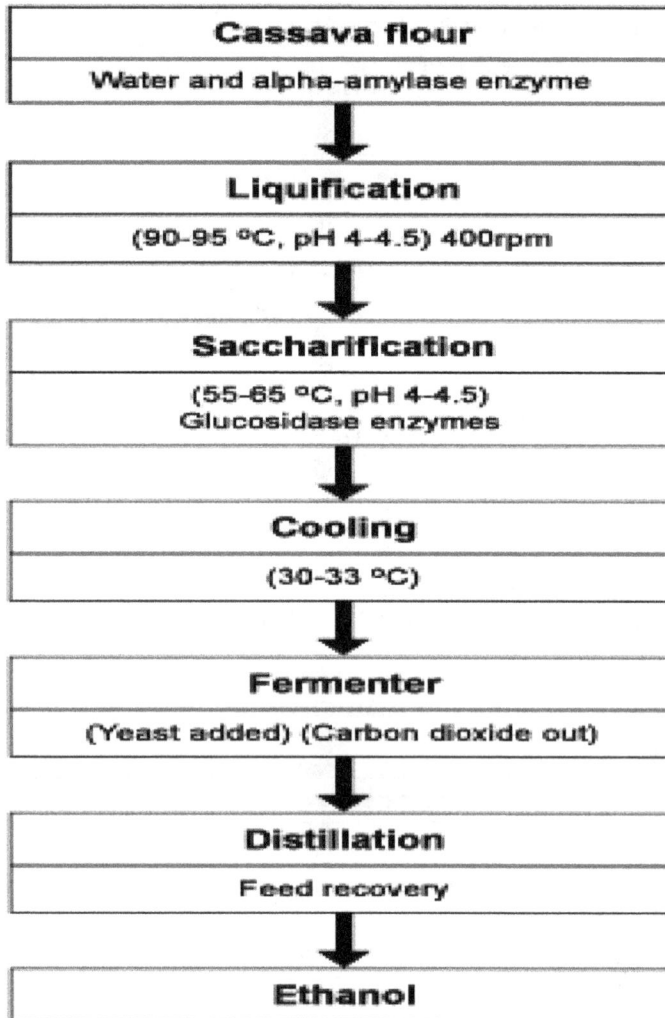

Figure 3. Flow chart showing the production of ethanol from cassava flour (Cassava Master Plan, 2006).

feed and some animal feed millers continued the practice until the late 90s when the price of cassava became too expensive vis-à-vis the price of maize. Presently, no major livestock feed mill uses cassava as a raw material, although smaller mills and large farms that blend their own feed use cassava chips or meal when these are locally available at low prices. The livestock sector in Nigeria is rapidly expanding and a continued demand for animal feed is predictable. In view of the relatively high-income elasticity for meat products, it is likely that this trend will continue during the remainder of this decade. Processing cassava chips into cassava pellets will further reduce transport costs and enhance product quality.

Cassava pellets for animal feed

Substituting maize with cassava in animal feed have been made using linear programming, saving of up 10%

in poultry feed costs and about 20% for pig feed (Cassava Master Plan, 2006). With the Nigerian livestock industry uses up to 1.2 metric tonnes of maize annually (Cassava Master Plan, 2006), substituting 10% of this figure with cassava would involve setting up of at least 200 cassava chip making factories processing about 10 tonnes of cassava roots per day (Cassava Master Plan, 2006), Pellets can be made either from cassava chips or flour. An indigenous Nigeria company, B & T Ventures, Ibadan, in collaboration with the cassava project at IITA, has designed and created a pelleting machine that can produce three different types of cassava pellets: hard, soft, and floating. The hard pellets are used for feeding poultry, the soft ones for feeding ruminants, and the floating ones for feeding fish. However, the machinery is still under R & D and is not as efficient as imported pelleting machines.

Ethanol

Ethanol is produced by the fermentation of sugar related materials such as molasses and sugar juice, or starchy materials. Cassava stands as one of the richest fermentable substances for the production of crude alcohol/ethanol, with dry chips containing up to 80% of fermentable substances (starch and sugars) (Cassava Master Plan, 2006). The process of cassava based ethanol production is described in Figure 3.

Tapioca

In Nigeria, this is made from partly gelatinised cassava starch, (although the cassava crop itself is called tapioca in some places, heat treated to a moist mash in shallow pans. Its shapes are irregular lumps called flakes, or perfectly ground beads. It is consumed in many parts of West Africa, soaked or cooked in water with sugar and/or milk added. High labour processing steps make it quite expensive (Sanni et al., 1992).

High quality garri

With a share of 70% of all cassava fresh roots harvested, gari will continue to dominate the cassava sector in the short term. The growth rate of gari has been put at least 4 to 6% per annum, primarily due to population growth and increasing urbanization, and export to the regional West African market. It already provides livelihoods to more than 5 million farmers and processors (often poor rural women) in Nigeria, as well as to numerous equipment manufacturers, wholesale and retail traders, and transporters. In addition, small-scale garri processing has gradually become the main source of non-farm rural employment in many countries. The process flow chart is described in Figure 3.

Sweeteners

Cassava starch and HQCF can be used as raw material for sweeteners, primarily high fructose syrup (HFS), glucose, and sorbitol. Sweeteners are obtained by hydrolysis of cassava starch or flour or wet cake, to produce glucose, which is further purified to produce HFS or hydrogenated to produce sorbitol. One ton of starch yields 900 kg of glucose, 550 tons of HFS (55% purity), and 1.1 ton of sorbitol (70% purity) (Cassava Master Plan, 2006). The annual demand for these sweeteners in Nigeria is: 150,000 tons of HFS, as part replacement for imported sugar in the soft drink and juice industry, 40,000 tons/year of glucose, and 14,000 tons of sorbitol (FAO, 2011). The sweetener industry is a strong market that is expected to grow by 50% over the next ten years. Ekha Agro Nigeria Limited is currently the only cassava processor, producing sweeteners, liquid glucose, from cassava for supply to Guinness Plc (FAO, 2011).

CONCLUSION

The research in cassava processing has established the fact that there is a lot more in cassava than starch. The nutritional quality content in cassava can be enhanced by developing new varieties by biofortificaton, cassava could be source of raw materials for a number of industrial products example include, the starch, flour and ethanol. The production of cassava is relatively easy as it is tolerant to the biotic and edaphic encumbrances that hamper the production of other crops. Cassava has a shelf life that is generally accepted to be of the order of 24 to 48 h after harvest. Rapid postharvest deterioration means that processing is more important than for any other root crops. Processing reduces food losses and stabilizes seasonal fluctuations in the supply of the crop. Processing cassava can affect the nutritional value of cassava roots through modification and losses in nutrients of high value. Although raw cassava root contains significant vitamin C, it is very sensitive to heat and easily leaches into water, and therefore almost all of the processing techniques seriously affect its content.

Conflict of Interest

The authors declare that there is no conflict of interest.

REFERENCES

Andrew W (2002). Cassava utilization, storage, and small scale-processing. Natural resource institute, Chatham maritime. UK. 14: 270-290.
Aristizabal J, Sanchez T (2007). Guía técnica para producción y análisis de almidón de yuca. Boletinde servicios agricolas de la FAO 130 FAO Rome., Italy P. 134.
Cassava Master Plan (2006). A strategic action plan for the development of the Nigeria cassava industry. UNIDO pp. 42-50.
Centro Internacial de Agricutura Tropical (CIAT), (2004). Dried cassava and its By - Products. Available at www.ciat.cgiar.org/agroempesas/ sistema_yaca/english/ dried cassava.htm.
Charles AL, Chang YH, Ko WC, Sriroth K, Huang TC (2004). Some physical and chemical properties of starch isolates of cassava genotypes. Starch/Starke 56:413-418.
Charles AL, Sriroth K, Huang TC. (2005). Proximate composition, mineral contents, hydrogen cyanide and phytic acid of 5 cassava genotypes. Food Chem. 92:615–620.
Crentsil D, Gallat S, Bancroft R (1995). Low cost fresh cassava root storage project achievement to date. In: Proceeding of the Workshop on Postharvest Experience in Africa, Accra 4-8 July, 1995. Edited by FAO, Rome.
El-Sharkawy MA (2003), Cassava biology and physiology. Plant Mol. Biol. 53:621-41.
EFDI-Technoserve (2005). Assessment of different models of cassava processing enterprises for the south and south-east of Nigeria, including the Niger Delta. Draft Final Report submitted to IITA-CEDP, March 2005.
Food and Agriculture Organization (FAO) (2011)."FAOSTAT: Production, Crops, Cassava, 2010 data".
FAO (2013)-Food and Agriculture Organization of the United Nations. Statistical Database _ FAOSTAT, http://faostat.fao.org/; 2013 [Accessed 14 July 2013].
Frederick DO (2008). Hog and Hominy: Soul Food from Africa to America. Columbia University Press. 1-2:1-24.
Gil JL, Buitrago AJA (2002). La yuca en la alimentacion animal. In: Ospina B, Ceballos H, editors. La yuca en el tercer milenio: Sistemas modernos de producción, procesamiento, utilizacio comercializacónn. Cali, Colombia. Centro Internacional de Agricultural Tropical. pp. 527–569. From: http://www.clayuca.org/PDF/libro_yuca/capitulo28.pdf. Accessed Jun 29, 2008.
Hahn SK (1994). An overview of traditional processing and utilization of cassava in Africa. pp. 2-8.
IITA (1990). Post Harvest Technology". In: Cassava in Tropical Africa A Reference Manual. edited by IITA Ibadan. pp. 82-120.
IITA (2010). Post Harvest Technology". Annual report. pp. 62-80.
IITA (2011). Annual report. pp. 34-38.
IITA (2012). An annual report on cassava production. pp. 4-6.
Julie AM, Christopher RD, Sherry AT (2009). Nutritional value of cassava for use as a staple food and recent advances for improvement. institute of food technologists. Comprehensive reviews in food science and food safety 8:181-192.
Kormawa P, Akoroda MO (2003). Cassava supply chain arrangement for industrial utilization in Nigeria. Ibadan: IITA.
Nahdy SM, Odong M (1995). Storage of fresh cassava tuber in plant based Storage Media. In: Proceeding of the Workshop on "Post-Harvest technology Experience in Africa". Accra, 4-8 July, 1994. Edited by FAO, Rome.
Oirschot Q, O'Brien GM, Dufour D, El-Sharkawy MA, Mesa E (2000). The effect of pre-harvest pruning of cassava upon root deterioration and quality characteristics. Journal of the Science of Food and Agriculture. 80:1866-1873.
Onabolu A, Abass A, Bokanga M (1998). New food products from cassava. International Institute of Tropical Agriculture, Ibadan, Nigeria. P. 40.
Osunde ZD, Fadeyibi A (2011). Storage methods and some uses of cassava in Nigeria. Continental J. Agric. Sci. 5(2):12-18.
Phillips TP, Taylor DS, Sanni L, Akoroda MO (2004). A cassava industrial revolution in Nigeria The potential for a new industrial crop. International Institute of Tropical Agriculture, Ibadan, Nigeria. International Fund for Agricultural Development, Food and Agriculture Organization of the United Nations, Rome, Italy.
Ravi V, Aked J, Balagopalan C (1996). Review on tropical root and tuber crops I. Storage methods and quality changes. Critical Rev. Food Sci. Nutr. 36:661-709.
Rawel HM, Kroll J (2003). Die Bedeutung von Cassava (Manihot esculenta, Crantz) als Hauptnahrungsmittel in tropischen L¨andern. Deutsche Lebensmittel-Rundschau 99:102-110.
Sánchez H, Ceballos F, Calle JC, Pérez C, Egesi CE, Cuambe AF,

Escobar DO, Chávez AL, Fregene M (2010). Tolerance to Postharvest Physiological Deterioration in Cassava Roots. Crop Sci. 50(4):1333-1338.

Sanni AI, Morlon-Guyot J, Guyot JP (2002). New efficient amylase-producing strains of *Lactobacillus plantarum* and I. fermentum isolated from different Nigerian traditional fermented foods. Int. J. Food Microbiol. 72:53–62.

Sanni LO, Charles A, Kuye A (1992). Moisture sorption isotherms of *fufu* and tapioca. J. Food Eng. 34:203-212.

Sanni LO (2005). Cassava Utilisation and Regulatory Framework in Nigeria, UNIDO.

Stone GD (2002). Both Sides Now. Current Anthropology 43(4):611-630.

Tewe OO, Lutaladio N (2004). Cassava for livestock feed in sub-Saharan Africa. Rome, Italy: FAO.

USDA (2009). USDA National Nutrient Database for Standard Reference.

Wenham JE (1995). Post-harvest deterioration of cassava. A biotechnology perspective. FAO Plant Production and Protection Paper 130.NRI/FAO. Rome. P. 90.

Westby A (2002). Cassava utilization, storage and small-scale processing. In: Hillocks, RJ. Thresh, J.M., Bellotti, A.C. (Eds.) Cassava Biology, Production and Utilization CABI Publishing Oxfordshire, UK. P. 480.

29

Insecticidal effect of *Jatropha curcas* L. seed oil on *Callosobruchus maculatus* Fab and *Bruchidius atrolineatus* Pic (Coleoptera: Bruchidae) on stored cowpea seeds (*Vigna unguiculata* L. Walp.) in Niger

Zakari ABDOUL HABOU[3] , Adamou HAOUGUI[3], Adamou BASSO[3] Toudou ADAM[2], Eric HAUBRUGE[1] and François J. VERHEGGEN[1]

[1]Unité d'Entomologie fonctionnelle et évolutive, Gembloux Agro-Bio Tech, Liege University, Passage des Déportés 2, B-5030 Gembloux, Belgium.
[2]Faculty of Agronomy, Abdou Moumouni University of Niamey, BP 10960, Niger.
[3]Institut National de Recherche Agronomique (INRAN) BP 429, Niger.

We report the insecticidal efficacy of *Jatropha curcas* seed oil against two bruchid beetle species, *Callosobruchus maculatus* Fab and *Bruchidius atrolineatus* Pic, devastating stored cowpea seeds (*Vigna unguiculata*). *J. curcas* oil concentrations ranging from 0.0, 0.25, 0.5, 1.5, 2.5, 5.0 and 7.5 ml were mixed with 200 g of cowpea seeds before introduction of 10 pairs (5 males and 5 females) of *C. maculatus* or *B. atrolineatus* as the case may be. Mortality, fecundity and rate of emergence were observed and compared with untreated control and a standard (Deltamethrin). *J. curcas* oil reduced adult survival in both species, *B. atrolineatus* being more sensitive than *C. maculatus*. Oviposition was also reduced by 85 to 90% in the females of both species after exposure to 2.5 ml of *J. curcas* oil solution. Only 9% of *C. maculatus* nymphs emerged as adults in seeds treated with 2.5 ml of oil. In *B. atrolineatus,* emergence was reduced to 12% in seeds treated with 1.5 ml of oil.

Key words: Natural insecticide, *Jatropha curcas* oil, pea beetle, cowpea seed.

INTRODUCTION

Cowpea (*Vigna unguiculata* L. Walp.) is among the principal world food leguminous plants (Pasquet and Baudoin, 1997). In Africa, it is appreciated for its green leaves, pods and seeds, which can be consumed (ISRA, ITA, CIRAD, 2005). In Niger, it is one of the principal cultivated crops, and has an important place in the rural population's food (Ibro and Bokar, 2001).

Cowpea seeds are attacked severally during storage by a variety of insect pests (Douma et al., 2002). In a study comparing the susceptibility of twenty varieties of cowpea to insect infestations in Niger, Douma et al. (2002) concluded that all varieties were subject to infestations by two Bruchid beetles, *Bruchidius atrolineatus* and *Callosobruchus maculatus* (Coleoptera: Bruchidae).

During the period of cowpea storage, losses associated with insect damages in stored cowpea could reach 30 to 100% (Singh and Allen, 1979).

The insecticidal effect of many locally harvested plants has been evaluated as protestants of stored cowpea seeds before the advent of synthetic insecticides. More than 2000 vegetable species have been prospected (Ngamo and Hance, 2007; Benayad, 2008). *Jatropha curcas*, locally named pourghère, has been the center of recent researches as its oil has been demonstrated to have insecticidal effect (Ratnadas et al., 1997; Abdoul Habou et al., 2011; Katoune et al., 2011). The seed oil of *J. curcas* has been shown to contain toxic substances called phorbol esters which exhibit insecticidal effects (Makkar and Becker, 1997; Solsoloy and Solsoloy 1997; Adebowale and Adedire, 2006). Boateng and Kusi (2008) and Abdoul Habou et al. (2011) have variously shown the effectiveness of *J. curcas* oil in controlling insects of cotton, rice and cowpea.

This study evaluates the insecticidal activity of *J. curcas* oil on two Bruchidae beetles *C. maculatus* Fabricius and *B. atrolineatus* Pic (Coleoptera: Bruchidae) on stored cowpea.

MATERIALS AND METHODS

Cowpea seeds

TN5-78 cowpea seeds, a variety susceptible to bruchid infestations, were conserved in the freezer for one week, in order to disinfest the seeds.

Breeding of the beetles

C. maculatus and *B. atrolineatus* were reared separately, but following the same procedure. 200 g of cowpea seeds were placed in large boxes (11.5 × 14 cm) and 20 beetle couples, previously collected from the field, introduced into them. After four days, the insects were removed and the seeds were left at 30°C for 21 days until adult emergence.

Evaluation of the insecticidal activity of *J. curcas* oil

An emulsifying concentrate (EC) was prepared using 50% of *J. curcas* oil, 30% of ethanol as a stabiliser and 20% arabic gum as an adjuvant in order to fix active molecules on the plant. Different dilutions of the EC were prepared in order to have seven solutions containing respectively 0.25, 0.5 1.5, 2.5, 5.0 and 7.5 ml of *J. curcas* oil solution. 200 g of treated cowpea seeds were then introduced inside small boxes (4.6 cm high with 4 cm of superior diameter and 3 cm of inferior diameter) before introducing twenty insects aged 24 h. In each box, 13.5 g of one of the seven above-mentioned solutions of *J. curcas* oil were applied, corresponding to the following doses: 0.25, 0.5, 1.5, 2.5, 5.0 and 7.5 ml per 200 g of seeds. A control box was also set up with 13.5 g of EC prepared without *J. curcas* oil (50% distilled water, 20% arabic gum and 30% ethanol). For each treatment, 5 boxes were set up as replicates. All boxes were placed in an incubator (30°C and 70% relative humidity. As a reference, deltamethrin (Decis 25 EC) treated seeds were also evaluated.

The number of dead insects was estimated, within one week of the experiment, in each box. The average mortality of the beetles (M0) was expressed as a corrected mortality (Mc), taking into account the natural mortality observed on the control (Mt) according to Abbott's formula (Abott, 1925):

$$Mc = \frac{M0 - Mt}{100 - Mt} \times 100$$

To calculate LC 50, a binary logistic regression analysis was used:

$$\ln\left(\frac{p_i}{1 - p_i}\right) = \beta_0 + \beta_1 X_i$$

Where: p_i = probability; β_0 and β_1 = predictors, and X_i = doses (Duyme and Claustriaux, 2006).

Statistical analysis

The data collected were subjected to analysis of variance and the means separated using Fisher's least significant differences.

Adult bruchid emergence

In another set of experiments, we evaluated the potential effect of *J. curcas* oil on adult emergence. Twenty couples of *C. maculatus* or *B. atrolineatus* adults, aged 48 h, were introduced in small boxes containing 50 treated cowpea seeds. The same doses of *J. curcas* oil as presented above were applied. Each treatment was evaluated 5 times per bruchid specie. Five days after, corresponding to the maximum period of egg laying (Sanon, 1997) adult insects (alive or dead) were removed. Seeds with eggs were transferred into Petri dishes and placed in an incubator at 30°C. After 10 days, the rate of emergence was evaluated.

RESULTS

Effect of *J. curcas* oil on adult bruchids' survival

The percent mortality of *C. maculatus* and *B. atrolineatus* adults in cowpea grains with different concentrations of *J. curcas'* seed oil is presented in Table 1. The oil was highly toxic to both *C. maculatus* and *B. atrolineatus* after 7 days. For each concentration tested, a significant death rate was observed (p ≤ 0.05), being higher in presence of high concentration of oil. Treatments with 10 and 15% (5.0 and 7.5 ml/200 g) *J. curcas* oil concentrations caused significant mortality on adult *C. maculatus* (42.0 and 48% respectively) after 72 h. This result was not different after 7 days of the introduction. Similarly, these treatments induced 65 and 84% mortality, respectively on *B. atrolineatus* after 3 days. The mortality rose to 95 and 98% respectively, after 7 days.

Effect of *J. curcas* oil on adult bruchid emergence

The oil of *J. curcas* seeds has an inhibitory effect on

Table 1. Effect of *J. curcas* oil concentrations on mean mortality of *C. maculatus* and *B. atrolineatus*.

J. curcas oil (ml)	*C. maculatus* (time of observation)							*B. atrolineatus* (time of observation)						
	1 days*	2 days	3 days*	4 days*	5 days*	6 days*	7 days*	1 day*	2 days*	3 days*	4 days*	5 days*	6 days*	7 days*
0	0±0.0[f]	0±0.0[f]	0±0.0[e]	0±0.0[e]	0±0.0[e]	0±0.0[e]	0±0.0[e]	0±0.0[e]	0±0.0[e]	0±0.0[e]	0±0.0[e]	0±0.0[e]	0±0.0[e]	0±0.0[e]
0.25	8±0[de]	13±1.1[ef]	18±0.5[ef]	29±1.0[d]	37±0.5[e]	48±1.5[d]	63±3.8[c]	17±0.8[de]	19±1.4[de]	23±1.6[d]	32±2.1[d]	45±1.4[d]	66±0.4[d]	75±3.0[d]
0.5	14±0.8[cd]	17±1.3[ef]	25±1.2[e]	33±1.3[d]	46±1.7[de]	62±1.8[d]	72±2.5[bc]	27±1.3[cd]	37±2.1[cd]	42±1.6[cd]	50±2.4[cd]	58±2.6[cd]	74±1.7[c]	81±0.8[c]
1.5	19±1.3[cd]	25±1.5[de]	33±0.8[de]	41±1.3[d]	58±2.6[cd]	72±1.8[cd]	84±0.8[ab]	26±1.0[cd]	36±1.6[cd]	47±3.6[cd]	61±4.1[bc]	74±2.9[bc]	83±2.6[bc]	86±1.7[bc]
2.5	27±1.8[c]	33±1.5[cd]	41±2.3[bc]	59±1.7[c]	73±0.8[bc]	82±1.1[bc]	85±1.4[ab]	51±2.1[c]	53±3.7[c]	55±3.0[c]	65±2.7[bc]	79±2.5[abc]	87±1.9[abc]	95±1.4[ab]
5.0	30±1.5[bc]	35±1.8[c]	48±2.3[bc]	60±1.0[c]	74±1.6[b]	85±0.7[b]	87±0.8[ab]	43±4.2[cd]	56±4.4[c]	65±4.3[ab]	80±2.4[ab]	88±2.3[ab]	94±1.6[ab]	98±0.8[ab]
7.5	32±1.3[bc]	38±1.8[c]	53±1.7[c]	61±3.1[c]	75±2.0[b]	84±1.4[b]	88±0.0[a]	67±3.2[b]	74±2.0[ab]	84±2.3[ab]	91±2.6[a]	92±2.6[ab]	96±1.3[ab]	100±0.0[a]
Decis	92±0.8[a]	93±1.1[a]	99±0.4[a]	100±0.0[a]	100±0.0[a]	100±0.0[a]	100±0.0[a]	95±1.0[a]	96±1.3[a]	99±0.4[a]	100±0.0[a]	100±0.0[a]	100±0.0[a]	100±0.0[a]

The averages which have the same letter are appreciably not different. Turkey test, α=5%.

oviposition and subsequent emergence of adult *C. maculatus* and *B. atrolineatus*. The average number of eggs laid by *C. maculatus* in the control was 64.8±13.9 eggs. Under exposure to *J. curcas* oil, the oviposition rate reduced as indicated in Table 2. Oviposition was significantly inhibited in both insect species with concentrations of 5.0 and 7.5 ml of *J. curcas* oil solution, after four days infestation.

In the control, the rates of emergence of *C. maculatus* and *B. atrolineatus* were 76.2 and 76.1%, respectively, indicating that exposure to *J. curcas* oil drastically reduced adult emergence in the two insect species as stated in Table 2.

The lethal concentration of *J. curcas* oil that effected 50% mortality of *C. maculatus* at the end of the 1st, 2nd and 3rd days were, respectively, 3.75, 3.10 and 2.75 ml. For *B. atrolineatus*, they were, 2.90, 2.45 and 2.25 ml, respectively (Table 3).

DISCUSSION

J. curcas oil contains phorbol esters with activate the protein kinase C (PKC) (Makkar et al., 1997). This phenomenon endues a cellular proliferation or apoptosis (Blumberg et al., 1987). The toxicity of phorbol esters are demonstrated by many studies on insects pest species (Solsoloy and Solsoloy; Ratnadas et al. 1997; Abdoul Habou et al., 2011). This study compared the sensitivity of two beetles pea bruchid

In this study, we have demonstrated the toxic effect of *J. curcas* oil on *C. maculatus* and *B atrolineatus* adult's survival and emergence. *J. curcas* oil reduced adult survival in both species, *B. atrolineatus* being more sensitive than *C. maculatus*. Oviposition is also reduced in the females of both species, reaching up to 85 and 90% fewer eggs deposited after exposure to a 2.5 ml of *J. curcas* oil solution. Only 9% of *C. maculatus* nymphs emerged to adults in the presence of a 5% solution of oil. In *B. atrolineatus*, emergence was reduced to 12% with the use of a 1.5 ml of oil solution. Similar result was found by Ravindra and Kshirsagar (2010) on the development of *C. maculatus*. In their work, these authors showed that the eggs of *C. maculatus* were the most susceptible at the developmental stage to *J. curcas* oil. Our results are comparable to those obtained by Udo (2011) and Singh et al. (1978) who used pure oil. Udo (2011) showed that palm oil has a significant effect on progeny development of *C. maculatus* as Singh et al. (1979) demonstrated that groundnut oil at 5 ml/kg completely protected cowpea seeds in storage for the same insect for up to 180 days. The oil treatment prevented emergence of progeny rather than affecting oviposition or mortality of the adult weevils.

Adebowale and Adedire (2006) conducted a similar experiment in the laboratory on *C. maculatus* Fabr devastating insects of cowpea in Nigeria. They observed a significant reduction in laying for all tested concentrations and a total inhibition of eggs and larvae. The number of eggs laid by *C. maculatus* females was also reduced. They also observed that the emergence rate of the adults is null for all the concentrations applied. These authors observed that the seeds are thus protected during 12 weeks and advance that the insecticidal effect could originate in sterols and terpenes alcool contained in *J. curcas* seeds' oil. Boateng and Kusi, 2008 highlighted the insecticidal effect of *J. curcas* oil on *C. maculatus* (Coleoptera: Bruchidae) and its parasitoid *Dinarmus basalis* (Hymenoptera: Pteromalidae). The adults of the two species have the same sensibility for the toxic effect of *J. curcas* oil. *C. maculatus*' eggs are more sensible than those of *D. basalis* because they are protected by the

Table 2. Effect of *J. curcas* oil concentrations on oviposition and rate of emergence of *C. maculatus* and *B. atrolineatus*.

J. curcas oil (ml)	*C. maculatus*			*B. atrolineatus*		
	number of eggs*	number of adults emergent's*	rate of emergent*	number of eggs*	Number of adults emergent's*	rate of emergent*
0	64.8±13.8^a	48.8±8.7^a	76.2±0.1^a	75.4±11.5^a	57±1.9^a	76.1±1^a
0.25	40.8±10.2^b	25.8±10.5^b	63.4±0.1^b	59.6±8.5^b	16.8±6.3^a	27±01^b
0.5	38.6±6.4^b	11.6±8.9^b	29.1±0.1^b	47.2±4.3^b	6.6±8.0^b	13.2±0.1^b
1.5	36.6±12.0^b	7.8±5.1^b	19.1±0.1^b	18.8±7.0^c	2.6±7.6^c	11.3±0.1^b
2.5	9.6±3.0^c	1.0±1.2^c	11.6±0.0^c	8.2±4.3^c	0±0.0^c	0±0.0^c
5.0	0±0.0^c	0±0.0^c	0±0.0^c	0±0.0^c	0±0.0^c	0±0.0^c
7.5	0±0.0^c	0±0.0^c	0±0.0^c	0±0.0^c	0±0.0^c	0±0.0^c
Delthametrin	0±0.0^c	0±0.0^c	0±0.0^c	0±0.0^c	0±0.0^c	0±0.0^c

Averages which have the same letter are not different, Turkey test α= 5%.

Table 3. Regression coefficients estimated by logistic binary analysis.

Parameter	observation time					
	C. maculatus			*B. atrolineatus*		
	1 day	2 days	3 days	1 day	2 days	3 days
Probability	P<0.001	P<0.001	P<0.001	P<0.001	P<0.001	P<0.001
Cstant (β0)	-2.51	-2.19	-1.65	-2.13	-1.86	1.62
Coef β1	0.033	0.033	0.029	0.036	0.047	0.049
DL 50 (ml)	3.75	3.10	2.75	2.90	2.45	2.25

seeds. Solsoloy and Solsoloy (1997) also showed the effectiveness of *J. curcas* seeds oil on *Sitophilus zeamais* Motschulsky and *Callosobruchus chinensis* L. insects devastating seeds of maize and *Phaseolus vulgaris* L.

Ogunleye et al. (2010) tested the properties of seed oil of three botanicals namely *J. curcas*, *Heliathus annus* (sunflower) and *Cocos nucifera* (coconut) on maize weevil, *Sitophilus zeamais* Mots. The results of insects treated with all dosage rates of *C. nucifera* showed a significantly higher mortality when compared with the control. The least rate of application of *J. curcas* produced 70% mortality after 24 h, while the dosage of 0.3 ml and 0.4 ml produced 80% mortality after 24 h. The control experiment remained at 0% level throughout the period of the experiments.

J. curcas seeds' oil has a toxic effect on the adults of *C. maculatus* and *B. atrolineatus*. This toxicity increases with the concentration. A reduction of laying of the females of these two species is observed for weak concentrations. The emergence rate of adults is more significant to *B. atrolineatus* than *C. maculatus*. In Niger, the peasants' cans used 7.5 ml/200g to detruded 100% of beetle's pea contained in the grains which will be preserving after 24 h. Germination test performed on treated seeds showed germination rate above 80%. *J. curcas* oil does not affect the germination of seeds.

Conflict of Interest

The authors have not declared any conflict of interest.

REFERENCES

Abbot WS (1925). A method of computing the effectiveness of an insecticide. J. Econ. Entomol. 18:265-267.

Abdoul Habou Z, Haougui A, Mergeai G, Haubruge E, Adam T, Verheggen FJ (2011). Insecticidal effect of *Jatropha curcas* oil on the aphid *Aphis fabae* (Hemiptera: Aphididae) and on the main insect pests associated with cowpeas (*Vigna unguiculata*) in Niger. Tropicultura 29:225-229.

Adebowale KO, Adedire CO (2006). Chemical composition and insecticidal properties of the underutilized *Jatropha curcas* seed oil. Afr. J. Biotechnol. 5:901-906.

Benayad N (2008). Les huiles essentielles extraites des plantes médicinales marocaines : moyen efficace de lutte contre les ravageurs des denrées alimentaires stockées. *Rapport d'étude*. Université Mohammed V–Agdal. Maroc, P. 61.

Boateng BA, Kusi F (2008). Toxicity of *Jatropha* Seed Oil to *Callosobruchus maculatus* (Coleoptera: Bruchidae) and its Parasitoid. *Dinarmus basalis* (Hymenoptera: Pteromalidae). J. Appl. Sci. Res. 4(8):945-951.

Blumberg PM, Nakadate T, Warren B, Dell Aquila M, Sako T, Pasti G, Sharkey NAB (1987). Phorbol esters as probes of the modulatory site on protein kinase C. an overview. Botanical J. linean Soc. 94(1-2):283-292.

Duyme F, Claustriaux JJ (2006). La régression logistique binaire. *Notes statistique et d'informatique*. Gembloux Agro-bio- Tech, Belgique, P. 24.

Douma A, Abbas IL, Adam T, Alzouma I (2002). Comportement de vingt variétés de niébé (*Vigna unguiculata* (L.) Walp.) vis-à-vis de *Bruchidius atrolineatus* (Pic.) et*Callosobruchus maculatus* (F.) (Coleoptera: Bruchidae. *Cahiers Agricultures* 15:187-193.

Ibro G, Bokar M (2001). Transfert de nouvelles technologies dans les systèmes de production des paysans au Niger, étude de cas: adoption des variétés améliorées et de nouvelles techniques de protection de la culture du niébé. *Rapport du projet PRONAF*, P. 17.

ISRA, ITA, CIRAD (2005). Bilan de la recherche agricole et agroalimentaire au Sénégal, *rapport*, P. 520.

Katoune HI, Malam Lafia D, Salha H, Doumma A, Yaye DA, Pasternak

D, Ratnadass A (2011). Physic nut (*Jatropha curcas*) oil as a protectant against field insect pests of cowpea in Sudano-Sahelian cropping systems. J. Sat Agric. Res. 9:1-6.

Makkar HPS, Becker K, Sporer F, Wink M (1997). Studies on nutritive potential and toxic constituents of different provenances of *Jatropha curcas* L. J. Agric. Food Chem. 45:3152-3157. http://dx.doi.org/10.1021/jf970036j

Ngamo LST, Hance TH (2007). Diversité des ravageurs des denrées et méthodes alternatives de lutte en milieu tropical. Tropicultura 25:215-220.

Ogunleye RF, Ogunkoya MO, Abulude FO (2010). Effect of the seed oil of three botanicals, *Jatropha curcas*, *Helianthus annus* and *Cocos nucifera* on the maize weevil, *Sitophilus zeamais* (Mots). Plant Product Res. J.14:14-18.

Pasquet RS, Baudoin JP (1997). Le niébé *In* Charrier A., Jacquot M., Hammon S. *biotechnology research* network. Proceedings of the Second International Scientific Meeting, 7–10 Sept. 1993, CIAT (Cali, Colombia), pp. 69-75.

Ratnadas A, Cissé B, Diarra A, Mengual L, Taneja SL, Thiéro CAT (1997). Perspectives de gestion bio intensive des foreurs des tiges de sorgho en Afrique de l'Ouest. *Insect Sci.* 17:227-233.

Ravindra V, Kshirsagar (2010). Insecticidal activity of *Jatropha* seed oil against *Callosobruchus maculatus* (Fabricius) infesting *Phaseolus* aconitifoliusJacq. Bioscan 5:415-418.

Sanon A (1997). Contribution à l'étude du contrôle biologique des populations de Bruchidae ravageurs des graines de niébé (*Vigna unguiculata* L. Walp) au cours de leur stockage au Burkina-Faso. Thèse de Doctorat, Université de Ouagadougou, P. 162.

Singh SR, Luse RA, Leuschner K, Nangju D. (1979). Groundnut oil treatment for the control of *Callosobruchus maculatus* (F.) during cowpea storage. J. Stored Prod. Res. 14(2-3):77-80. http://dx.doi.org/10.1016/0022-474X(78)90001-2

Singh SR, Allen DJ (1979). Les insectes et les maladies du niébé. IITA Ibadan, Nigeria, P. 113.

Solsoloy AD, Solsoloy TS (1997). Pesticidal efficacy of formulated product *Jatropha curcas* oil on pests of selected field crops, pp. 216-226 in: Gubitz G.M. Mithelbach M et TrabiM. Symposium on Biofuel and Industrial Products from *Jatropha curcus* and other Tropical Oil Seed Plants, February 23-27, Managua. Nicaragua,

Udo IO (2011). Protectant effect of plant oils against cowpea weevil (*Callosobruchus maculatus*) on stored cowpea (*Vigna unguiculata.*) ARPN J. Agric. Boil. Sci. 6(20):1-4.

Permissions

All chapters in this book were first published in AJAR, by Academic Journals; hereby published with permission under the Creative Commons Attribution License or equivalent. Every chapter published in this book has been scrutinized by our experts. Their significance has been extensively debated. The topics covered herein carry significant findings which will fuel the growth of the discipline. They may even be implemented as practical applications or may be referred to as a beginning point for another development.

The contributors of this book come from diverse backgrounds, making this book a truly international effort. This book will bring forth new frontiers with its revolutionizing research information and detailed analysis of the nascent developments around the world.

We would like to thank all the contributing authors for lending their expertise to make the book truly unique. They have played a crucial role in the development of this book. Without their invaluable contributions this book wouldn't have been possible. They have made vital efforts to compile up to date information on the varied aspects of this subject to make this book a valuable addition to the collection of many professionals and students.

This book was conceptualized with the vision of imparting up-to-date information and advanced data in this field. To ensure the same, a matchless editorial board was set up. Every individual on the board went through rigorous rounds of assessment to prove their worth. After which they invested a large part of their time researching and compiling the most relevant data for our readers.

The editorial board has been involved in producing this book since its inception. They have spent rigorous hours researching and exploring the diverse topics which have resulted in the successful publishing of this book. They have passed on their knowledge of decades through this book. To expedite this challenging task, the publisher supported the team at every step. A small team of assistant editors was also appointed to further simplify the editing procedure and attain best results for the readers.

Apart from the editorial board, the designing team has also invested a significant amount of their time in understanding the subject and creating the most relevant covers. They scrutinized every image to scout for the most suitable representation of the subject and create an appropriate cover for the book.

The publishing team has been an ardent support to the editorial, designing and production team. Their endless efforts to recruit the best for this project, has resulted in the accomplishment of this book. They are a veteran in the field of academics and their pool of knowledge is as vast as their experience in printing. Their expertise and guidance has proved useful at every step. Their uncompromising quality standards have made this book an exceptional effort. Their encouragement from time to time has been an inspiration for everyone.

The publisher and the editorial board hope that this book will prove to be a valuable piece of knowledge for researchers, students, practitioners and scholars across the globe.

List of Contributors

R. A. Wani
Department of Horticulture, SHIATS Allahabad U.P-211007, India

V. M. Prasad
Department of Horticulture, SHIATS Allahabad U.P-211007, India

S. A. Hakeem
Department of Horticulture, SHIATS Allahabad U.P-211007, India

S. Sheema
Department of Horticulture, SHIATS Allahabad U.P-211007, India

S. Angchuk
Department of Horticulture, SHIATS Allahabad U.P-211007, India

A. Dixit
Department of Horticulture, SHIATS Allahabad U.P-211007, India

Umezuruike Linus Opara
South African Research Chair in Postharvest Technology, Faculty of AgriSciences, Stellenbosch University, Stellenbosch 7600, South Africa

Asanda Mditshwa
South African Research Chair in Postharvest Technology, Faculty of AgriSciences, Stellenbosch University, Stellenbosch 7600, South Africa

Rewati Raman Bhattarai
Tribhuvan University, Central Campus of Technology, Hattisar, Dharan, Nepal

Raj Kumar Rijal
Food Research Officer, Regional Food Technology and Quality Control, Hetauda, Nepal

Pashupati Mishra
Tribhuvan University, Central Campus of Technology, Hattisar, Dharan, Nepal

Samira Nair
Laboratory Research on Biological Systems and Geomatics, Faculty of Nature and Life, University of Mascara, Algeria

Boumedienne Meddah
Laboratory Research on Biological Systems and Geomatics, Faculty of Nature and Life, University of Mascara, Algeria

Abdelkader Aoues
Laboratory of Experimental Biotoxicology, Biodepollution and Phytoremediation, University of Es-Senia, Oran, Algeria

Khaled A. Alqadi
Ecosystem Management, School of Environmental and Rural Science, University of New England, Armidale, NSW, 2351, Australia

Lalit Kumar
Water Resources and Environment Management, Al Balqa Applied University, Amman, Jordan

Al-Zu'bi Jarrah

S. O. Agbeniyi
Cocoa Research Institute of Nigeria, P. M. B. 5244, Ibadan, Oyo State, Nigeria

M. S. Ayodele
Department of Biological Sciences, University of Agriculture, Abeokuta, Ogun State, Nigeria

S. N. Kirmani
Division of fruit Sciences, Sher-E-Kashmir University of Agricultural Sciences and Technology of Kashmir, Shalimar-191121 Srinagar Kashmir-India

G. M. Wani
Division of fruit Sciences, Sher-E-Kashmir University of Agricultural Sciences and Technology of Kashmir, Shalimar-191121 Srinagar Kashmir-India

M. S. Wani
Division of fruit Sciences, Sher-E-Kashmir University of Agricultural Sciences and Technology of Kashmir, Shalimar-191121 Srinagar Kashmir-India

M. Y. Ghani
Department of Plant Pathology, Sher-E-Kashmir University of Agricultural Sciences and Technology of Kashmir, Shalimar-191121 Srinagar Kashmir-India

M. Abid
Central Institute of Temperate Horticulture, Old Air Field Rangreth Srinagar-190007, Jammu and Kashmir-India

S. Muzamil
Central Institute of Temperate Horticulture, Old Air Field Rangreth Srinagar-190007, Jammu and Kashmir-India

Hadin Raja
Division of fruit Sciences, Sher-E-Kashmir University of Agricultural Sciences and Technology of Kashmir, Shalimar-191121 Srinagar Kashmir-India

A. R. Malik
Division of fruit Sciences, Sher-E-Kashmir University of Agricultural Sciences and Technology of Kashmir, Shalimar-191121 Srinagar Kashmir-India

Lílian Moreira Costa
Federal Institute of Education, Science and Technology of Goiás (Instituto Federal de Educação, Ciência e Tecnologia Goiano – IF Goiano) – Rio Verde Câmpus, GO, Rodovia Sul Goiana, Km 01 - Zona Rural - CEP: 75901-97, Brazil

Osvaldo Resende
Board of Undergraduate Studies, IF Goiano – Rio Verde Câmpus, GO, Rodovia Sul Goiana, Km 01 - Zona Rural - CEP: 75901-970, Brazil

Douglas Nascimento Gonçalves
PIBIC/CNPq scholar, IF Goiano – Rio Verde Câmpus, GO, Brazil

Anderson Dinis Rigo
PIBIC/CNPq scholar, IF Goiano – Rio Verde Câmpus, GO, Brazil

Meire C. N. Andrade
Universidade Estadual Paulista, UNESP, Faculdade de Ciências Agronômicas, FCA, Departamento de Produção Vegetal/Defesa Fitossanitária, Módulo de Cogumelos. Rua José Barbosa de Barros, 1780 - Fazenda Lageado. Caixa Postal 237, 18610-307. Botucatu, SP, Brasil

João P. F. Jesus
Universidade Estadual Paulista, UNESP, Faculdade de Ciências Agronômicas, FCA, Departamento de Produção Vegetal/Defesa Fitossanitária, Módulo de Cogumelos. Rua José Barbosa de Barros, 1780 - Fazenda Lageado. Caixa Postal 237, 18610-307. Botucatu, SP, Brasil

Fabrício R. Vieira
Universidade Estadual Paulista, UNESP, Faculdade de Ciências Agronômicas, FCA, Departamento de Produção Vegetal/Defesa Fitossanitária, Módulo de Cogumelos. Rua José Barbosa de Barros, 1780 - Fazenda Lageado. Caixa Postal 237, 18610-307. Botucatu, SP, Brasil

Sthefany R. Viana
Universidade Estadual Paulista, UNESP, Faculdade de Ciências Agronômicas, FCA, Departamento de Produção Vegetal/Defesa Fitossanitária, Módulo de Cogumelos. Rua José Barbosa de Barros, 1780 - Fazenda Lageado. Caixa Postal 237, 18610-307. Botucatu, SP, Brasil

Marta H. F. Spoto
Universidade de São Paulo, USP, Escola Superior de Agricultura, ESALQ, Departamento de Agroindústria, Alimentos e Nutrição. Av. Pádua Dias, 11, Caixa Postal 9, 13418-900. Piracicaba, SP, Brasil

Marli T. A. Minhoni
Universidade Estadual Paulista, UNESP, Faculdade de Ciências Agronômicas, FCA, Departamento de Produção Vegetal/Defesa Fitossanitária, Módulo de Cogumelos. Rua José Barbosa de Barros, 1780 - Fazenda Lageado. Caixa Postal 237, 18610-307. Botucatu, SP, Brasil

Y. Mijinyawa
Department of Agricultural and Environmental Engineering, Faculty of Technology, University of Ibadan, Oyo State, Nigeria

John O. Alaba
Department of Agricultural and Environmental Engineering, Faculty of Technology, University of Ibadan, Oyo State, Nigeria

B. Patil Atul
ACHF Farm, Navsari Agriculture University, Navsari-396 450, Gujarat, India

D. T. Desai
ACHF Farm, Navsari Agriculture University, Navsari-396 450, Gujarat, India

A. Patil Sandip
ACHF Farm, Navsari Agriculture University, Navsari-396 450, Gujarat, India

R. Ghodke Umesh
ACHF Farm, Navsari Agriculture University, Navsari-396 450, Gujarat, India

K. Wojtkowiak
Department of Fundamentals of Safety, University of Warmia and Mazury in Olsztyn, Poland

A. Stępień
Department of Agriculture Systems, University of Warmia and Mazury in Olsztyn, Poland

M. Tańska
Department of Food Plant Chemistry and Processing, University of Warmia and Mazury in Olsztyn, Poland

I. Konopka
Department of Food Plant Chemistry and Processing, University of Warmia and Mazury in Olsztyn, Poland

S. Konopka
Department of Working Machine and Methodology of Research, University of Warmia and Mazury in Olsztyn, Poland

Fekadu Gemechu
Department of Plant Sciences and Horticulture, Jimma University College of Agriculture and Veterinary Medicine, P. O. Box-307, Jimma, Ethiopia

Dante R. Santiago
Department of Environmental Sciences and Technology, Jimma University College of Public Health and Medical Sciences, P. O. Box-378, Jimma, Ethiopia

Waktole Sori
Department of Plant Sciences and Horticulture, Jimma University College of Agriculture and Veterinary Medicine, P. O. Box-307, Jimma, Ethiopia

T. Krishnakumar
Department of Food and Agricultural Process Engineering, Tamil Nadu Agricultural University, Coimbatore-3, Tamil Nadu, India

C. Thamilselvi
Department of Food and Agricultural Process Engineering, Tamil Nadu Agricultural University, Coimbatore-3, Tamil Nadu, India

C. T. Devadas
Department of Food and Agricultural Process Engineering, Tamil Nadu Agricultural University, Coimbatore-3, Tamil Nadu, India

C. S. Ugwu
Department of Agricultural Extension, University of Nigeria Nsukka, Enugu State, Nigeria

J. C. Iwuchukwu
Department of Agricultural Extension, University of Nigeria Nsukka, Enugu State, Nigeria

H. Ibrahim
Department of Crop Production, School of Agriculture and Agricultural Technology, Federal University of Technology, Minna, Nigeria

J. A. Oladiran
Department of Crop Production, School of Agriculture and Agricultural Technology, Federal University of Technology, Minna, Nigeria

H. Mohammed
Department of Crop Production, School of Agriculture and Agricultural Technology, Federal University of Technology, Minna, Nigeria

Kai Ying Chiu
Department of Post-Modern Agriculture, Mingdao University, Peetow, Changhwa County, 523, Taiwan

Jih Min Sung
Department of Food Science and Technology, Hung Kuang University, 34 Chung-Chie Rd, Sha Lu, Taichung City, 43302, Taiwan

M. O. OKE
Bioresources Engineering, School of Engineering, University of Kwazulu Natal, Private Bag X01, Scottsville, Pietermaritzburg, Kwazulu Natal, 3209 South Africa

T. S. WORKNEH
Bioresources Engineering, School of Engineering, University of Kwazulu Natal, Private Bag X01, Scottsville, Pietermaritzburg, Kwazulu Natal, 3209 South Africa

U. Zakka
Department of Crop and Soil Science, Faculty of Agriculture, University of Port-Harcourt, Choba, P. M. B. 5005, Rivers State, Nigeria

N. E. S. Lale
Department of Crop and Soil Science, Faculty of Agriculture, University of Port-Harcourt, Choba, P. M. B. 5005, Rivers State, Nigeria

N. M. Duru
Department of Crop and Soil Science, Faculty of Agriculture, University of Port-Harcourt, Choba, P. M. B. 5005, Rivers State, Nigeria

C. N. Ehisianya
National Root Crops and Research Institute, Umudike, Abia State, Nigeria

Girish Sharma
Department of Fruit Breeding and Genetic Resourses, Dr. YSP UHF, Solan, HP, India

Nirmal Sharma
Division of Fruit Science, SKUAST-J, Jammu and Kashmir, India

Rubiqua Bashir
Department of Fruit Breeding and Genetic Resourses, Dr. YSP UHF, Solan, HP, India

Amit Khokhar
Division of Fruit Science, SKUAST-J, Jammu and Kashmir, India

Mahital Jamwal
Division of Fruit Science, SKUAST-J, Jammu and Kashmir, India

I. P. Dike
Department of Biological Sciences, Covenant University, Ota, Ogun State, Nigeria

J. Ekou
Department of Animal Production and Management, Faculty of Agriculture and Animal Sciences, Busitema University, P. O. Box 236, Tororo, Uganda

Franciele Caixeta
Department of Agriculture, Federal University of Lavras, C.P. 3037, 37200-000, Lavras, MG, Brazil

Édila Vilela De Resende Von Pinho
Department of Agriculture, Federal University of Lavras, C.P. 3037, 37200-000, Lavras, MG, Brazil

Renato Mendes Guimarães
Department of Agriculture, Federal University of Lavras, C.P. 3037, 37200-000, Lavras, MG, Brazil

Pedro Henrique Andrade Rezende Pereira
Department of Agriculture, Federal University of Lavras, C.P. 3037, 37200-000, Lavras, MG, Brazil

Hugo Cesar Rodrigues Moreira Catão
Department of Agriculture, Federal University of Lavras, C.P. 3037, 37200-000, Lavras, MG, Brazil

M. K. Yadav
Department of Horticulture, N. M. College of Agriculture, Navsari Agricultural University, Dandi Road, Navsari-396450, India

N. L. Patel
ASPEE College of Horticulture and Forestry, Navsari Agricultural University, Dandi Road, Navsari -396450, India

Patrícia de Lima Martins
Universidade Estadual da Paraíba – UEPB, Pós-graduação em Ciências Agrárias, Rua Baraúnas n. 351, CEP: 58429-500 Bairro Universitário, Campina Grande, PB, Brazil

Roseane Cavalcanti dos Santos
Empresa Brasileira de Pesquisa Agropecuária, Embrapa Algodão. Rua Oswaldo Cruz n. 1143, CEP: 58428-095, Centenário, Campina Grande, PB, Brazil

João Luis da Silva Filho
Empresa Brasileira de Pesquisa Agropecuária, Embrapa Algodão. Rua Oswaldo Cruz n. 1143, CEP: 58428-095, Centenário, Campina Grande, PB, Brazil

FábiaSuelly Lima Pinto
Empresa Brasileira de Pesquisa Agropecuária, Embrapa Algodão. Rua Oswaldo Cruz n. 1143, CEP: 58428-095, Centenário, Campina Grande, PB, Brazil

Liziane Maria de Lima
Universidade Estadual da Paraíba – UEPB, Pós-graduação em Ciências Agrárias, Rua Baraúnas n. 351, CEP: 58429-500 Bairro Universitário, Campina Grande, PB, Brazil
Empresa Brasileira de Pesquisa Agropecuária, Embrapa Algodão. Rua Oswaldo Cruz n. 1143, CEP: 58428-095, Centenário, Campina Grande, PB, Brazil

Matteo Barbari
Department of Agricultural, Food and Forestry Systems(GESAAF) -University of Firenze Via San Bonaventura, 13 – 50145 Firenze - Italy

Massimo Monti
Department of Agricultural, Food and Forestry Systems(GESAAF) -University of Firenze Via San Bonaventura, 13 – 50145 Firenze - Italy

Giuseppe Rossi
Department of Agricultural, Food and Forestry Systems(GESAAF) -University of Firenze Via San Bonaventura, 13 – 50145 Firenze - Italy

Stefano Simonini
Department of Agricultural, Food and Forestry Systems(GESAAF) -University of Firenze Via San Bonaventura, 13 – 50145 Firenze - Italy

Francesco Sorbetti Guerri
Department of Agricultural, Food and Forestry Systems(GESAAF) -University of Firenze Via San Bonaventura, 13 – 50145 Firenze - Italy

G. H. M. Biego
Laboratory of Biochemistry and Food Sciences, UFR Biosciences, University of Abidjan-Cocody, 25 BP 313 Abidjan 25, Cote d'Ivoire
Department of Public Health, Hydrology and Toxicology, Faculty of Pharmaceutical and Biological Sciences, University of Abidjan-Cocody, BP 34 Abidjan, Cote d'Ivoire

K. D. Yao
Laboratory of Biochemistry and Food Sciences, UFR Biosciences, University of Abidjan-Cocody, 25 BP 313 Abidjan 25, Cote d'Ivoire

K. M. Koffi
Laboratory of Biochemistry and Food Sciences, UFR Biosciences, University of Abidjan-Cocody, 25 BP 313 Abidjan 25, Cote d'Ivoire
Central Laboratory for Food, Hygiene and Agro-Industry (LCHAI), National Laboratory for Agriculture Development Support (LANADA), Department of Animal Production and Fish Resources, Abidjan, Cote d'Ivoire

P. Ezoua
Laboratory of Biochemistry and Food Sciences, UFR Biosciences, University of Abidjan-Cocody, 25 BP 313 Abidjan 25, Cote d'Ivoire

L. P. Kouadio
Department of Public Health, Hydrology and Toxicology, Faculty of Pharmaceutical and Biological Sciences, University of Abidjan-Cocody, BP 34 Abidjan, Cote d'Ivoire

C. A. Onyenwoke
Agricultural and Bioresources Engineering Department, Michael Okpara University of Agriculture, Umudike, P. M. B. 7267, Umuahia, Abia State, Nigeria

K. J. Simonyan
Agricultural and Bioresources Engineering Department, Michael Okpara University of Agriculture, Umudike, P. M. B. 7267, Umuahia, Abia State, Nigeria

Zakari ABDOUL HABOU
Institut National de Recherche Agronomique (INRAN) BP 429, Niger

Adamou HAOUGUI
Institut National de Recherche Agronomique (INRAN) BP 429, Niger

Adamou BASSO
Institut National de Recherche Agronomique (INRAN) BP 429, Niger

Toudou ADAM
Faculty of Agronomy, Abdou Moumouni University of Niamey, BP 10960, Niger

Eric HAUBRUGE
Unité d'Entomologie fonctionnelle et évolutive, Gembloux Agro-Bio Tech, Liege University, Passage des Déportés 2, B- 5030 Gembloux, Belgium

François J. VERHEGGEN
Unité d'Entomologie fonctionnelle et évolutive, Gembloux Agro-Bio Tech, Liege University, Passage des Déportés 2, B- 5030 Gembloux, Belgium

www.ingramcontent.com/pod-product-compliance
Lightning Source LLC
Chambersburg PA
CBHW080657200326
41458CB00013B/4885